The Anatomy of Aging in Man and Animals

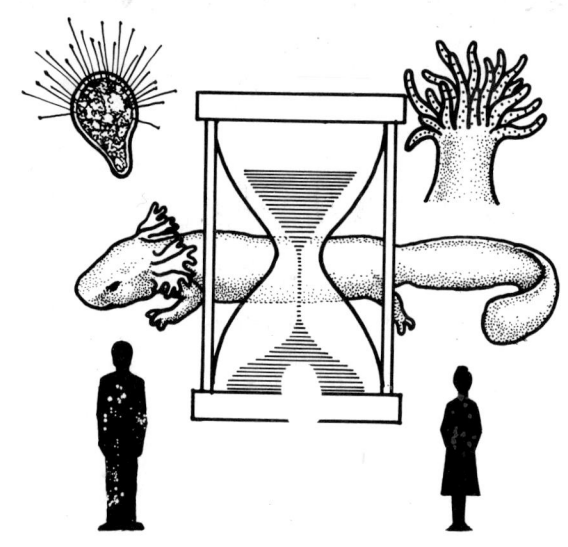

The Anatomy of Aging in Man and Animals

Warren Andrew, Ph.D., M.D.

Chairman, Department of Anatomy
Indiana University Medical Center

GRUNE & STRATTON
New York and London

©1971 by Grune & Stratton, Inc.
All rights reserved. No part of this publication may be reproduced or transmitted in any form or by any means, electronic or mechanical, including photocopy, recording, or any information storage and retrieval system, without permission in writing from the publisher.

Grune & Stratton, Inc.
757 Third Avenue
New York, New York   10017

Library of Congress Catalog Card Number 73-92018

International Standard Book Number 0-8089-0640-2

Printed in the United States of America (P-B)

Contents

Preface vii

PART I THE INVERTEBRATES 1
    Chapter 1 Protozoa 3
    Chapter 2 Porifera, Coelenterata, Rotifera 11
    Chapter 3 Mollusca and Annelida 20
    Chapter 4 Arthropoda 24
    Chapter 5 Other Phyla 35

PART II VERTEBRATES OTHER THAN MAMMALS 41
    Chapter 6 Fishes 43
    Chapter 7 Amphibians 50
    Chapter 8 Reptiles and Birds 59

PART III MAN AND MAMMALS 65
    Chapter 9 Skin and Fascia 71
    Chapter 10 The Skeleton 94
    Chapter 11 The Muscular System 101
    Chapter 12 The Blood Vascular System 110
    Chapter 13 The Lymphocyte and Lymphoid Tissue 123
    Chapter 14 The Respiratory System 144
    Chapter 15 The Digestive System 148
    Chapter 16 The Urinary System 172
    Chapter 17 The Reproductive System 183
    Chapter 18 The Endocrine Glands 194
    Chapter 19 The Sense Organs 206
    Chapter 20 The Nervous System 218

PART IV CONCLUSION 239
    Chapter 21 An Overview 241

Index 243

Preface

The title of this book is intended as a generally accurate description of what one may expect to find in it. It deals primarily with changes in the structure of the human and animal body in chronological aging. The structure with which it deals is visible structure—that which can be seen with the naked eye or which the eye can see with the aid of the light microscope or the electron microscope. Such structure, while it lies at vastly different levels of organization—from organism to organ, tissue, cell, and cell organelle—still constitutes the subject matter of modern anatomy. Fascinating as the study of structure which lies beyond this level may be, in the realm of the molecule, the atom, and those far smaller particles which have been made known to us in recent decades, the methods of research and integrative interpretation still remain largely different from those of anatomical investigation. The fact that the borderline is not always definite will not be denied, but it is not our purpose in this work to pass beyond the realm of the visible unless such a step is required in possible explanation of anatomical changes.

Does the title of our book preclude a discussion of function? Certainly not, no more than any rational study of anatomy will preclude such discussion. Structure and function, organization and the activity made possible by that organization, go together in the great realm of animal life and in the human body.

The title does indicate, however, that as in anatomical studies in general, it is structure and structural change on which the emphasis is placed. The research career of the author has been concerned primarily with investigations on this subject, and he has become increasingly aware of the many avenues which are open for work in this field and the wide vistas that are opened by proceeding down these individual avenues.

Interest in the aging process has been combined for the author with a curiosity about comparative structures of animals. Because of this and because of the very probable value of the comparative approach in studies of senescence, we discuss in Chapters 1 to 5 the phyla of the Invertebrata; in Chapters 6 to 8 we discuss the classes of vertebrates other than mammals. While far less is known for many of these groups than for mammals and for man, we have considered it worthwhile to discuss such anatomical data as are available, for a number of careful studies have been made. We have given some attention also in the various groups to the relationship of growth, of reproduction, and of regenerative capacity to the longevity or natural life span.

A very important part of any book of which the theme is anatomical must be its illustrations. We wish to express our sincere appreciation to the many research workers who have given us permission to reproduce figures from their books and papers.

We wish to express also our gratitude for the understanding, the patience, the interest, and the unfailing courtesy shown to us by the Publishers.

October 1970   W. A.

The Anatomy of Aging
in Man and Animals

part I

The Invertebrates

1
Protozoa

At a recent symposium on the Biology of Aging (Krohn, 1966), the following statements were made concerning bacteria and the aging process. "You can certainly say that there is no possible comparison between the death of an individual and the death of a bacterial culture which divides by binary fission" (Orgel, p. 39), and "I do not claim that bacteria age, because they go on by binary fission, and in a sense the whole culture keeps going forever, if an environment is maintained constantly."

The one-celled animals, or Protozoa, the great majority of which reproduce by binary fission, might well be thought to be comparable to the bacterial cells in a culture.

Indeed, Weismann (1882) held that Protozoa are "immortal" and present no phenomenon of natural death such as seen in multicellular animals. The cycles of growth followed by fission would form, then, a part of an endless stream of nuclear and cytoplasmic continuity. Vast numbers of individuals are lost through predation by other organisms and by environmental changes but they would be rescued from "aging" by the simple act of cellular reproduction. Whether there is a beginning of senescence at some time prior to reproduction, a change in metabolism which even might "trigger" cell division, is a subject for interesting speculation, but at present one on which we do not have firm data.

It is true, however, that even at a fairly early period in the history of Protozoology, some investigators had found evidence of a process of deterioration in a *clone*, the series of generations arising from a single individual.

Because of their abundance and ease of cultivation, the ciliates have been used most frequently in these studies. Clones are reared on depression slides and daughter cells are transferred at intervals to fresh slides. Studies by R. Hertwig (1903) and Maupas (1888) in which a number of species were used led them to conclude that after a certain number of *asexual* generations a clone does show a decline both in vitality and in rate of division. Calkins (1906) obtained similar results. These investigators indicated, therefore, that Protozoa, like Metazoa, undergo a senescence and death unless some act of rejuvenation occurs. For the Protozoa this is, according to them, the process of conjugation of two cells with an interchange of nuclear (genetic) material. The comparison with the process of fertilization for multicellular animals is obvious and the whole story, as thus told, would have been a rather satisfying one. Other workers, however, particularly Woodruff (1926) and Enriques (1907) presented evidence that the results of the earlier workers may have been due to inadequate cultural conditions. Thus Woodruff maintained a clone of *Paramecium aurelia* which went for over 15,000 generations without conjugation and which showed no decline of fission rate or vitality. He did find, however, a complicated process of nuclear reorganization taking place in individual cells—*endomixis*. The question naturally arose as to whether this process is substituting for that of conjugation to maintain the vigor of the clone.

The careful studies of Sonneborn (1954, 1955) indicate that for *Paramecium aurelia* there is indeed a definite cycle of clonal aging and that continued existence of a clone is not possible without fertilization, either by conjugation or autogamy. Clones which did not undergo fertilization lived at most for 303 generations. Sonneborn also described anatomical evidence of degeneration in the later generations of a waning clone. These include abnormalities of the micronuclei which are conspicuous at the time of cell division.

Sonneborn states that the "life-cycle of Paramecium, like that of higher organisms, begins with fertilization (conjugation or autogamy), shows progressive changes in vitality and ends by death (of the clone) after 125 days or less." (personal communication, 1969) He investigated the effects of "parental" age on various clones, beginning with a parent from a clone 7 days old and going on by steps to one 80 days old. The frequency of sexually produced offspring clones which he classified as "vigorous" declined from 99.4 per cent for the offspring of the 7 day parents to *zero* when the parents are 80 days and older. The frequency of "subnormal" offspring rose from 0.2 per cent for the 7 day parents to 28 per cent in 67 day parents. The frequency of "lethal" clones (those which die out within 4 days) rose from 0.4 per cent in the offspring of the youngest parents to *100 per cent* in that of the eldest.

It appeared that effects of parental age were transmitted through subnormal offspring to succeeding generations but were not transmitted by vigorous offspring. Even very old individuals, those which paired with others of the same age would give rise only to lethals, can conjugate with young animals and help to produce not only the subnormal but even the vigorous type of offspring. Breeding analysis shows that old-by-young hybrids do have abnormal nuclei. Nuclear change can be seen in the chromosomal aberrations that appear at the time of division.

Sonneborn also (1960) described some very great differences in "length of life" of closely related ciliates, indicating that some clones appear to be practically immortal in the sense of not showing a necessity for the rejuvenative phenomena of either conjugation or endomixis. Such clones, however, have lost their micronuclei and are unable to conjugate, hence they may be said to be sexually and genetically dead (Sonneborn, 1969, personal communication).

The genetic mechanism of the rejuvenative effect of conjugation or endomixis may be in part related to the greater occurrence of inheritable variations after these phenomena. Jennings (1920, 1929) demonstrated this greater variation in his comparison of clones reared from conjugants and nonconjugants in Paramecium. Rejuvenating effects are found in exconjugants of Uroleptus if they are artificially separated even *before* exchange of micronuclei has occurred. The effects then would seem to be due to renewal of the macronucleus, a concept which is in accord with the results of endomixis.

Difference between "parent" and progeny among Protozoa is found in certain groups only. Korschelt (1922) observed such a difference to exist particularly in the Suctoria, as Spirochona, Podophrya, Acanthocystis, and Tokophrya.

There is one group of ciliates, the suctorians, in which a good opportunity is presented for study of age changes in the *individual* protozoan cell.

Tokophrya infusionum (Fig. 1–1), one of the group of suctorians, offers peculiarly favorable material for the study of aging of a single-celled organism. In this animal the cell retains its individuality in the reproductive process, for reproduction is accomplished by endogenous budding, an *embryo* forming within the parent organism and being liberated from it. The parent may survive for 2 or 3 weeks and reproduce a number of times. The aging process in this species has been studied in some detail (Rudzinska, 1958, 1961).

Like other ciliates, Tokophrya has two types of nuclei, a minute micronucleus and a large macronucleus. These both arise from the fusion of two micronuclei

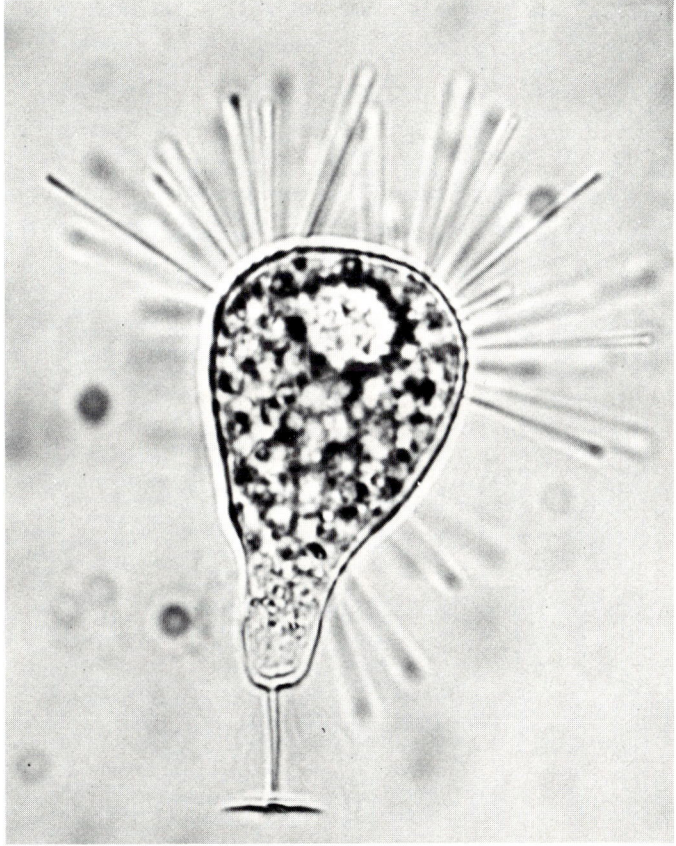

Fig. 1–1. Living organism of adult Tokophrya showing a basal disc, a slender stalk, radiating tentacles, and a large, clear, contractile vacuole in the cytoplasm. ×200 (Courtesy of M. Rudzinska.)

from different individuals to form the synkarya during conjugation. While at the outset they probably are of similar makeup, the macronucleus undergoes rapid growth and changes in structure. The micronucleus decreases in size. During reproductive budding of the organism, the macronucleus divides amitotically, the micronucleus mitotically. Only a small "bud" of the macronucleus goes to the embryo. In conjugation, the macronucleus disintegrates. Our interest, in relation to the description of senescent change, is in the macronucleus.

The macronucleus is spheroid in the young but oblong to irregular (loboid) in the old organism. It contains chromatin bodies each about half a micron in diameter. The number of these bodies is from 40 to 50 in young animals and up to over 300 in old ones. They are highly refractile and Feulgen-positive. They are evenly distributed except along the central part of the macronucleus, forming rows there lateral to elongated areas which are free of granules.

The chromatin bodies of the young animal show a dense, spongy character, homogeneous throughout. In older animals it is not uncommon to discover some of these granules with central cavities. Such granules generally are enlarged. Also, irregularly shaped masses of Feulgen-positive material make their appearance. The hollow granules, with adequate resolution by the electron microscope, frequently show an internal organization, with regularly spaced lines or a honeycomb appearance (Fig. 1–2). The lines are about 120 Å in diameter and the spaces between them measure 230 Å. It is thought that the longitudinally lined appearance represents longitudinal sections; the honeycomb appearance, cross-sections. In the old organism the macronucleus also shows with the electron microscope irregular, wispy masses, with vague boundaries, which show again the same ordered structure as do the cavity-containing chromatin bodies.

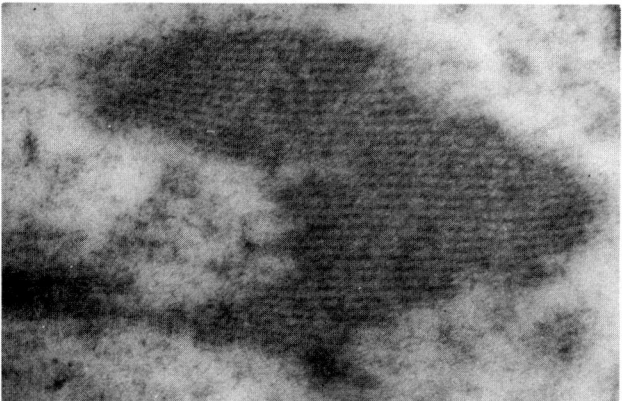

Fig. 1–2. Electron micrograph of one of the chromatin bodies in the nucleus of a senile individual of Tokophyra. In old organisms some of these bodies show a complex pattern, a type of honeycomb arrangement of their substance not seen in any young individuals. ×67000 (Courtesy of M. Rudzinska.)

The macronucleus increases from a diameter of 5μ in young to 20μ in old. Narrow, deep invaginations of the nuclear membrane may be seen in old age.

The macronucleus of the old Tokophrya differs not only in structure but also in behavior from that of the young one, for one often sees a peculiar process of *hemixis* occurring, in which the macronucleus undergoes division one or more times without accompanying division of the cell, so that there may be two to four macronuclei in an old cell. This type of process in various kinds of Protozoa has been considered as probably resulting in a revitalization of the cell, as in autogamy of organisms in old or aging clones (Fauré-Fremiet, 1953). Rudzinska (1955) has compared this type of nuclear division with the amitotic division which we have described in the cells of some tissues of metazoans as a probable means of self-preservation of cells (Andrew, 1955).

In addition to the changes in the macronucleus, the organism changes from pyriform to an irregular shape in old age. The body averages 30μ in diameter in the young and 60μ in diameter in the old *Tokophrya*. The young organism is able to produce an embryo every 2 to 4 hours, while the old one is unable to reproduce. The number of tentacles decreases from 60 in the young animals to 15 or less in the very old.

The cytoplasmic organelles of Tokophrya also show changes with age. In young cells the mitochrondria have many "cristae," in this case actually villous projections from the membrane into the matrix (Fig. 1-3). In old animals (8-10 days) many of the mitochondria appear unaltered but show a greatly reduced number of villous projections. The distribution of mitochondria also changes. In young cells they are distributed rather evenly through the cytoplasm, while in old cells they tend to aggregate near the periphery of the cell.

Of much interest is the fact that "old" individuals of *Tokophrya*, even though they have had less than two weeks for the process, accumulate many pigment granules of a lipid-like material (Fig. 1-4). There is often a close spatial relationship of these granules to individual mitochondria and the pigment material may even occur within mitochondria. The distribution of the pigment, like that of the mitochondria in the old cells, generally is peripheral. As we shall see later, a similar close relation between pigment and mitochondria has been described in cells of higher organisms (Hess, 1955; Duncan *et al.*, 1960).

Ribosomes, about 150 Å in diameter, are abundant in young individuals of Tokophrya, scanty in old ones. A type of particle of greater size, about 300 Å in diameter, occurs in small numbers in young cells, while in old cells it often occurs in large numbers, forming round or oval groupings.

In *Tokophrya infusionum* the most common form of the endoplasmic reticulum is vesicular rather than as cisternae or canaliculi. Vesicles, single or in stacks, are abundant in the cytoplasm of young individuals but undergo a marked decrease with age.

Thus, Tokophrya has added a considerable amount of knowledge to our study of the aging process, has shown that individual single-celled animals show evidences of senescence, and has made possible comparison of certain aspects of aging of single-celled organisms with aging of single cells of many-celled organisms.

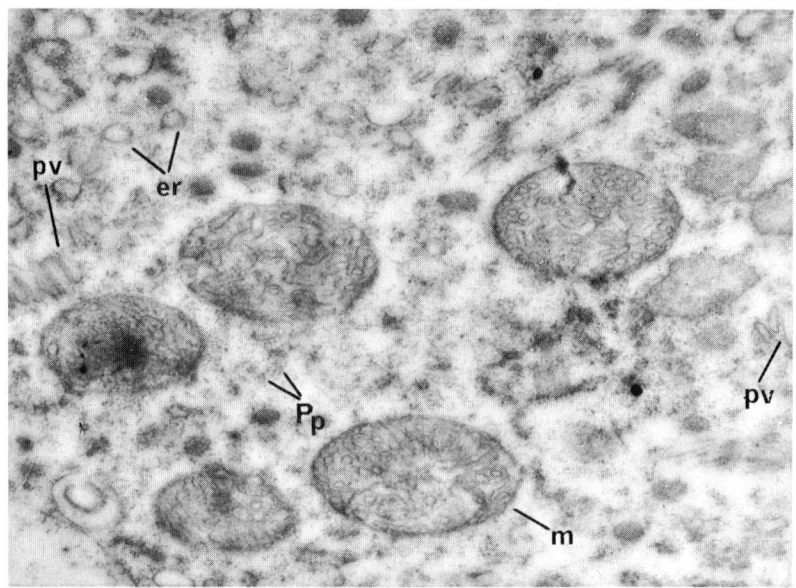

Fig. 1-3. Electron micrograph of portion of young Tokophyra, with mitochondria (*m*), vesicles of endoplasmic reticulum (*er*), rows of vesicles (*pv*), and Palade's particles (ribosomes) (*Pp*). ×31,000 (Courtesy of M. Rudzinska.)

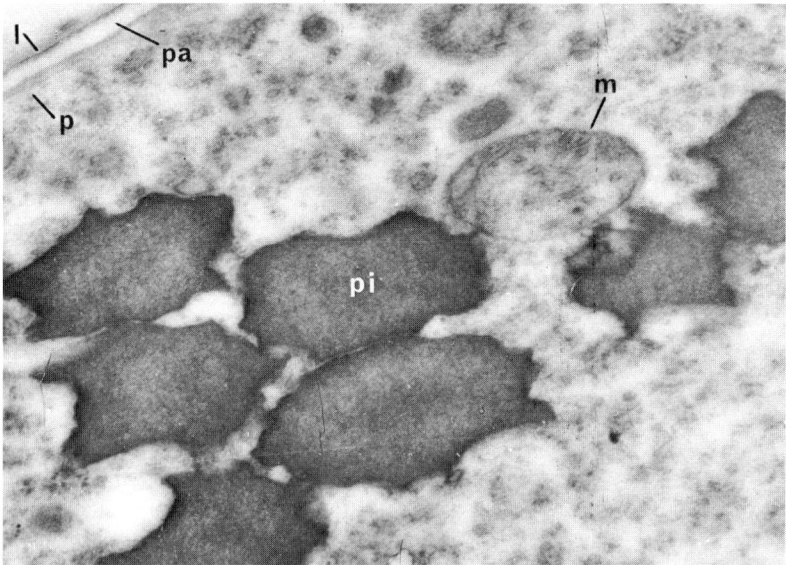

Fig. 1-4. Electron micrograph of portion of old Tokophrya, with one mitochondrion (*m*), a number of pigment bodies (*Pi*), pellicle (*p*), plasma membrane (*pa*), and a layer of homogeneous substance (*l*) just outside the pellicle. ×31,000 (Courtesy of M. Rudzinska.)

Reorganization and Possible Rejuvenation

There is evidence of reorganization processes occurring in the protozoan cell body and nucleus at or near the time of fission. In Ancistruma, a portion of the macronucleus is detached during division of that organ and this portion soon disappears, apparently by absorption into the cytoplasm of the dividing cell. In Aspidisca, a visible reorganization of the interior of the macronucleus occurs just before its division.

The cell body itself shows extensive changes, especially if it possesses complicated sets of appendages, for these disappear during fission (Wallengren, 1901).

These changes in nucleus and cytoplasm are intimately bound up with reproduction. Do they also indicate a change from a condition of senility to one of youth? If we consider senility as a period of decline and of lost powers, this hardly seems to be the case. Calkins (1911) discovered that in Uronychia the young protozoans just after fission have little power to regenerate lost parts and that only in old protozoans close to fission is this power well-developed, an increase in growth energy apparently preceding division. But such an increase does not seem to be present in all species (Moore, 1924), and further studies on this subject are indicated. The increase noted, however, may lead us to speculate as to how closely the morphological features of dedifferentiation are related to rejuvenation or increase of growth potential, or whether such increase actually may precede and help to bring about the morphological changes.

In some cases, at least, reproduction can be delayed almost indefinitely by removing a part of the individual at repeated intervals, as has been shown by Hartmann (1928) who used Stentor and Ameba. The reorganization brought about by loss of part of the cytoplasm seems to have somewhat the same effect as cell division and to enable the animal to continue living. One ameba was thus maintained for 4 months without division.

REFERENCES

Calkins, G. N., 1906. The protozoan life-cycle. New York and Lancaster, Pa.: Science, n.s., vol. 23, pp. 367–370.

―――― 1911. Regeneration and cell division in Uronychia. J. Exper. Zool., vol. 10, pp. 95–116.

―――― 1933. The Biology of the Protozoa. 2d. ed., Philadelphia: Lea and Febiger, 607 pp.

Curtis, W. C., 1928. Old problems and a new technique. Science, vol. 67, pp. 141–149.

Dhar, N. R., 1932. Senescence, an inherent property of animal cells. Quart. Rev. Biol., vol. 7, pp. 68–76.

Enriques, P., 1907. La coniugazione e il differenziamento sessuale negli infusori. Arch. f. Protistenk., Jena, vol. 9, pp. 195–296.

Hartmann, M., 1928. Über experimentelle Unsterblichkeit von Protozoen-Individuen. Ersatz der Fortpflanzung von Amoeba proteus durch fortgesetzte Regenerationen. Zool. Jahrb. Abt. f. allg. Zool. u. Physiol., vol. 45, pp. 973–987.

Hertwig, R., 1903. Über Korrelation von Zell- und Kerngrösse und ihre Bedeutung für die geschlechtliche Differenzierung und die Teilung der Zelle. Centralbl., Leipz., vol. 23, p. 108.

Jennings, H. S., 1920. Life and Death, Heredity and Evolution in Unicellular Organisms. Boston: Richard G. Badger, 233 pp.

——— 1929. Genetics of Protozoa. Bibliog. Genet., vol. 5, pp. 105–330.
Korschelt, E. 1922. Lebensdauer, Altern und Tod. Fischer, Jena.
Krohn, P. L. (Editor), 1966. Topics in the Biology of Aging (The Salk Institute for Biological Studies). New York: Interscience Publishers, 177 pp.
Maupas, E., 1888. Recherches expérimentales sur la multiplication des infusoires ciliés. Arch. de Zool. Exper. Gen., Ser. 2., vol. 6, pp. 165–277.
Moore, E., 1924. Regeneration at various phases in the life history of the infusorians Spathidium spathula and Blepharisma undulans. J. Exper. Zool., vol. 39, pp. 249–316.
Rudzinska, M. A., 1955. Differences between young and old organisms in Tokophrya infusionum. (Abstract) J. Geront., vol. 10, p. 469.
——— and K. R. Porter, 1955. Observations on the fine structure of the macronucleus of Tokophrya infusionum. J. Biophys. Biochem. Cytol., vol. 1, pp. 421–428.
——— 1958. An electron microscope study of the contractile vacuole in Tokophrya infusionum. J. Biophys. Biochem. Cytol., vol. 4, pp. 195–202.
——— 1961a. The use of a protozoan for studies on aging. I. Differences between young and old organisms of Tokophrya infusionum as revealed by light and electron microscopy. J. Geront., vol. 16, pp. 213–224.
——— 1961b. The use of a protozoan for studies on aging. II. The macronucleus in young and old organisms of Tokophrya infusionum: light and electron microscope observations. J. Geront., vol. 16, pp. 326–334.
Sonneborn, T. M., 1930. Genetic studies on Stenostomum incaudatum n. sp. I. The nature and origin of differences in individuals formed during vegetative reproduction. Z. Exp. Zool., vol. 57, p. 57.
——— 1938. The delayed occurrence and total omission of endomixis in selected lines of Paramecium aurelia. Biol. Bull., Wood's Hole, vol. 74, pp. 76–82.
——— 1959. In Advances in Virus Research, vol. 6, Smith, K. M. and M. A. Lauffer, Eds., New York: Academic Press, pp. 229–256.
——— 1960. The human early foetal death rate in relation to the age of father. The Biology of Aging, AIBS, Washington, no. 6, p. 288.
——— 1963. In The Nature of Biological Diversity. J. M. Allen, Ed., New York: McGraw-Hill, pp. 165–221.
Sonneborn, T. M. and M. Schneller, 1955. The basis of aging in Variety 4 of *Paramecium aurelia*. J. Protozool., 2 (suppl.), p. 6.
——— and M. Schneller, 1960a. Physiological basis of aging in Paramecium. Publ. No. 6, *The Biology of Aging*, AIBS, Washington, pp. 283–284.
——— and M. Schneller, 1960b. Age induced mutations in Paramecium. Publ. No. 6, *The Biology of Aging*, AIBS, Washington, pp. 286–287.
Wallengren, H., 1901. Zur Kenntniss des Neubildungs- und Resorptions-processes bei der Theilung der hypotrichen Infusorien. Zool. Jahrb., vol. 15, pp. 1–38.
Woodruff, L. L., 1926. Eleven thousand generations of Paramecium. Quart. Rev. Biol., vol. 1, No. 3, pp. 436–438.

2
Porifera, Coelenterata, Rotifera

Porifera

Sponges have so many distinctive features in comparison with other Metazoa that the original proposals by Huxley (1875) and Sollas (1884) that they should be considered in a separate isolated branch, now have been well accepted (Hyman, 1940) and they are considered as "Parazoa," after Sollas.

Bidder (1932) includes the sponges along with giant trees, cultures of chick cells, and other living things which show that indefinite growth is natural and that limitation on such growth at a definite size is a later evolutionary development.

Sponges grow by spreading and branching in a plant-like manner. Their powers of regeneration are high. Their level of organization is low but the forms and characteristics of the various species, with their specific types of skeletal elements and the cooperation of the types of cells, even though these cells show relatively slight degrees of differentiation, indicate a grade somewhat above that of colonial protozoans.

The cells of sponges, except for the collar-cells or choanocytes which line the gastrovascular cavities, appear to represent only slight modifications of a primitive ameboid cell which corresponds to the undifferentiated connective tissue or mesenchymal cell of higher forms. Even the choanocytes may be considered as little differentiated cells. They have rounded or oval bodies with their bases resting on the mesenchyme and the distal end showing a transparent, contractile collar lying at the base of a single long flagellum similar to those of the flagellate protozoans. The nucleus is located either near the base or the apex of the cell. It is thought that they can differentiate (or de-differentiate) into ameboid cells, sex cells, and perhaps other varieties. Sponges are devoid of nerve cells and indeed of sensory cells, so that responses are direct reactions to stimuli.

The question of aging in organisms such as sponges is naturally a somewhat different one from that in other Metazoa and, due to the manner of growth and regeneration, somewhat allied to that in the plants.

Some data are available on life span. Individuals of the smaller species live from a few months to several years. Fresh-water sponges have been kept in aquariums for a year or more. In nature, they generally die out in the autumn, while new individuals develop in the spring from resistant bodies called gemmules. Certain types of large marine sponges are believed to live for 50 years or more. It is probable, however, that no individual cells in these sponges have a life span even approaching such a duration, since none of them corresponds to the long-lived nerve cells or special sensory cells of higher metazoans.

The longest records of sponges in captivity are those of Arndt (1941) and are for *Adocia alba* (9 years) and *Suberites carnosus* (15 years).

Arndt (1928) expressed the opinion that senescence does not occur in Porifera.

Coelenterata

The phylum Coelenterata, or Cnidaria, contains the first members of what we may call the Metazoa Proper or Eumetazoa. These are animals with definite symmetry and individuality. Their tissues are well developed and they possess more or less definite organs. The body surface is continuous rather than showing pores, and the digestive tube is lined by a regular epithelium of endodermal origin. Members of the phylum frequently show a polymorphism or alternation of generations, from a sessile polyp which reproduces asexually to a free-swimming, sexually reproducing medusa. Where only the polyp form occurs, as in *Hydra*, it may reproduce either asexually or sexually.

A number of authors have discussed whether the coelenterates show a true process of aging. Pearl and Miner (1935) constructed a life table for Hydra, as they had done for Drosophila. For this purpose they used data from Hase (1909).

Comfort (1956) mentions the controversy concerning whether or not Hydra actually ages. Early authors, including Hertwig (1906), Berninger (1910), and Boecker (1914) found that the individuals in their cultures underwent a depression after awhile, and that anatomical changes accompanied the depression. Goetsch, however (1922), with close attention to conditions of culture, was able to keep individuals alive for longer periods—up to 27 months in the case of *Pelmatohydra oligactis*. It was the experience of Gross (1925) that individuals of this species did not live longer than about 1 year, and he spoke of "senile" alterations as early as the fourth month. David (1925) concluded that individuals of *Pelmatohydra oligactis* had a life span of between 20 and 28 months.

Schlottke (1930) felt that David's animals were heavily parasitized. Various kinds of unicellular organisms are found as intracellular parasites in Hydra. Nevertheless, Schlottke did believe that there are age changes, which included, according to him, pycnosis of nuclei of the ectodermal cells, a removal of some of these cells, and an accumulation of *guanine* deposits, apparently the remains of cells. He noted also that degenerating nematocytes, as well as ordinary ectodermal cells, tend to be moved into the endoderm.

Schlottke believed that all of the cell types in Hydra are replaced continuously from a reserve of interstitial cells, an opinion which seems to receive support from the work of Brien (1953).

The case of the hydranths of colonial hydroids, however, appears to show clearly an involution and a resorption (Huxley and de Beer, 1923).

Whether or not the coelenterates may be considered as metazoan organisms, the bodies of which are potentially immortal, is a subject that has been discussed both by earlier and more recent investigators. Ashworth and Annandale (1904) made a series of observations on the anemone, *Cereus pedunculatus* (earlier classified as *Sagartia troglodytes*). There were 16 of the animals, and they were at least 50 years old at that time. They were transferred later from the care of an individual to the Edinburgh Zoo where they died, apparently all in one night, in 1942. Hence they had reached an age of almost 90 years!

Brien (1953), used marking techniques and described what he considered a continued formation of new cells in the hypostomal region of Hydra, followed by slow migration of cells over the body surface of the animal to its foot where death and resorption occur. This would be a process reminiscent of the migration of intestinal epithelial cells up the sides of a villus, in that case to be discarded at the apex.

The aging process that occurs in many protozoan clones, where some rejuvenative process such as conjugation or endomixis must supervene if the clone is to survive, does not seem to be present in the lines of coelenterates in which the only form of propagation is asexual budding, since such lines of Hydra survive in the laboratory for *decades* without sexual reproduction. Studies by Crowell (1953, 1957) and Crowell and Wyttenbach (1957) and by Strehler and Crowell (1961) indicate that the colonial hydrozoan *Campanularia flexuosa* is able to propagate indefinitely without sexual reproduction. This is not to say, however, that the individual hydranths do not "age." Indeed, Strehler (1961) describes a definite series of steps in senescence of the hydranth (Figs. 2–1, 2–2). First there is a slight decrease in length of the tentacles, with an appearance of "knobiness" and a change in the refractive index. The tentacles continue to shorten until they have drawn in far toward their bases. After complete contraction of the tentacles, phenomena occur which seem to represent a breakdown of intercellular cement and an actual autolysis of cells. As a final stage, the whole tissue of the degenerated animals is passed back into the major stem and the hydrotheca or skeletal cup is left empty.

In this sort of "senile" change, it would seem that an approach to the explanation of the basic mechanism of degeneration might be more feasible than in the process of senescence of the far more complicated body of the higher organisms. Strehler (1961) found that the enzyme, acid phosphatase, which often is associated with lysosomes (Essner and Novikoff, 1960) increases greatly in quantity in the regressing animal. In young hydranths of Campanularia it is practically confined to the nuclei. In the senescent hydranth there are great numbers of acid phosphatase-positive granules in the cytoplasm as well. It may be that an activation of lysosomes, rich in enzymes, is the major process involved in regression of the hydranth, but further investigation of this subject is needed.

Another enzyme which shows a quantitative change with age in the hydranths is adenosine triphosphate (ATP). The ATP level decreases in 5-day hydranths to about one-third of that found in 1-day hydranths. The value was calculated per hydranth rather than on a dry weight basis and it is not known whether there is a change in dry weight of Campanularia during these days.

Thus, the coelenterates are of considerable interest in studies on the aging process, both from the evidence that some of them have a sort of potential immortality and from the opportunities to study in these relatively simple organisms what may be some fundamental processes in senescence.

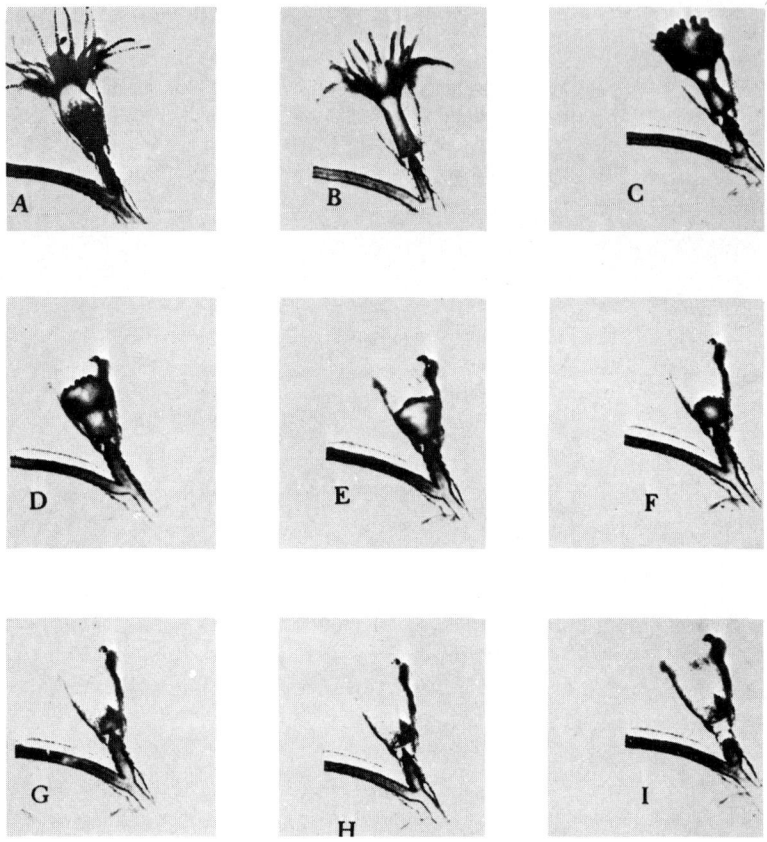

Fig. 2-1. Senescent regression of hydranth of Campanularia (taken from a time lapse sequence). Progressive shortening of tentacles prior to their resorption is seen. Time period covered is approximately 6 hours. (After Strehler, 1961.) Low magnification.

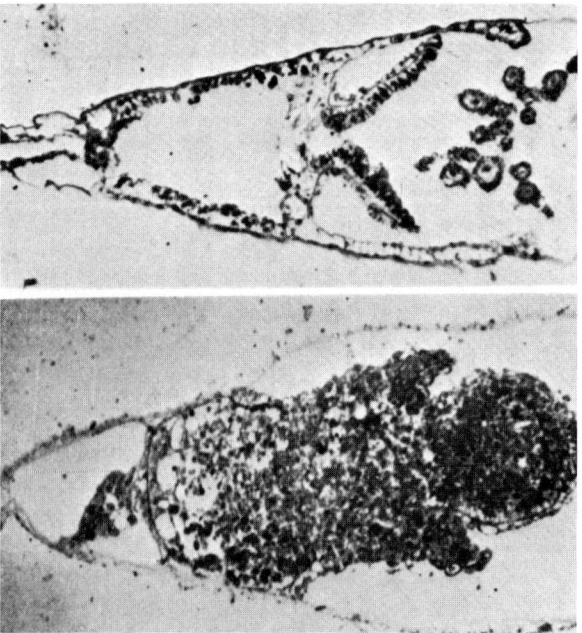

Fig. 2-2. Regression of Campanularia hydranth as seen in sections. A. Normal young hydranth. B. Senile hydranth. Magnification of both A and B is ×500. (After Strehler, 1961.)

Rotifera (Phylum Aschelminthes)

The rotifers are small multicellular animals with a rather closely determined total number of cells, or rather nuclei, and even a definite number for each organ and system. They belong to the pseudocoelomates in which no peritoneal epithelium separates the cells of the visceral organs from the fluid of the false body cavity or pseudocoelom. The anterior end consists of a ciliary apparatus, the corona, with a differentiated pharynx with movable pieces or "jaws." The protonephridia or excretory organs are of the typical flame-bulb type. In spite of their considerable degree of complexity, they are smaller than many of the Protozoa. Their minute size seems to preclude the need for a circulatory or respiratory system.

The majority of rotifers show a syncytial condition as adults. In those which have been studied, the total number of nuclei is from 900 to 1000. In a common species, *Epiphanes senta* (Martin, 1912), the numbers according to organs and parts are as follows: brain, 183; peripheral nervous system, 63; mastax (jaw) nerve cells, 34; mastax musculature, 42; epithelium of mastax, 91; esophagus, 15; stomach, 39; each of two gastric glands, 6; intestine, 14; each protonephridium, 14; oviduct, 3; vitellarium, 8; pedal glands, 19; body wall circular muscle, 22; retractor muscle, 40; coronal epidermis, 192; and trunk and foot epidermis, 108.

Powers of regeneration appear to be lacking, or almost so, in rotifers. From the theoretical standpoint, it would seem probable that animals with constant numbers of cells or nuclei would be unable to regenerate parts since the nuclei have ceased to divide.

Differences in rotifers at different ages can be seen in the living animals. In very old animals the cilia heat feebly, the gastrointestinal tract becomes less active, and the whole appearance of the organism is one of decline of vigor.

The accumulation of calcium in aging rotifers was demonstrated by Lansing (1942); and he actually succeeded in prolonging their life by putting them in dilute solutions of citrate.

Signs of aging in rotifers have been described by a number of authors (Metchnikoff, 1907; Spemann, 1924; Jennings and Lynch, 1928; Szabo, 1935). The individual rotifers grow sluggish. Speed of ciliary beat is diminished. They crawl rather than swim, their responses to stimuli are slow, and they become opaque and granular in appearance. Accumulation of pigment occurs in the mastax, gut, and digestive gland.

The study of rotifers has made a very considerable contribution to the general knowledge of the aging process. Lansing (1948) demonstrated the presence of a transmissible, cumulative, and reversible factor in the rotifer ovum which acts to accelerate the aging process. This factor does not exist, or at least does not express itself, in growing, adolescent rotifers but makes its appearance as full growth is reached. Thus, the older the parent, in general, the less the longevity of the offspring.

Miner (1954) has reviewed the important subject of the general effects of maternal age on offspring in various groups of animals.

The evidence for an accumulative factor in senescence of the rotifer is fairly strong. Whether accumulation, such as occurs in many cells and tissues of higher organisms in relation to pigment and to calcium (Lansing, 1951) is itself a primary cause of aging or a consequence, which itself may be reversible, is an unanswered question.

In recent anatomical studies, Lansing (1964) has used the species *Philodina citrina*, which contains approximately 1,000 cells, has a life-span of 30 days, and reproduces by parthenogenesis, the eggs hatching in 1 day. Growth of the newly hatched rotifer is by rapid increase in cell size. The individual attains its maximum size and sexual maturity between days 5 and 6. The period of adult vigor extends from this time to days 10 and 12. Then a gradual decline sets in and death generally occurs between 24 and 30 days.

The animals for study of age changes in fine structure were raised in artificial pond water at 20°C and were fed the alga *Chlorella vulgaris*.

Fixation was in cold, unbuffered osmium tetroxide, and embedment in Vestopal.

Sections were stained with lead by the Karnovsky technique.

Sections of young and old animals showed significant age changes in ultrastructural features (Figs. 2–4, 2–5). The epithelial cells of the stomach of rotifers 1 to 6 days old show an abundance of ribosomes distributed through the cytoplasm singly or in clusters. Rough endoplasmic reticulum is distributed irregularly through the cell. Lipoid bodies of moderate electron density are present and appear gray in electron micrographs.

The epithelial cells of the stomach of an old rotifer, 21 days or more, show very few free ribosomes and a sparse rough endoplasmic reticulum. The lipoid bodies show a very great electron density and appear black in electron micrographs.

Stomach cells of both young and old rotifers show food vacuoles in which intracellular digestion is proceeding (Fig. 2–5). These are seen to begin as pockets at the luminal surface.

A striking difference between young and old rotifers is seen also in the cortical region of the stomach cells. In very young rotifers single cortical membranes frequently are found just subjacent to the cell membrane. With rare exceptions, such membranes have ribosomes in association with them. Between 3 and 6 days there is a progressive increase in frequency of smooth-surfaced and stacked cortical membranes near the cell surface with a corresponding decrease of the long unstacked ribosome-associated cortical membranes. Thus, from 7 to 21 days there is a steady increase in the numbers of stacked, smooth cortical membranes and a steady decrease proceeding to a virtual disappearance of the ribosome-associated single cortical membranes.

Lansing says that although the extensions of the cell membrane into the cortical cytoplasm of the epithelial cells of the rotifer gut do not fit into the strict definition of endoplasmic reticulum, he is referring to such extensions as cortical membranes continuous with the cell membrane. The smooth-surfaced, trilaminar membranes stacked in approximately parallel arrangement in adult rotifers appear to originate from invaginations of the cell membrane bordering the pseudocoelom. Lansing believes that the frequent occurrence of cortical membranes with a smooth distal wall and a rough proximal wall (distal and proximal with respect to the cell interior) warrants consideration of the possibility that the rough reticulum is not only continuous with the cell membrane, but also may be derived from appropriately modified cell membranes.

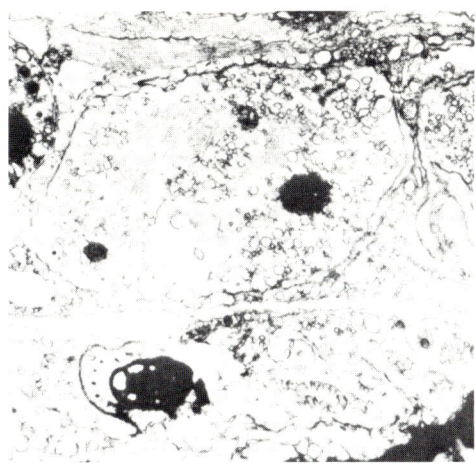

Fig. 2–3. Electron micrograph of regressing Campanularia hydranth. Breakdown of cytoplasmic organization and a marked vacuolation are seen. Magnification about ×24,000. (After Strehler, 1961.)

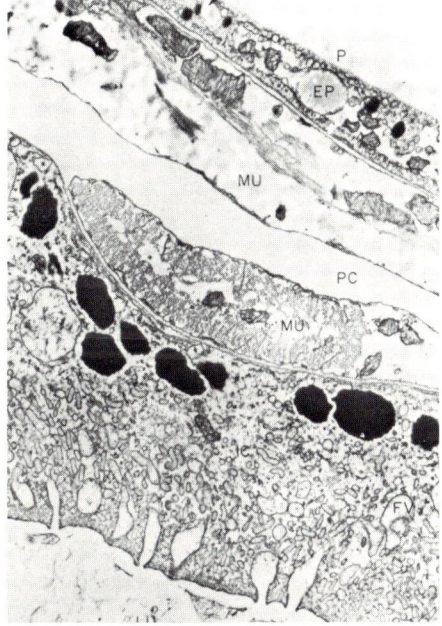

Fig. 2–4. Portion of a cross section through the body of an old rotifer, an individual 21 days of age. The outermost layer is the pellicle, P. Beneath it is the integumentary epithelium, EP, and deep to that the circular muscle, MU, which shows mitochondria and much amorphous material considered to be glycogen. Between body wall muscle and alimentary tract muscle is the pseudocoelom, P.C. The lowest layer is the epithelium of the stomach, conspicuous by the presence of many dark, electron-dense lipid bodies which appear in old rotifers. Stomach epithelium here differs from that of young rotifers in showing very little rough endoplasmic reticulum and a general paucity of ribosomes. ×24,000. (After Lansing, 1964.)

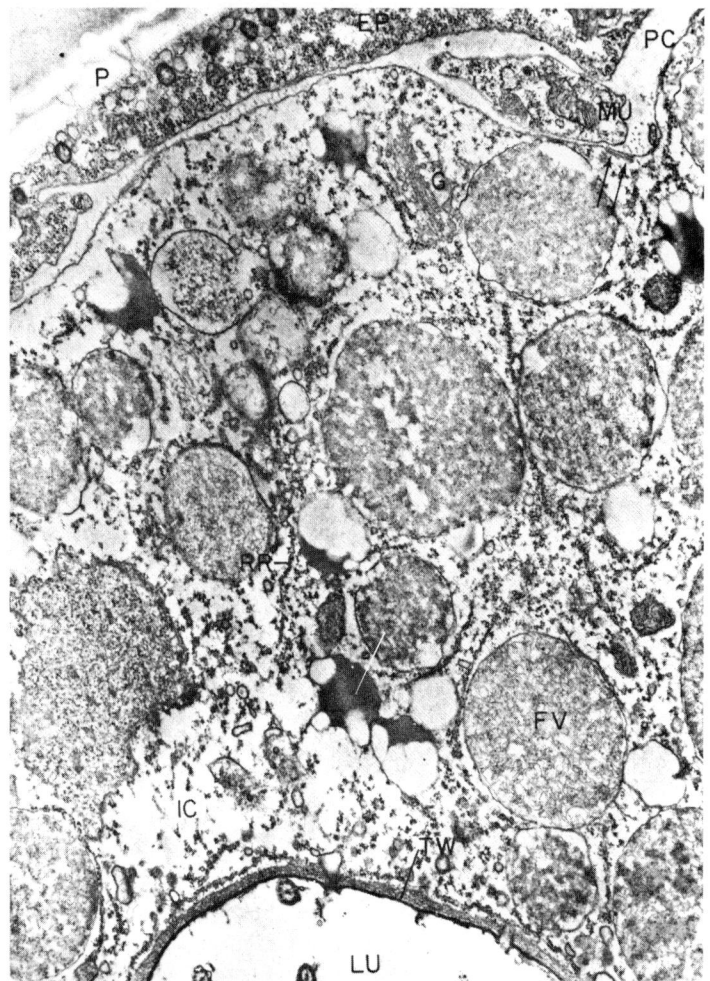

Fig. 2–5. Cross section through a portion of a young rotifer, an individual two days old. The pellicle, P, is seen at upper left. Beneath it is the integumentary epithelium, EP. A small portion of muscle, MU is shown, as well as a narrow section of pseudocoelom, PC. Most of the field is occupied by stomach (intestinal) epithelium, IC, containing large food vacuoles (FV). Rough endoplasmic reticulum (RR) is abundant. Golgi apparatus, G, and a terminal web, TW, are shown. The double arrow points to a smooth cortical membrane, not common in rotifers this young but common in adult rotifers. ×24,000. (After Lansing, 1964.)

Lansing exphasizes the fact that very young and actively growing rotifers show a much less elaborate pattern of cortical membranes than do adults. In the young animal the cortical membranes stain poorly with lead and it is very uncommon to find the type of system of well-defined, smooth-surfaced, and trilaminar membranes which is common in the adult and older rotifers. Lansing believes that protein synthesis is occurring at a higher rate in the young; that the infoldings of the cell membrane are being converted rapidly into rough endoplasmic reticulum by the attachment of ribosomes, and that chiefly for this reason, only single (as opposed to stacked) cortical membranes are found just beneath the cell membrane. In the adult and older rotifer the infoldings of the cell membrane probably continue to occur but they, and the synthesis of protein, proceed at a slower rate. Thus ribosome attachment would be retarded and the smooth cortical membranes would pile up or "stack," while only those lying well within the cell would show attached ribosomes. Whether the fact that fewer ribosomes are available in the older rotifer, or that the rate of attachment of individual ribosomes is the chief factor in the slow rate of formation of rough endoplasmic reticulum, remains a question.

The study by Lansing seems a significant one in demonstrating changes in cellular synthetic activities with increasing age.

REFERENCES

Arndt, W., 1941. Lebendbeobachtungen an Kiesel- und Horn-Schwammen des Berliner Aquarium. Zool. Gart., Lpz., vol. 13, p. 140.

Ashworth, J. H. and N. Annandale, 1904. Observations on some aged specimens of Sagartia troglodytes and on the duration of life in Coelenterates. Proc. Roy. Soc. Edin., vol. 25, p. 295.

Brien, P., 1953. La perennité somatique. Biol. Rev., vol. 28, pp. 308–349.

Crowell, S., 1953. The regression-replacement cycle of hydranths of Obelia and Campanularia. Physiol. Zool., vol. 26, pp. 319–327.

——— 1957. Differential responses of growth zones to nutritive level, age, and temperature in the colonial hydroid, Campanularia. J. Exp. Zool., vol. 134, pp. 63–90.

——— and C. Wyttenbach, 1957. Factors affecting terminal growth in the hydroid Campanularia. Biol. Bull., vol. 113, pp. 233–244.

David, K., 1925. Zur Frage der potentiellen Unsterblichkeit der Metazoen. Zool. Anz., Leipz., vol. 64, pp. 126–134.

Goetsch, W., 1922. Beiträge zum Unsterblichkeit der Metazoen; ii, Lebensdauer und geschlechtliche Fortpflanzung bei Hydren. Biol. Zbl., Leipz., vol. 42, pp. 231–240.

Gross, J., 1925. Versuche und Beobachtungen über die Biologie der Hydriden. Biol. Zbl., Leipz., vol. 45, pp. 385–417.

Hase, A., 1909. Uber die deutschen Susswasser-polypen Hydra fusca. Arch. für Rassen- und Gesellschafts-biologie, vol. 6, pp. 721–753.

Hertwig, R., 1906. Über Knospung und Geschlectsentwicklung von Hydra fusca. Biol. Zlb., vol. 26, pp. 469–508.

Huxley, J. S. and G. R. DeBeer, 1923. Studies in dedifferentiation-IV. Resorption and differential inhibition in Obelia and Campanularia. Quart. J. Micr. Sci., vol. 67, p. 473.

Jennings, H. S. and R. S. Lynch, 1928. Age, mortality, fertility and individual diversity in the Rotifer Proales sordida Gorse. II. J. Exp. Zool., vol. 51, p. 339.

Lansing, A. I., 1948. The influence of parental age on longevity in rotifers. J. Geront., vol. 3, p. 6.

——— 1964. Age variations in cortical membranes of rotifers. J. Cell Biol., vol. 23, pp. 403–422.

Miner, R. W. (Ed.), 1954. Parental age and characteristics of the offspring. Ann. N.Y. Acad. Sci., vol. 57, p. 451.

Pearl, R. and J. R. Miner, 1935. Experimental Studies on the Duration of Life. XIV. The comparative mortality of certain lower organisms. Quart. Rev. Biol., vol. 10, pp. 60–79.

Schlottke, E., 1930. Zellstudien an Hydra. I. Altern und abbau von Zellen und Kernen. Z. Mikr. Anat. Forsch., vol. 22, pp. 493–532.

Spemann, F. W., 1924. Über Lebensdauer, Altern und andere Fragen der Rotatorienbiologie. Z. wiss. Zool., vol. 123, p. 1.

Strehler, B., 1961. Aging in Coelenterates. In The Biology of Hydra, Coral Gables, University of Miami Press. H. M. Lenhoff and W. F. Loomis (eds.). pp. 373–398.

3
Mollusca and Annelida

Mollusca

The molluscs present a great diversity of form, size, and way of life. There are some 70,000 existing species. In numbers of individuals some species, such as those of the oyster, run into many billions while others appear to be very rare. Most of the molluscs have a shell, but many groups lack one or show only an inconspicuous remnant.

Molluscs in the wild state may be divided into annual species, which include many small freshwater species and nudibranchs, plurennial species with a short reproductive life, and plurennial species with a long reproductive life, some of the latter probably having an open-ended or indeterminate life-span.

Some molluscs show in their shell the passage of time. Such is the Chambered Nautilus (Nautilus pompileus) of the Indian and Pacific Oceans.

The animal lives in a large chamber at the aperture, or opening. Behind that chamber a number of partitions have divided the cavity of the shell into other chambers, now tenantless and empty. These empty chambers are earlier dwelling places of the Nautilus, onto which it has successively built larger and larger chambers. Such a manner of life inspired Oliver Wendell Holmes to make a comparison with human aspirations when he wrote:

> "Build thee more stately mansions, O my
>   soul
> As the swift seasons roll!
>   Leave thy low-vaulted past!
> Let each new temple, nobler than the last
>   Shut thee from heaven with a dome more vast,
> 'Til thou at length art free
>   Leaving thy outgrown shell by life's
> unresting sea!"

Perhaps the scientist too may at some time be inspired by this life in clear-cut stages to make a further study of the tissues of the Nautilus as it grows older!

There are cases among the *Mollusca* where reproductive exhaustion appears to bring on senescence, somewhat as in the salmon among the fishes. Pelseneer (1932) has described a method of determining age in molluscs by using sections of statoliths and he has commented on exhaustion and death in relation to reproduction.

There has been much disagreement on the reliability of growth rings of the shell as determining age. Annual rings as laid down in a particular species in one locality may fail to be well defined at another. It appears also that any check to growth that leads to a drawing back of the edge of the mantle may produce rings.

According to Haskins (1955), the data from such rings in pelecypods can be accepted as a measure of chronological age if studies are confined to places and species where the yearly production either of a single ring or a definite pattern of rings can be confirmed by experiment.

Specimens of the American oyster (Crassostrea virginica Gmelin) which are of marketable size generally measure 10–15 cm. (4–6 in.) in their greatest dimension which is known as the "height" because of the normally upright position of the shell. Depending on the place of origin, oysters of such size are considered as 3, 4, or 5 years of age.

The oyster generally does not stop growing after attaining a certain size. Rather, growth continues in all directions and the shell may reach considerable proportions. The very old oysters are found on oyster grounds not disturbed by commercial fishing. Galtsoff (1964) says that the largest oyster in his collection had a height of 20.6 cm. (8.1 in.). The total weight was 1230 g., the shell weighing 1175 g., the "meat" 35.8 g., and the balance of 19.2 g. being the sea water retained between the valves. Figure 3–1 gives a profile view of one valve of the shell of an old specimen of C. virginica. In some species, shallow cavities or chambers may be formed (Fig. 3–2). It is thought by a number of authors that these are caused by a gradual shrinkage of the body of the oyster, but this is by no means certain.

Because of the difficulty of determination of age and of knowing whether one is dealing with really "senile" animals, it is not easy to assess the value of anatomical findings in molluscs. Szabo and Szabo (1931a,b; 1934) have studied histological changes in the nervous system and some other parts of some species of the slug, Agrolimax, which may offer interesting material for future investigators working with carefully controlled aquarium material.

Among the changes mentioned by them is deposition of pigment in nerve cells of Agrolimax and Helix.

Annelida

Although the annelid worms are not very conspicuous animals, living either under the ground or in the sea, there are some 7,000 existing species and they probably represent some of the basic types from which the very conspicuous and successful arthropods arose.

The annelid worms, which include the earthworms, many marine worms, and the leeches, are notable for the highly segmental nature of their organization. Organs which usually are single or paired in higher animals, (such as heart and kidneys) in this phylum show several or many repeating units, one or a pair for each segment of the body.

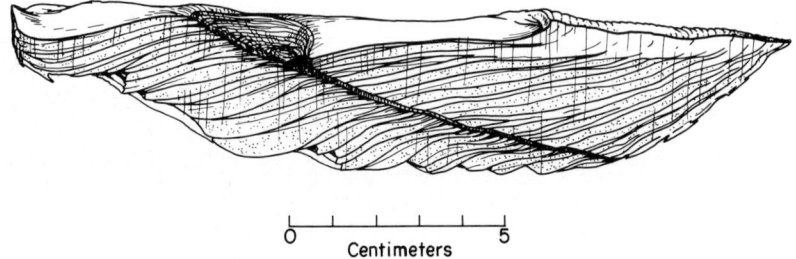

Fig. 3–1. Left valve of the shell of an old individual of *Crassostrea virginica*, the American Oyster, cut along the principal axis of growth. Chalky areas on either side of the hypostracum (the dark platform for attachment of the adductor muscle) are enclosed in thin layers of a hard crystalline material. Hinge would be at the right side of figure. Natural size. (Courtesy of Paul S. Galtsoff, from *The American Oyster*, 1964.)

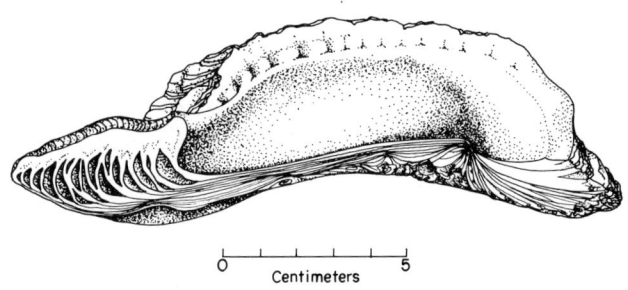

Fig. 3–2. Shell of *Ostrea iridescens* cut at right angles to the hinge. A series of empty chambers is seen at the hinge area on the left. (Courtesy of Paul S. Galtsoff, from *The American Oyster*, 1964.)

Along with the segmental arrangement goes a high power of regeneration, such that individual segments with their component parts are easily replaced.

Stolte (1924, 1927) studied the microscopic changes with age in Nais. He described a degeneration of visceral ganglia, an apparent termination of the production of the primitive reserve cells, and a loss of the normal zonation.

A description of certain degenerative changes of wide distribution in the body of the common earthworm, *Lumbricus terrestris*, has been made by Cooper and Baculi (1968). These changes occurred in animals in captivity, at least 18–20 weeks having passed while they were on an empirically devised medium. They were judged to be about 3 years or older when obtained from a commercial source.

In this "syndrome" the worms lost weight, decreased in length, and were sluggish.

Conspicuous alterations occurred in the body wall and in the organs, including atrophy of muscle, fibrosis, and deposits of a substance in the large blood vessels and the nephridia. The hyaline material stained with congo-red and appears similar to the amyloid of higher forms. Further work is indicated. The study seems to us to emphasize two things: (1) the great gaps in our knowledge of what, chronologically, constitutes aging in some very common invertebrate animals, and (2) the interesting possibilities for anatomical study in a form such as Lumbricus which has many tissues bearing a close resemblance to those of the vertebrates.

There is a great lack of literature on age changes in the annelids and with the great powers of regeneration, these might be thought to be minimal. There are, however, a number of tissues, such as those of the nervous system and in marine worms of the well-developed eyes, in which the cells have a relatively long life span and which might prove fruitful objects of study. The probable phylogenetic relationship of annelid worms to the arthropods would add to the interest of such studies.

REFERENCES

Cooper, Edwin L. and Buena S. Baculi 1968 Degenerative changes in the annelid, *Lumbricus terrestris*. J. Geront., vol. 23, pp. 375–381.

Galtsoff, P. S., 1964. The American Oyster Crassostrea virginica Gmelin. Fishery Bulletin, vol. 64, 480 pp.

Haskin, H. H., 1955. Age determination in molluscs. Trans. N.Y. Acad. Sci., vol. 16, pp. 300–304

Stolte, H. A.,1924. Altersveränderung bei limnicolen Oligochäten.Verhand. Deutsch. Zool. Ges., vol. 29, p. 43.

——— 1927. Studien zur Histologie des Altersprozesses. Z. Wiss. Zool., vol. 129, p. 1.

Szabo, I., 1935. Senescence and death in invertebrate animals. Riv. Biol., vol. 19, p. 377.

——— and M. Szabo, 1931a. Todesursachen und pathologische Erscheinungen bei Pulmonaten. II. Hautkrankheiten bei Nacktschnecken. Arch. Molluskenk, vol. 63, pp. 156–160.

——— 1931b. Histologische Studien über den Zusammenhang der verschiedenen Alterserscheinungen bei Schnecken. I and II. Z. vgl. Physiol., vol. 15, pp. 329, 345.

——— 1934a. Alterserscheinungen und Alterstod bei Nacktschnecken. Biol. Zbl., vol. 54, pp. 471–477.

4
Arthropoda

Not all of the arthropods are short-lived creatures. Some large beetles are said to live over 10 years (Labitte, 1916), the Tarantula as much as 20 years (Baerg, 1945), and the lobster Homarus perhaps 50 years (Herrick, 1911).

Crustacea

A very large amount of information has been published concerning the physiology of the Crustacea (Waterman, 1960).

It is among the lower Crustacea, in particular the Entomostraca, that important information on life span and on various factors which affect it are available. Earlier workers (Anderson and Jenkins, 1942) gathered information on the life history of the water-flea, *Daphnia magna*. They state that in the female the period of youth or juvenescence ends when the ovaries become visible. Adulthood then starts in the fifth, sixth, or seventh instar.

If the animal lives long enough, there is a period of senility, marked by an end of egg laying even when the animal is living under optimal conditions (Banta, 1939).

Schulze-Röbbecke (1951) described microscopic changes in the muscles, intestinal epithelium, and fat body of Daphnia. This author found that Daphnia continued to produce young, although in decreased numbers, until death, which seemed to come finally as the result of nutritional failure caused by alterations of the digestive system.

According to the studies of Meijering (1958) and Fritsch (1958) heart beats, and only heart beats, lend themselves as quantitative means to measure aging in Daphnia. The moults and the heart beats follow different time "places" in the male and female Daphnia, but instars of the same number, i.e., corresponding instars in the same sex, are equal in their number of heart beats. Although the males reach only about half of the female life span (23 days with 8 instars as compared with 62.5 ± 8.8 days with 20 instars) the male life span appears to be proportional to that of the females as the position of the mortality maximum is concerned. The maximum duration of life in the males was 45 days, comprising 14 instars and 30.3 million heart bears. In the female the maximum duration of life was 30 instars with 47.2 million heart beats (Meijering, 1958).

The present author has carried out studies on the relationship between the parthenogenic production of young and the longevity in *Daphnia longispina*. The reproductive lives of 391 females were studied. The animals were observed daily and the young counted and transferred, by pipetting, to fresh medium. The inference was made that the number of young produced in a unit of time, or a particular portion of the life span, as compared with the number produced throughout life, is a fair measure of the "vital capacity" of the dynamics of the animal. The above inference seems to be justified during that unit of time, since reproduction in these femaes (after maturity is reached) is not complicated by the onset of any definite period of sterility.

The animals were divided into 11 groups according to the duration of life in days: (1) 12–17, (2) 18–23, (3) 24–29, (4) 30–35, (5) 36–41, (6) 42–47, (7) 48–53, (8) 54–59, (9) 60–65, (10) 66–71, and (11) 72–77 days. For the individuals in each group the life was then divided into six equal periods. In the group living 18–23 days, for instance, each period or sixth of the life is 3 days, any excess days being automatically dropped from the time before reproduction begins. In the 72–77 day groups each of the six periods is 12 days in length. This division of the life was designed to show whether in the animals which attained or came close to attaining a "normal" life span, a process of decline or senescence could be detected.

The nature of the results is seen in Fig. 4–1. We can see that smaller percentages were produced in the last periods of life in the long-lived animals as compared with the short-lived ones, i.e., that there is a physiological as well as a chronological aging.

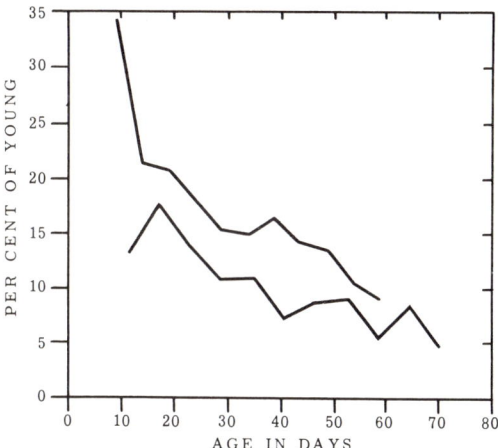

Fig. 4–1. The two lines of this graph show the difference in numbers of young produced in the last two periods of life among the different longevity groups of *Daphnia longispina*. Each point on the graph indicates a longevity group on the abscissa and a per cent of young (total young produced during the lifetime) on the ordinate. The upper line represents the reproductive rate during the fifth period for the different longevity groups and the lower line the reproductive rate during the sixth and last period for each group.

The shortest-lived mothers, whatever the cause of death, still showed a high productivity in the last two periods of their lives, while the longest-lived mothers, those which "had time" to senesce, showed a very low productivity in these last two periods.

26   *The Invertebrates*

Study of the oil globules in the bodies of a separate series of 25 animals through their complete life cycles showed these appearing first on the second day of life as very small structures along the intestine, increasing in size and already beginning to concentrate in the ovary on the third day (Fig. 4–2). The next day the ovary assumed a dark color due to deposition of yolk which obscured the globules. On the fifth day eggs or embryos were seen in the brood-chambers. Whenever eggs had been laid or delivered into the brood-pouch, the ovary contained few or even no globules.

In the later days of life of these 25 animals, of which the longest-lived continued for 55 days, there were marked disturbances in the pattern of globule accumulation in the ovary and their appearance in the embryos. It may be that such disturbances are a primary cause of the decline in reproductive capacity.

Studies on Drosophila

Investigators in the broad biological field have recognized that the insects offer a number of advantages for gerontological research. They have a relatively short life-span, measurable in months, weeks, or even days; their small size makes it possible to raise and maintain large numbers of specimens at small cost; genetically well-known strains are available; and the wide variety of habits and habitats

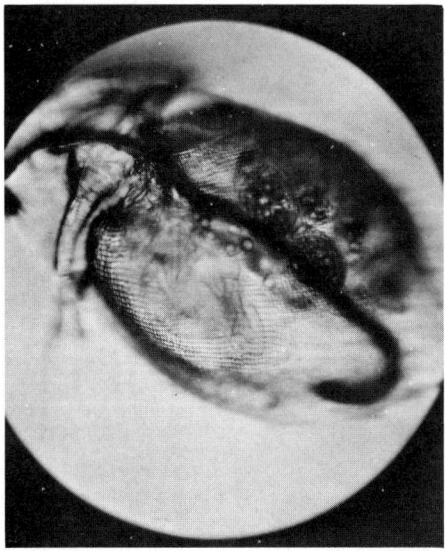

Fig. 4–2. A young female specimen of *Daphnia longispina*. Fat droplets have begun to appear along the intestinal tract and are seen within each developing embryo in the brood pouch. In senile individuals disturbances of the accumulation and transport of this food material may be a cause of the decrease in fertility. Low magnification.

makes important comparisons of types and rates of age change feasible. An insect such as Drosophila about which so many data have been compiled, would seem to be especially favorable.

The fat body of Drosophila has been the subject of study by Miquel *et al.* (1969). Oregon R wild type male flies, after emerging as adults, were kept in groups of 10 per vial under standard conditions. Mortality data were obtained by counts of the numbers decreased at various intervals in a group of 150 animals. Histological study was made of a very large number of specimens, with sacrifice of 30 flies being made at the following ages after emergence from the pupal case: 2, 6, 15, 19, 30, 50, 70, 84, 96, 100, and 104 days. Sections of the flies were prepared in various ways to show the content and histochemical reactions of the fat bodies as studied with the light microscope. In addition, electron microscope studies were made on flies of 5, 15, 76, and 91 days. For these specimens, buffered glutaraldehyde was used in fixation and epon for embedding, with staining by the double method of uranyl acetate and lead hydroxide.

In very young flies (2-15 days old) the cells of the fat body were stained almost homogeneously with amido black, indicating an abundance of basic protein. At 30 days of age the cells of the fat body showed a honeycombed structure which the investigators interpreted as indicating the presence of globules of fat separated by narrow partitions of cytoplasm. Beyond 30 days in the abdominal fat bodies and beyond 50 days in the cephalic fat bodies the cytoplasm of the cells showed a granular appearance, the granularity seeming to increase with age. There appeared to be a gradual disorganization of the tissue with increase in extracellular space, weakening of the tinctorial affinities of the cell membranes for amido black, and probable loss of some cells by lysis.

Study of the glycogen content of the cells, using the dimedone-PAS method, showed a great abundance of this substance in flies of 2 days. Some diminution was seen in flies 30-70 days of age and the glycogen was not spread through the entire cell.

In some of the 100-day-old flies a striking loss was seen, with some areas completely lacking in glycogen (Fig. 4-3).

The lipid content of the cells included some large fat droplets in very young adults. In 15-day-old flies all of the droplets are small in size but abundant. Finally, in old flies, there were again some large droplets but these were unlike those in young flies in being irregular in form and staining more weakly.

Study with the electron microscope showed many large lipid bodies and considerable numbers of lysosomes in the old flies. These investigators (Miquel, 1968) believe that the large numbers of lysosomes are due to an accumulation through the life of the animal. In both young and old specimens, there were peculiar virus-like structures within the nucleus. If these represent intracellular parasites, they apparently are not very harmful, since they are found in specimens that have attained almost the maximum life span.

In young flies glycogen is seen well in electron micrographs (Fig. 4-4). Rough endoplasmic reticulum with its ribosomes, is not abundant and seems confined chiefly to the perinuclear areas or as chain-like formations associated with mitochondria. In old flies, ribosomes are very numerous and are widely distributed

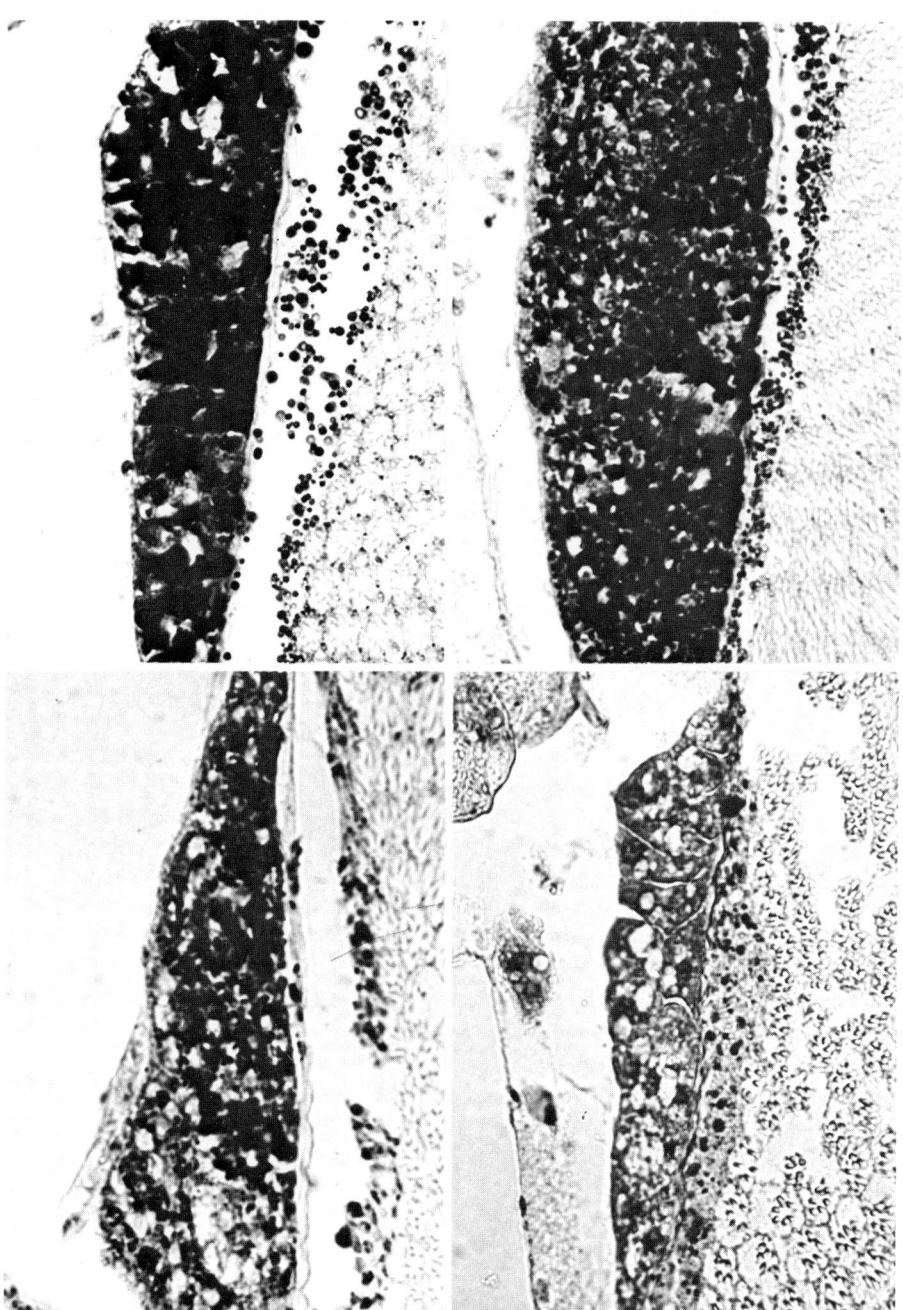

Fig. 4–3. Glycogen decrease in the fat bodies of male *Drosophila melanogaster*, induced by aging and by starvation. Age: (a) 2 days; (b) 30 days; (c) 100 days; (d) 67 days, fasted during the 24 hours preceding sacrifice. Dimedone-PAS-Gallocyanin stain. Digestion by amylase of parallel sections completely prevented the staining by dimedone-PAS. ×400. (Courtesy of Dr. J. Miquel.)

through the cytoplasm. A surprising number of degenerating mitochondria are in evidence in fat body cells from old animals (Fig. 4–5).

The definite changes with age in the fat body of Drosophila, an important organ of intermediary metabolism, would lead one to believe that its alterations

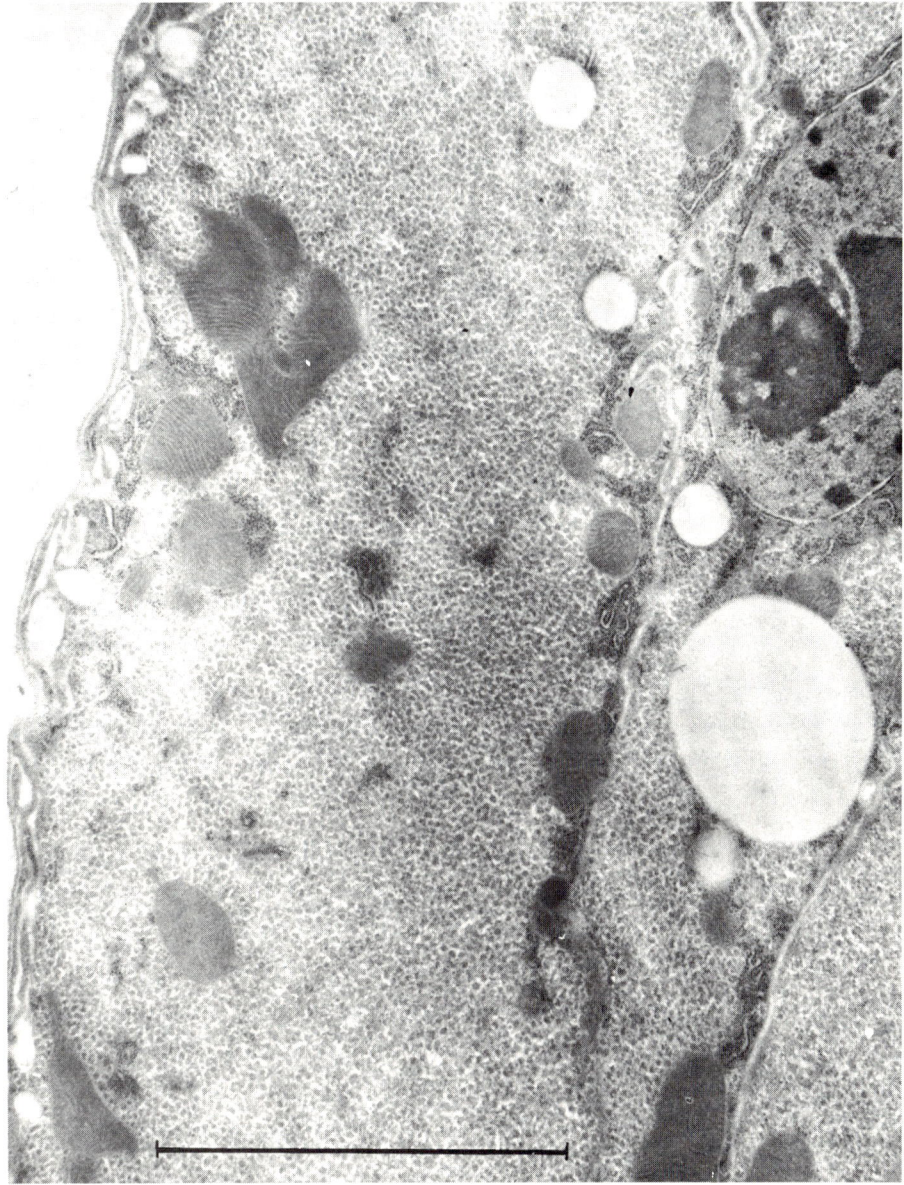

Fig. 4–4. Electron micrograph of abdominal fat body of a 5-day-old male *Drosophila melanogaster* showing abundant glycogen corpuscles, normal mitochondria, and empty round spaces, which presumably were occupied *in vivo* by neutral fats. The mark measures 1μ. Tissue from both young and old Drosophila was fixed in glutaraldehyde, followed by osmium and stained with uranyl acetate and lead hydroxide. (Courtesy of Dr. J. Miquel.)

play an important role in aging of the organism as a whole. It is true, however, that age changes have been found also in the nervous system and alimentary tract; the relative roles of these organs in the degenerative changes of senescence have yet to be worked out.

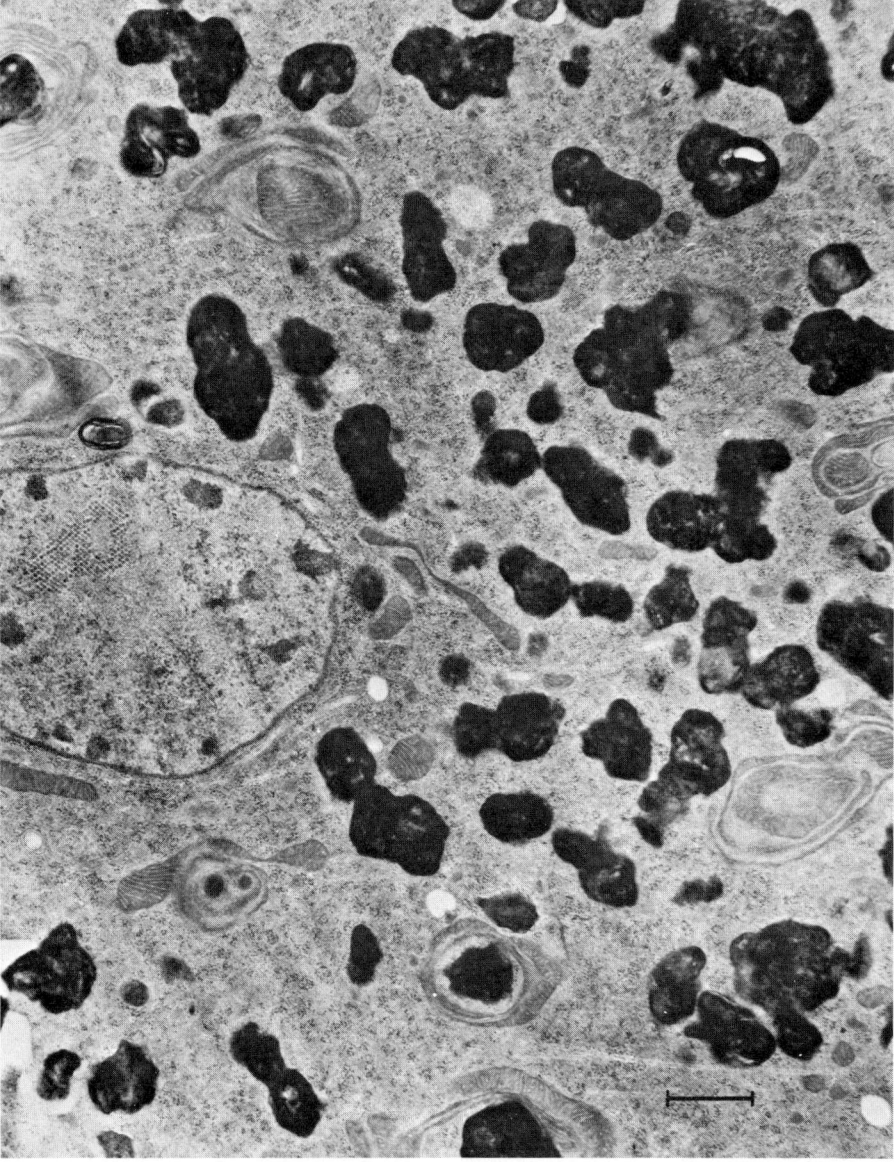

Fig. 4–5. Electron micrograph of abdominal fat body of a 76-day-old Drosophila showing a great increase in the number of ribosomes, mitochondria with splitting of the lamellae and dark cores, and abundant osmiophilic bodies. Virus-like particles are present in the nucleus. The mark measures 1μ. Tissue from both young and old Drosophila was fixed in glutaraldehyde, followed by osmium and stained with uranyl acetate and lead hydroxide. (Courtesy of Dr. J. Miquel.)

Studies on Musca domestica

Rockstein has made a number of studies on the common housefly, *Musca domestica*, reared under controlled environmental conditions. (Rockstein, 1957; Rockstein and Bhatnagar, 1964).

A Life Table has been constructed for both male and female flies (Rockstein and Lieberman, 1959) (see Fig. 4-6). The life span of the females is considerably greater than that of the males, mean length for males, about 17 days; for females, about 29 days. Fifty per cent mortality occurred at 16 days for the males and at 30 days for the females. In an investigation of the giant mitochondria or sarcosomes of the flight muscles, it was found (Rockstein and Bhatnagar, 1964) that variations in mitochondrial size and number are both sex-related and age-related. In the male fly both the total number and the mean size increase up to the beginning of the second week and then begin to decline by the end of the second week. In the female fly maximum number and size are reached by the third day, remaining unchanged until the end of the third week, when both size and number show a marked decline. Observations on specific changes in populations of differently sized sarcosomes lead these authors to believe that the multiplication, growth, and loss of these organelles actually are continuous processes throughout the life of the fly but that their *rates* change with age. Male flies lose their wings as they age. In them the changes in numbers and size of the sarcosomes are preceded by this loss of the wings (de-alation) and by a decline in ATP-ase and in alpha-glycerophosphate dehydrogenase activity. In the female fly the changes in the mitochondria appear to occur with the decline in biochemical and physical functions necessary for flight, although de-alation occurs in both sexes.

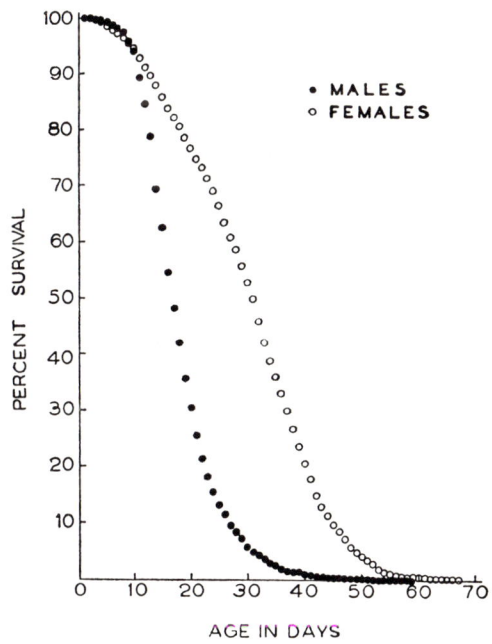

Fig. 4-6. Survival curves for male and female house flies (Musca domestica). (After Rockstein and Lieberman, 1959.)

Nutrition, Temperature, and Aging

Maurizio (1961) states that in caged young bees life span and physiological condition are closely related to the protein content of the food. An increase in life span and a fuller development of pharyngeal glands, ovaries, and fat bodies is seen when a diet of casein is used in place of a pure sugar diet. While the addition of the vitamins B_1, B_2, B_6, nicotinic, and panthothenic acids does not increase the life span significantly, it does insure the vitality of the next generation which, in the absence of vitamins, is impaired as seen by bad rearing in this second generation. In vitamin-starved bees, however, addition of pantothenic acid or a mixture of vitamins to the diet will restore normal brood-rearing within 24 hours!

The idea that life at an increased temperature leads to a more rapid decrease in something which is vital to the organism seems to be opposed by the findings of Clarke and Maynard Smith (1961b) on Drosophila. They kept adult flies at high temperatures, then transferred them back to a cooler environment. When they tried to measure an expected decrease in life span, they found no subtraction from the survival time. The factor which varies with temperature appears to be the "vitality required for survival" (Maynard Smith, 1958a; 1959c; Clarke and Maynard Smith, 1961b). There seems not to be a simple thermal acceleration of life processes.

Nervous System

Among anatomical changes which have been described for the aging process in insects are accumulation of urates (Metchinikoff, 1915) and hardening and weakening of chitin (Blunck, 1924); depletion of reserves of food (Bilewicz, 1953) and degenerative changes in nervous tissue (Pixell-Goodrich, 1920; Schmidt, 1923; Weyer, 1931; Rockstein, 1959).

Bee

In a series of experiments on the Italian honey-bee (*Apis mellifica*), Rockstein (1950, 1959) compared changes in cell number in the nervous system with alterations in cholinesterase activity with age, the time of final age of sampling of specimens depending upon the availability of sufficient numbers of animals for the two kinds of studies. This time occurred when 1 per cent or less of the original number of bees employed were still living and represented values close to maximum lifespan, varying under different conditions from 51 days to about 70 days for worker bees.

The number of neurons at two distinct levels of the brain was used as an anatomical criterion for biological old age. There was a loss of about 35 per cent of the original number of nerve cells, a figure which Rockstein points out as rather closely comparable to what had been found by earlier authors studying the brains of rodents and of man. The degree of loss appeared to be virtually identical for indoor and outdoor (hive) senescent bees.

Metamorphosis and Aging

Comfort (1964) raises the question as to how long, if metamorphosis could be postponed indefinitely, the insect larva would continue to live. Presumably, he says (p. 197), it might do so with no apparent limit or it might undergo a "specialized type of senescence due to the suppression of development or to imbalance between continued, divergent growth processes, or it might nevertheless undergo senescence from the same cause, whatever that cause may be, which limits the life of the imago." Recent work, particularly that of Hadorn (1966), seems to indicate that life of larvae or of their tissues may be prolonged indefinitely.

Transplantation Studies

Transplantation studies on Arthropoda have recently contributed in an important way to the fundamental question as to whether aging of tissues or cell-lines is an inevitable process.

Hayflick (1966) states that most diploid cell strains show limited growth span when they are serially substituted *in vitro*. He feels that this fact may reflect a limitation to cell division and replacement which is *intrinsic*, especially since it has been widely observed even when considerable alterations are made in the technical methods of culture. He states that considerable evidence indicates that such intrinsic change may be due primarily to alterations in permeability which in turn lead to decrease in cell proliferation, with onset of some degree of nuclear instability.

This apparent experimental evidence has been the first actually to support strongly the theoretical concept of intrinsic aging of cells. It is of great interest that at this same time the transplantation studies of Hadorn (1966), on *Drosophila* larvae, offer excellent evidence *against* such intrinsic aging.

REFERENCES

Baerg, W. J., 1945. The black widow and the tarantula. Trans. Conn. Acad. Arts Sci., vol. 36, p. 99.

Bilewicz, S., 1953. Doswiadczenia nad wplywem czynnosci rozrodczych na dlugosc zycia u muchy owocowej Drosophila melanogaster. (Influence of mating on the longevity of D. m.) Folia Biol. Warsaw, vol. 1, p. 175.

Blunck, H., 1924. Lebensdauer, Fortpflanzungsvermögen und Alterserscheinungen beim Gelbrand (Dytiscus marginalis L.). Zool. Anz., vol. 58, p. 163.

Clarke, J. M. and J. Maynard Smith, 1961b. Independence of temperature of the rate of aging in Drosophila subobscura. Nature, London, vol. 190, pp. 1027–1028.

Hatai, S., 1902. Number and size of the spinal ganglion cells and dorsal root fibers in the white rat at different ages. J. Comp. Neurol., vol. 12, pp. 107–124.

Hadorn, E., 1966. Konstanz, Wechsel und Typus der Determination und Differenzierung in Zellen aus Mannlichen Genitalanlagen von Drosophila Melanogaster nach Dauerkultur in vivo. Develop. Biol., vol. 13, no. 3, pp. 424–509.

Hayflick, L. 1966. Senescence and cultured cells. *In* Perspectives in Experimental Gerontology, Chap. 14, pp. 195–211 (Nathan W. Shock, ed.). Springfield, Ill.: Charles C Thomas.

Herrick, F. H., 1911. Natural history of the American lobster. Bull. U.S. Bureau of Fisheries, vol. 29, p. 149.

Labitte, A., 1916. Longevité de quelques insectes en captivité. Bull. Mus. Hist. Nat. Paris, vol. 22, p. 105.

Maynard Smith, J., 1958b. Prolongation of the life of Drosophila subobscura by a brief exposure of adults to a high temperature. Nature, London, vol. 181, pp. 496–497.

────── 1959c. Rate of aging in Drosophila subobscura. CIBA Foundation Colloquia on Aging, vol. 5, pp. 269–281.

Metchnikoff, E., 1915. La mort du papillon du murier—un chapitre de thanatologie. Ann. Inst. Pasteur, p. 477.

Miquel, J., 1968. Personal communication.

────── 1970. Advances in Gerontological Research. New York: Academic Press (in press).

Pixell-Goodrich, H., 1920. Determination of age in honey bees. Quart. J. Micr. Sci., vol. 64, p. 191.

Rockstein, M., 1950. The relation of cholinesterase activity to change in cell number with age in the brain of the adult worker honey bee. J. Cell. Comp. Physiol., vol. 35, pp. 11–23.

────── 1959. Biology of Aging in Insects. CIBA Foundation Symposium on The Lifespan of Animals, pp. 247–264.

Schmidt, H., 1923. Über den Alterstod der Biene. Jena Z. f. Naturwiss., vol. 59, p. 343.

Schulze-Röbbecke, G., 1951. Untersuchungen über Lebensdauer, Altern, und Tod bei Anthropoden. Zool. Jb., vol. 62, p. 366.

Weyer, F., 1931. Cytologische Untersuchungen am Gehirn alternden Alterstod. Z. Zellforsch., vol. 14, pp. 1–54.

5
Other Phyla

Phyla of invertebrates which we have not discussed include the worms other than annelids, namely the flatworms (Platyhelminthes) and the roundworms (Nemathelminthes), and the great group of Echinodermata the "spiny-skinned" animals which include the starfishes, sea urchins, brittle stars, and sea cucumbers. Unfortunately, very little has been written about aging in these groups.

Flatworms

The flatworms present remarkable powers of regeneration. One of the great early works relating to senescence (Child, 1915) was based to a large extent on studies of Planaria (now "Dugesia") a common flatworm (Fig. 5-1).

As many as 16 fragments of Planaria from one individual will regenerate with antero-posterior polarity, i.e., a new head will develop from the anterior cut end and a new tail from the posterior one on each fragment. The polarity here seems to be maintained primarily by means of quantitative differences between cells according to their position in the organism. Child has suggested that the physico-chemical basis of the antero-posterior gradient may be graded rates of oxidation

Fig. 5-1. Whole mount of *Planaria maculata* with the alimentary tract injected to show its complex branching. The pharynx (light) is centrally located while the three branches of the intestine, with their ramifications, are dark. Planaria (now "Dugesia") has been the subject of many fundamental studies on regeneration, senescence and rejuvenescence.

which would result in graded rates of energy production. Gradients of various kinds, however, have been found or suggested in different organisms: sulfhydryl gradients are seen in a number of forms and new gradients of ribonucleoprotein have been described for the amphibian egg, making it seem probable that quantitative differences in such substances may play a role in polarity.

The effect of cutting of flatworms into portions may be thought of, to some degree, as a rejuvenative, stimulating growth and calling on the cells to demonstrate their potentials for differentiation anew.

The studies of Child (1915) gave evidence that differentiation and senescence are reversible processes. The decline of oxygen uptake shown in aging is reversed when need for regeneration brings about a rejuvenative process. The resistance to cyanide poisoning, which rises steadily through normal life in planarians, falls when "rejuvenation" occurs. Planarians maintained in an environment in which they did not grow did not show a rise in resistance and starved planarians which actually were growing smaller showed a reversal of the normal trend. Rejuvenation, in part or in whole, appears to occur in regenerating fragments of planarians.

Roundworms

The roundworms, Class Nematomorpha, belong to the Aschelminthes, the same phylum in which we find the rotifers. As with the rotifers, *constancy* in cell numbers is a prominent feature. Cell division generally ceases at the time of hatching, except in the reproductive system, and usually there is no further increase in the number of nuclei. There is often, however, an increase in size of many cells. The cells of muscle, in particular, show elongation and it is thought that some of these cells, in larger nematodes are as much as 10 mm. ($2/5$ in.) in length!

Regenerative power in the roundworms apparently is limited. Study of possible age changes in them would seem to offer a fruitful field of work.

Echinodermata

The determination of age in the echinoderms, except for those which have been hatched and reared in aquaria, is a difficult matter. One important factor related to size, other than age, appears to be depth. Thus a common crinoid of western European coasts, *Antedon bifida* shows a spread of 120 mm. in shallow water but one of 220 mm. at greater depths. Another species, *Heliometra glacialis* is relatively small in waters close to northern Europe, reaches a spread of 500 mm. near Greenland, and in the Sea of Japan and the Sea of Okhotsk may be 700 mm. across. The chief determining factor here seems to be temperature—the colder the environment, the larger the individual echinoderm.

Among the holothurians, or sea cucumbers, the species in the ocean depths generally are different from those found in the littoral zones and the same is true when temperate, arctic, and antarctic waters are compared. The size relations according to depth and geographic region are not as easily studied as with the crinoids.

Some data are available on the longevity of sea cucumbers. Mitsukuri (1903) found that Stichopus japonicus attains sexual maturity in its third year and lives at least 5 years. It reaches a maximum length of 40 cm. (16 in.). Noll (1881) found that a specimen in his indoor aquarium was in healthy condition after 3 years and 4 months.

Edwards (1908) raised the young of *Holothuria floridana*. He found a slow growth, from 0.33 mm. length at 5 days of age to 4.00 mm. at 75 days. This perhaps would be slower than under natural conditions but was taken to mean that several years would be required for this large species to reach full size. Tao (1930) felt that 3 or 4 years would be required for the much smaller species, *Paracaudina chilensis*, to attain full growth.

Among the great group of *Asteroidea*, or starfishes, some interesting relationships of size to geographical distribution again are seen. Thus Fisher (1940) mentions a number of species of asteroids of the Antarctic region as being notable for their large size, either in absolute terms or in relation to other species of the same genus, a number having individuals over a foot in diameter and some over 2 feet. A considerable degree of similarity exists between asteroids of high southern and high northern latitudes. Whether the relatively great size of some asteroids in cold waters may be due actually to a *greater longevity* is an interesting question and would seem to be subject to study either in the aquarium or in nature.

Growth processes have been studied in several species of asteroids. Spawning of *Asterias forbesi* off the New England coast takes place in June and by the end of that month great numbers of the tiny starfishes, 1 mm. or less in diameter, are found among the seaweeds (Galtsoff and Loosanoff, 1939).

By the end of July the young animals have attained a diameter of 10–11 mm., by the end of September 35 mm., and by the end of October 50–80 mm. Growth is slowed during the winter but is speeded again as summer approaches. Spawning of specimens a year old seems to be the rule, whatever size they have reached. According to Smith (1940) *Asterias vulgaris* arrives at a diameter of 35 mm. at the end of 1 year and 60 mm. only at the end of the second year. Mortensen (1927) believed that full size is reached by *Ctenodiscus crispatus* in 3 years, by *Psilaster andromeda* in 4 years, and *Pseudarchaster parelii* in 4 years or more. According to several authors (see Barnes and Powell, 1951) *Asterias rubens* reaches a diameter of 40–100 mm. during the first year and spawns at 1 year. Growth occurs chiefly in the summer. According to this rate of more or less seasonal growth, 6 or 7 years would be needed for individuals of this species to attain the maximum size of some 70 cm. (28 in.) in diameter.

Echinoidea

The echinoids include the sea-urchins and sand-dollars. Some information is available on the growth and longevity of urchins and sand-dollars. According to Bull (1938) newly metamorphosed specimens of *Psammechinus miliaris*, under laboratory conditions, are somewhat over 1 mm. in diameter; at 8 months about 10 mm. (the "test"); at 1 year, 20; at 2 years, 26.2; at 3 years, 29.2; at 4 years, 30.3; at 5 years, 37; and at 6 years, 38.7 mm. Grieg (1928) studied specimens of *Strongylo-*

centrotus drobachinesis from the coast of Norway. From the rate of growth which he found, he estimated the largest specimens, 78 mm. in diameter, to be about 8 years of age. *Echinus esculentus*, which is thought to live about 4 years, grows relatively rapidly and reaches a diameter of 90 to 110 mm. in that span. The sand dollar, *Mellita sexiesperforata* appears to live about 4 years at Bermuda and attains a diammeter of about 120–130 mm.

Orton (1923) found urchins to grow at comparable rates at Spitsbergen where the water has a year-round temperature of 4°C with those of the same species seen along the shores of Great Britain where the temperature is far higher.

The echinoids thus appear to have life spans varying generally from 4 to 8 years and to grow for all or almost all of this span.

The ophiuroids, or brittle-stars, are not as conspicuous nor easy to collect as are many of the echinoderms because of their secluded habits and generally rather small size. As with other classes of echinoderms, however, they seem to be concentrated in population in the Indian-West Pacific Ocean area, although again like other classes of echinoderms there are some species to be found in almost every part of the world, including the Arctic and Antarctic regions.

Data on growth and life-span of brittle stars are scanty. It would appear that in some well known species several years would be needed to reach the maximum size. Mortensen (1929) says that *Ophiura texturata* probably requires 5 to 6 years to reach its maximum disk diameter of 30 to 35 mm. This species reaches sexual maturity in its third year at a disk diameter of only 7 to 11 mm. Several species from northern Europe (*Ophiopholis aculeata*, *Ophiocten sericeum*, and *Ophiura sarsi*) show a similar relationship in their life history, reaching sexual maturity long before full growth is attained (Grieg, 1928).

Regeneration in Echinoderms

The echinoderms in general show remarkable powers of regeneration, not only of small parts such as spines and pedicellariae but of large structures that comprise major portions of the body. A starfish torn in two by the irate oysterman who considers it an enemy, will produce two new individuals by regeneration. In the brittle stars, a disc deprived of all arms at the base soon dies. If, however, even one arm is left attached, complete regeneration of the others will occur (Dawydoff, 1901; Zeleny, 1903, 1905). In these animals, and apparently in echinoderms in general, the wandering cells, or coelomocytes, seem to play an important role in regeneration, aggregating at the site of injury in enormous numbers. Their role seems to be partly phagocytic, for the removal of injured and degenerating tissue, partly nutritive, for many of them are laden with spherules of stored material (Fig. 5–2), and partly one of differentiation into cells of different type. Cells of coelomic nature, according to Dawydoff, differentiate into muscle fibers.

While the regenerative powers in this phylum are so very great, there are cells which probably have a long span of life. With the knowledge already on hand concerning growth and life span of the organism, studies on possible age differences in the tissues of echinoderms would seem to warrant investigation.

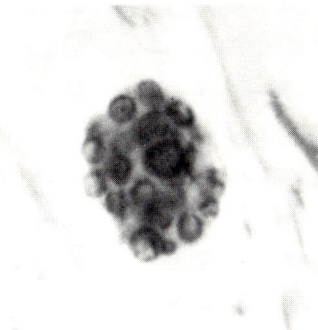

Fig. 5–2. Spherule cell in connective tissue of the body wall of the sea cucumber, *Stichopus badionotus*. Among the invertebrates, many types of motile cells are found, some of which transport nourishment to various tissues, others of which serve as replacement cells by their ability to differentiate into other cell types. The abundant spherule cells of the echinoderms are thought to be of the former type. ×1,000.

REFERENCES

Barnes, H. and H. T. Powell, 1951. The growth rate of juvenile Asterias rubens L. Jour. Marine Biol. Assoc., United Kingdom, vol. 30, pp. 381–385.
Bull, H. O., 1938. Growth of Psammechinus miliaris. Durham, England, University. King's College. Rept. Dove Marine Lab., Cullercoats, ser. 3, no. 6, p. 39.
Child, C. M. Senescence and Rejuvenescence. University of Chicago Press, 1915.
Dawydoff, C., 1901. Beiträge zu Kenntnis der Regenerationserscheinungen den Ophiuren. Ztschr. f. Wiss. Zool., Leipz., vol. 69, pp. 202–234.
Edwards, C. L., 1908. Variation, development and growth in Holothuria floridana Pourtalès and in Holothuria atra Jäger. Biometrika, Cambridge, vol. 6, pp. 236–301.
Galtsoff, P. S. and V. L. Loosanoff, 1939. Natural history and method of controlling the starfish (Asterias forbesi, Desor). Bull. U.S. Bur. Fish., vol. 49, pp. 73–132.
Grieg, J. A., 1928. The Folden Fiord Echinodermata. Tromsö Mus. Skrifter, vol. I, pp. 1–12.
Mitsukuri, K., 1903. Habits and life-history of Stichopus japonicus. Annotationes Zoologicae Japonenses, vol. 5, pp. 1–21.
Mortensen, T. H., 1927. Two new Ctenophores. Vidensk. Meddel. Dansk Naturhist. Foren., vol. 83, pp. 277–288.
——— 1927. Handbook of the echinoderms of the British Isles. London: Oxford University Press, 471 pp.
Nöll, F., 1881. Mein Seewasser-Zimmeraquarium. Zool. Garten, vol. 22, pp. 8–19, 33–42, 71–79, 137–147, 168–177, 194–206 (pub. in diff. issues).
Orton, J. H., 1923. Some experiments on rate of growth in a polar region (Spitsbergen) and in England. Nature, London, vol. III, No. 2779, pp. 146–148.
Smith, G. F., 1940. Factors limiting distribution and size in the starfish. Jour. Fisheries Research Bd., Canada, vol. 5, pp. 84–103.
Tao, L., 1930. The ecology and physiology of Caudina. Proc. 4th Pacific Sci. Congr., No. 3.
Zeleny, C., 1903–1904. A study of the rate of regeneration of the arms in the brittlestar, Ophioglypha lacertosa. Biol. Bull., Woods Hole, Mass., vol. 6, pp. 12–17.
——— 1905. Compensatory regulation. J. Exper. Zool., Baltimore, vol. 2, pp. 1–102.

part II
The Vertebrates Other than Mammals

6
Fishes

The cold-blooded vertebrates (fishes, amphibians, and reptiles) generally are considered as examples of indeterminate growth (Backman, 1938). Bidder (1925), regarding determinate size in vertebrates as a characteristic which had evolved with the development of life on dry land, emphasized the many cases among the fishes where continued growth and no evidence of declining vigor nor a definite life-span are seen. Because of the commercial importance of many kinds of fishes, extensive studies have been made by the fishery boards or similar organizations in various countries. These studies have shown that fishes in general have a prolonged period of growth (there are definite exceptions) and that growth goes on throughout life, slowing down to some extent as the fish grows older (Fig. 6-1). The use of the scales in determination of age of fishes in waters of the temperate zones has made possible a number of interesting studies. A rather surprising finding is that the fertility of the female fish does not decrease with age but that the longer and older the female, the greater is its capacity for egg production.

For the pike, the opercular bones and the scales have been used in age determination; for the carp, (Jhingran, 1957) and the salmon (Hartley, 1958), the scales; for the tuna, (Galtsoff, 1952), the shark (Haskell, 1948–1949), and for the ray (Ishiyama, 1951), the vertebrae; for the sturgeon (Chugunov, 1925), the fin rays.

Raitt (1932) gave the following figures for the haddocks of the North Sea: second year of life, 31,000 eggs; third year, 100,000; fourth year, 159,000; fifth year, 224,000; and sixth year, 278,000. It is true, however, that the limits of life in the haddock are not too well known, since human depredation (fishing) accounts for termination of life so frequently; and where this does not occur, natural depredation and parasitism intervene. It may be said rather generally of the wild populations of any of the vertebrates that the percentage of deaths that can be attributed to senility or even to degenerative diseases, is low.

The longevity of the carp is relatively well known because in the monasteries and gardens of the Orient these fishes are well cared for and often the object of close attention as individuals. Flower (1925, 1935) has written extensively on the longevity of fishes, as well as of other vertebrates. Gerking (1957) gave important evidence for a process of aging in populations of fish in the wild. According to his critical analysis, the oldest carp known as living in captivity was 47 years of age. A few other fishes are known to have attained a greater age: a specimen of eel *Anguilla anguilla*, 55 years, and of Wels (European catfish), *Silurus glanis*, 60 years.

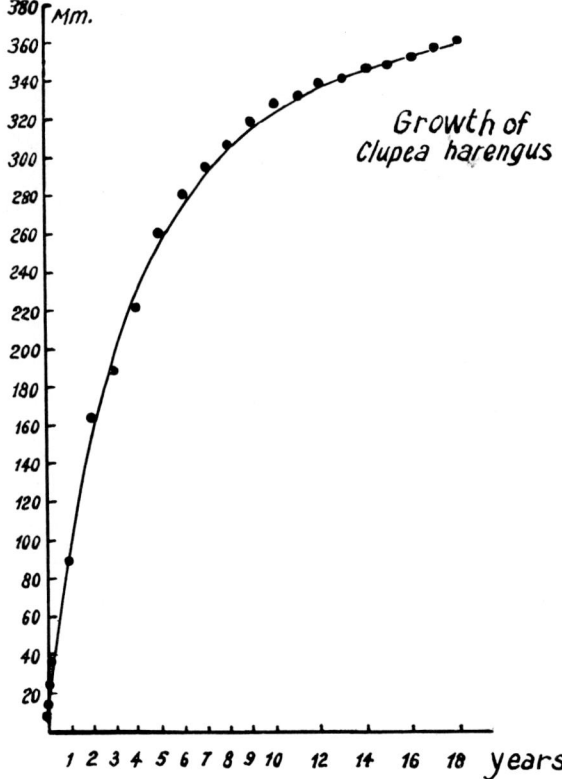

Fig. 6-1. The growth in length of the Atlantic Herring, *Clupea harengus*. This fish, one of the most abundant of all vertebrates, may live for 20 years but usually falls prey to other animals or to man at an earlier age. As the curve shows, growth slows down but is continuing at an ad-advanced age. (After Bachman, 1938.)

The early literature contains numerous accounts of fish said to have lived 100, 150, and even more years, but these are very poorly substantiated.

In fishes in general the life of the male is shorter than the female (Wimpenny, 1953), a feature which is true of spiders, houseflies, fruitflies, many beetles, and man! Minnows, however, seem to show a longer life for the males, at least in the wild (Van Cleave and Markus, 1929).

The tremendous range in adult size of the fishes is noteworthy. At one end of the scale is a tiny Goby, *Mistichthys luzonensis*, found only in one of the lakes of Luzon in the Philippine Islands, the fully mature individuals being ½ in. long; while at the other is the Whale Shark, *Rhineodon typicus*, which has a length of 50 ft. or more and a weight of several tons. In general, longevity is greater in larger fish. One of the shortest-lived of fishes is another small Goby, the White or Transparent Goby, *Latrunculus pellucidus*, which lives only 1 year and thus may be called an "annual vertebrate." There are a few other examples of annual fish.

In the case of fishes which migrate from the sea up the rivers and streams, and over considerable cataracts and long stretches of rough water, death may occur after spawning. However, some species of salmon return to the sea and come back for a second, and even (rarely) a third spawning. The scales of such salmon show a spawning mark, the result of an erosion or absorption of the scale margin and sometimes of a part of its surface also (Fig. 6-2). It has been said that the scale erosion is due to a calcium deficiency in the period of fasting when all of the efforts of the fish are directed toward the reproductive function (Crichton, 1935). In extreme cases, especially in large early-running male salmon, the erosion may be so marked that a large part of the scale disappears, leaving a small triangular remnant.

In at least some of the fishes which undergo a rapid "senescence" in relation to migration and spawning, nonmigratory populations of the same species show no such senescence in relation to reproduction and indeed may spawn a number of times (Robertson and Wexler, 1959, 1962). Castrated salmon show in actual senescence changes similar to "normal" salmon after the migratory spawning activity.

In the salmon the organ systems may progress from a stage of vigorous sexual maturity to death from physiological "aging" within a few weeks (Figs. 6-3 and 6-4). Thus these fish undergo in that short span changes comparable in many ways to those which take place over a 30–40 year span for the human organism. During

Fig. 6-2. Scale of a salmon, showing a "spawning mark." The erosion of the scale apparently is due to a reversal of activity of the cells which lay down the scale and probably is brought about by calcium deficiency. (From J. W. Jones, 1959. *The Salmon*. Harper & Row.)

the fortnight in the salmon, then, processes such as impairment of peripheral circulation, thickening of arterial walls, liver disorders, glandular hypertrophy, malnutrition, and the development of infections by bacteria and fungi, all form part of the complicated pattern of degeneration of the body of the fish.

In salmon that have reached the spawning grounds, marked changes have occurred in the tissues. Much of the fat and the protein appear to be used up for the development of the gonads and the body of the fish, due to this and to the exhaustion from the physical effort of the arduous journey, is in poor condition. Muscle protein was reduced to about 70 per cent of that seen in fish fresh from the sea and that fat is down to 20 per cent to 2 per cent of that seen in such fish. Meanwhile, the gonads have increased in weight by as much as five times.

Recently a research team of 37 scientists made a special 6-week study of the degenerative changes in the salmon, bringing, as it were, the laboratory into the field by following them in their arduous migration and making scientific studies of individuals at the various stages (Lobanov-Rostovsky, 1969). They studied the pink, *Oncorhynchus gorbuscha;* the chum, *Oncorhyncus keta;* the coho, *Oncorhynchus kisutch;* and the sockeye, *Oncorhynchus nerka.*

The gross changes in aspect of the head of the salmon are striking indeed. The following vivid and concise description is given by Benson (1969): "The spawned salmon is a miserable shadow of the beautiful silvery, deep-ocean marine animal. He has developed his hump and hooked jaw. His bones have become cartilaginous: his skin is peeling off. We even saw many with their tails falling off. His liver is a vivid olive green, from the decomposition products of his hemoglobin.

"Only the heart of the salmon remains in good condition—and even this suffers with thickening of the coronary artery walls. The demise of the salmon is indeed a dramatic model for impairment of biological function."

A somewhat surprising feature of these rapid degenerative changes is that the salmon does not seem to suffer from heart failure or from "stroke."

The bright orange fat serves as storage food during the strenuous migratory phase. The fish begins migration with a *six times* greater concentration of lipoprotein in the blood than is normally found in man.

During migration the fat is consumed and the fish sinks to a starvation level (Figs. 6-3, 6-4). It appears that in this case fasting and exercise are excellent preventatives of atherosclerosis.

A tremendous growth of the pituitary gland occurs when the salmon enters fresh water, according to Trams (1969). The liver soon begins to undergo a marked deterioration.

There are many aspects of the changes in the salmon, which, when carefully studied, may yield important new data on vertebrate aging in general.

One of the very interesting consequences of the weakened condition of the kelts (both male and female salmon in the post-spawning condition) is the greater susceptibility which they show to bacterial and fungus diseases. Such conditions take a heavy toll and many salmon never reach the sea. Thus, in one sense, a rapid aging has occurred and disease and death have followed.

Of probably more significance in relation to the nature of the normal aging process are descriptions of changes in the gonads in the later part of life. Rasquin and Hafter (1951) have described alterations in the testis of the teleost, *Astyanax mexicanus.* They studied the gonads of 53 specimens, 49 ranging from 3 months to

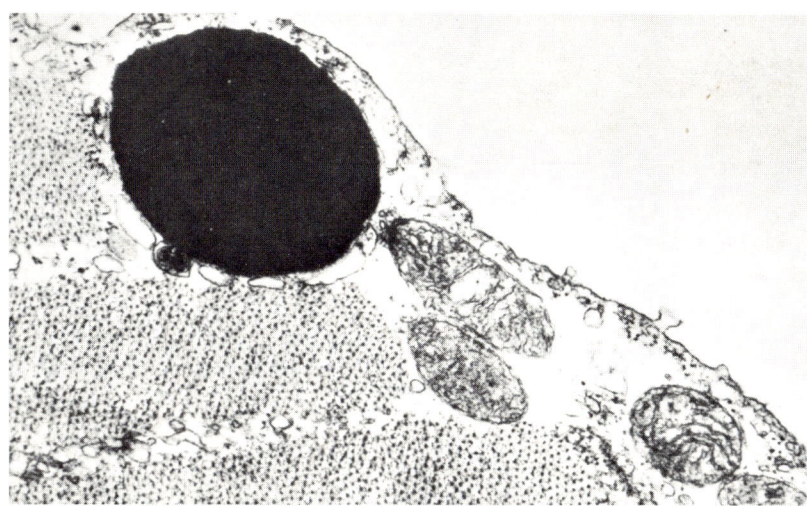

Fig. 6–3. Electron microgrpah of muscle tissue of pink salmon. Pre-spawning stage. One large ovoid fat storage body (black) and several mitochondria are seen in the subsarcolemmal position, while deep to them are several myofibrils with their numerous myofilaments. (Tissue prepared and picture taken aboard the ship, *Alpha Helix*, by Professor T. Bisalputra.)

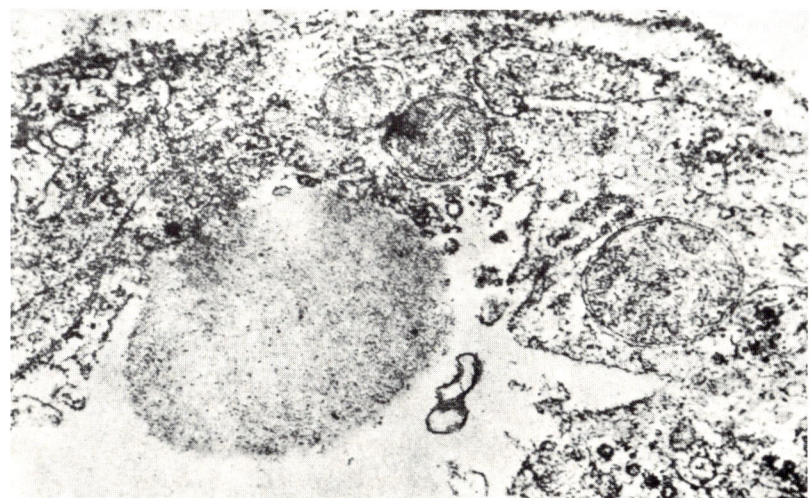

Fig. 6–4. Electron micrograph of muscle tissue of pink salmon. Post-spawning stage. A "ghost" of a fat storage body, now depleted of lipid, and several mitochondria are shown (Tissue prepared and picture taken aboard the ship, *Alpha Helix*, by Professor T. Bisalputra.)

3 years old, and 4 being 6-year-old fish. There was a decrease in production of spermatozoa and an occurrence of concretions in the tubules.

A further conspicuous age change in the Astyanax testis is an increase in amount of interstitial tissue. There is very little such tissue in the younger fishes between the lobules and among the coils of the tubules. In the older fish the connective tissue fibers and cells seem to penetrate from the capsule into the gland. The interstitial tissue contains fibroblasts, macrophages, and many reticular fibers. Collagenous fibers are associated only with blood vessels and the basement membrane of the sperm duct epithelium. A large amount of adipose tissue also is present. In some places the epithelium of the tubules has degenerated but it is not possible to tell whether spermatogenesis ever ceases completely.

Rasquin and Hafter (1951) state that the similarity between the concretions in the testis of the fish Astyanax and those in the aging prostate of man and other mammals is a striking one. In both cases they seem to be formed by condensation of fluid around desquamated cells or other debris. In both, they increase in number with age and frequently become calcified. In mammals, they may appear in the ejaculated seminal fluid and it is believed that many concretions·may leave the body of the older fish during spawning.

Somewhat similar changes have been shown by Krumholz (1948) in the western mosquito fish *Gambusia affinis affinis*.

The small fishes of aquaria frequently show a decline of reproductive capacity and even become completely infertile; yet after this change, the fish may be in apparently better condition than during the reproductive life, a fact true of Carassius, the common goldfish (Comfort, 1964).

It remains to be demonstrated whether fish of continuous growth, if they survive long enough, show any decrease in fertility, any decline in vigor, or any of the anatomical changes associated with senescence in higher vertebrates.

There is, however, a tendency for the mortality rate to increase with age in a number of species of fish, as found by Kennedy (1954) for the Lake Trout (*Cristivomer namaycush*), by Wohlschlag (1954), for the Alaskan whitefish (*Leucichthys sardeinella*), and by Ricker (1949) on several species.

In many of the smaller fishes continued growth after a certain age is no more evident than in, say, the male laboratory rat; yet the fish may survive for several times its growth period.

REFERENCES

Backman, G., 1938. Wachstumzyklen und phylogenetische Entwicklung. Lunds Univ. Arsskrift, N. F. Avd. 2, vol. 34, Nr. 5, pp. 3–142.

Benson, A. E., 1969. Cited from Lobanov-Rostovsky.

Bidder, G. P., 1925. The mortality of plaice. Nature, London, vol. 115, p. 495.

———, 1932. Senescence. Brit. Med. J., vol. ii, p. 5831.

Chugunov, N. I., 1925. On the methods of age determination in sturgeons. Bull. Fish. Econ. U.S.S.R., vol. 11, p. 33.

Flower, S. S., 1925. Contributions to our knowledge of the duration of life in vertebrate animals. I. Fishes. Proc. Zool. Soc., London, vol. 1, pp. 247–268.

———, 1935. Further notes on the duration of life in animals. I. Fishes; as determined by otolith and scale-readings and direct observations on living animals. Proc. Zool. Soc., London, vol. 1, pp. 265–304.

Galtsoff, P. S., 1952. Staining of growth rings in the vertebrae of Tuna (Thynnus thynnus). Copeia, p. 103.

Gerking, S. D., 1957. Evidence of aging in natural populations of fishes. Gerontologia, vol. 1, pp. 287–305.

Hartley, W. G., 1958. The microscopical study of salmon scales. J. Queckett micr. Club., vol. 28, pp. 95–98.

Haskell, W. L., 1949. An investigation of the possibility of determining the age of sharks through annuli as shown in cross sections of vertebrae. Annual Report of the Marine Laboratory of the Texas Game, Fish, and Oyster Commission, pp. 212–217.

Ishiyama, R., 1951. Studies on the rays and skates belonging to the family Rajidae found in Japan and adjacent regions. (Japanese-English summary). Bull. Jap. Soc. Sci. Fish., vol. 16, pp. 112–118, 119–124.

Jhingran, V. C., 1957. Age determination of the Indian major carp (Cirrhina mrigala Ham) by means of scales. Nature, London, vol. 179, pp. 468–469.

Krumholz, L. A., 1948. Reproduction in the western mosquitofish Gambusia affinis affinis (Baird and Girard), and its use in mosquito control. Ecolog. Monographs, vol. 18, pp. 1–43.

Lobanov-Rostovsky, I., 1969. Aging—Oceanography and Geriatrics. Oceans, vol. 1, no. 2, pp. 4–17.

Raitt, D. S., 1932. The fecundity of the haddock. Fishery Board for Scotland, Scientific Investigations, No. 1.

Rasquin, P. and E. Hafter, 1951. Age changes in the testis of the teleost, *Astyanax americanus*. J. Morph., vol. 89, pp. 397–404.

Robertson, O. H. and B. C. Wexler, 1959. Hyperplasia of the adrenal cortical tissue in Pacific salmon and Rainbow trout accompanying sexual maturation and spawning. Endocrinol., vol. 65, pp. 225–238.

———, 1962. Histological changes in the organs and tissues of senile castrated Kokanee Salmon. Gen. Comp. Endocr., vol. 2, pp. 458–472.

Trams, E., 1969. Cited from Lobanov-Rostovsky, I.

Van Cleave, H. J. and H. C. Markus, 1929. Studies on the life cycle of the blunt-nosed minnow. Amer. Nat., vol. 63, p. 530.

Wimpenny, R. S., 1953. The plaice. London: Arnold.

7
Amphibians

Until very recently, there has been almost nothing in the literature about age changes in amphibians. Brocas and Verzár (1961) studied the property of collagen contractility in the South African clawed frog Xenopus. They found that it changes steadily with age, as in mammals. Up to the age of 13 years, there was no sex difference in rate of aging demonstrable.

Some of the amphibians have surprisingly great powers of regeneration when compared with members of other classes. Thus a salamander can grow or regenerate lost limbs, although frogs and toads, also amphibians, cannot do so. The salamander also can regenerate a lost tail.

Flower (1925) gives the age attained by some species of amphibians under good conditions at the London Zoo as follows:

Salamandra maculosa	6 years
Molge crista	8 years
Molge pyrrhogaster	8 years
Cryptobranchus alleghaniensis	7 years

For some families of the Anura, he gives the following estimates, which he states (p. 288) are "for the larger species only."

Discoglossidae	6 years
Ranidae	7¾ years
Cystignathidae	7¾ years
Hylidae	8⅙ years
Bufonidae	11½ years

The life span for the amphibians, as indeed for the other classes of vertebrates, shows a great range: many species of frogs live 10–12 years at least, while the small *Triturus pyrrhogaster* has been said to live 25 years (Walterstorff, 1928) and the great Megalobatrachus over 65 years (Schneider, 1932).

Haploid amphibians have a shorter life than diploid ones; but as Comfort (1964) picturesquely puts it, they are, after all, "poor things."

Noble, in his comprehensive book on the Amphibia (1931) says that many species seem to have a determined absolute size, soon attained by the males, but that this is not true for many of the salamanders nor for some frogs of northern regions.

Senning (1940) has described an interesting method of age determination for the well-known Necturus, using the characteristics of the parasphenoid bones.

Some salamanders have remarkable ability to live at low temperatures. *Hydromantes platycephalus* lives at high altitudes in the Sierras of California and often is found active in water formed from melting snow banks. Some salamanders can survive when frozen in a block of ice, and many salamanders in the course of evolution, have retired to cool, underground waters in caves and wells.

The opportunity to study changes with age in the Axolotl, *Ambystoma mexicanum* was made possible for us by the availability of specimens of this animal from Dr. Rufus R. Humphrey of the Department of Zoology of Indiana University. Animals ranging from under 1 year up to 8 years 4 months have been studied.

Our attention has been concentrated on the liver. This organ shows conspicuous changes with age.

The livers of many (if not all) species of amphibians, when specimens are chosen at random, show scattered groups of cells that contain pigment. It seems to be generally presumed that these cells are of hepatic parenchymal type and that the pigment is derived from the breakdown of red blood corpuscles, probably being carried from the spleen, in finely dispersed form, through the portal vein and its branches to the liver cells.

Jordan (1931) made some extensive comparative studies on this pigment in various amphibians and discusses its occurrence and distribution.

In our series of *Ambystoma mexicanum*, all females, we find in the liver of young animals 7 months old, an almost complete lack of pigment (Fig. 7-1). In animals of 1 year and 8 months the pigment is conspicuous, but scattered in rather small groups of cells and not very densely aggregated in individual cells. In animals of 3 years, larger groups of cells are involved and there is a dense aggregation of pigment in the cytoplasm of the individual cells.

In our oldest specimen, an animal of 8 years and 4 months, the pigment is present in massive amounts (Fig. 7-2) and the nuclei of the pigment-containing cells generally are obscured.

In young animals the areas around the veins lack pigment (Fig. 7-3). In all of the old animals pigment occurs in these areas and there is a type of infiltration by lymphocytes and plasma cells (Fig. 7-4). In the older animals the association of this "ectopic" lymphoid tissue with the pigment-bearing cells is a conspicuous phenomenon.

Electron microscope studies of the pigmented areas indicate that the pigment is generally, if not always, confined to one type of cell and, although in close spatial relationship to parenchymal cells, is not taken in by them (Figs. 7-5, 7-6).

The pigment is present in discrete, electron-dense bodies which vary considerably in size and to a lesser degree in density (Fig. 7-7). Some of these bodies have a less dense interior and a circular rim of very dense material. In some areas, under the electron microscope, the pigment has an appearance somewhat similar to that found for lipofuscin, the "age pigment" (Andrew, 1961). It does not appear similar to hemosiderin. In fact, the dense bodies are morphologically more similar to the melanosomes of melanin than to any of the other types of animal pigment with which we are familiar. Earlier studies on liver pigment in amphibians, such as those of Okamoto (1925) and Jordan (1931) were done before the advent of

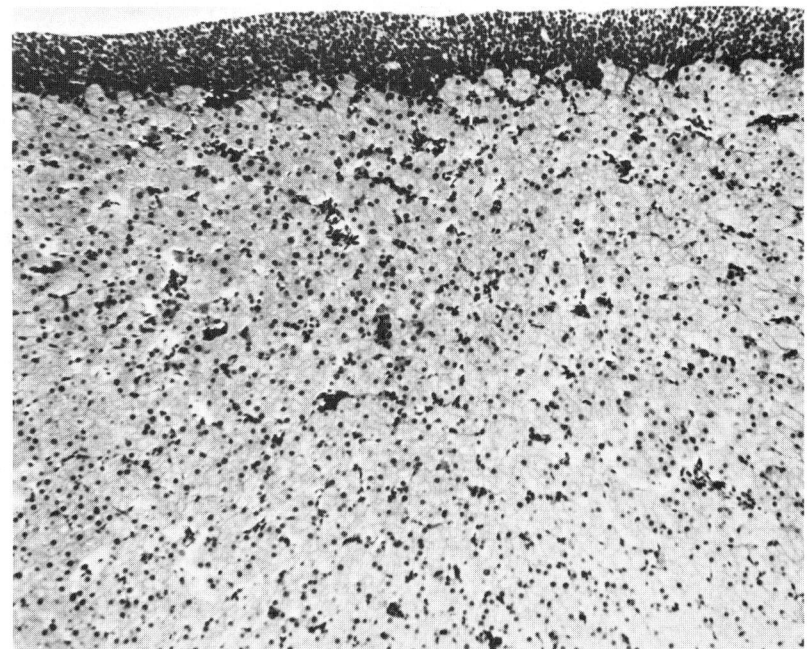

Fig. 7–1. Liver of a 7-month-old female *Ambystoma mexicanum*. The subcapsular hemopoietic tissue is a normal finding. Masson's trichrome. ×35.

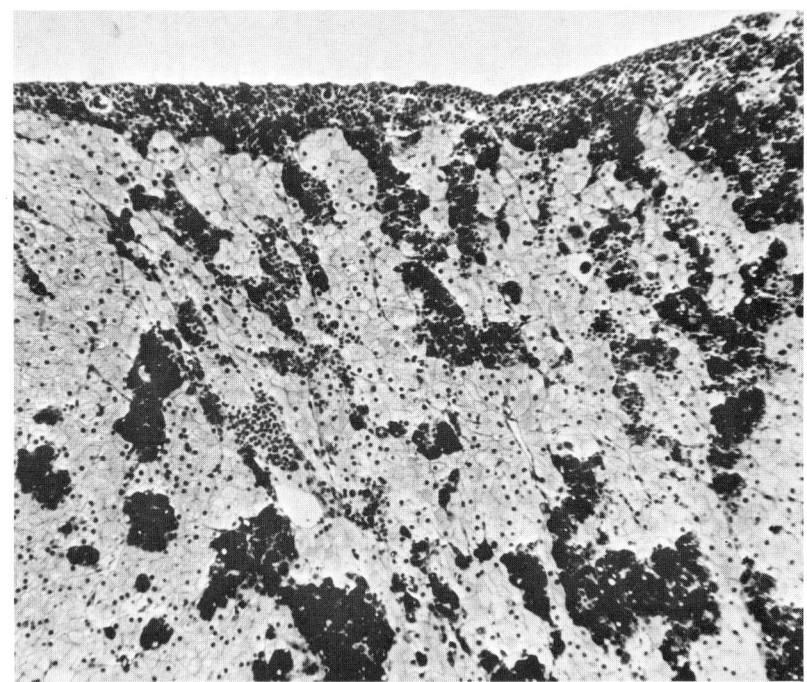

Fig. 7–2. Liver of a very old female *Ambystoma mexicanum*, 8 years and 4 months of age. Hemopoietic tissue extends widely in the liver parenchyma and pigment cells are abundant just beneath the subcapsular hemopoietic tissue and accompanying the intrahepatic masses. Masson's trichrome. ×35.

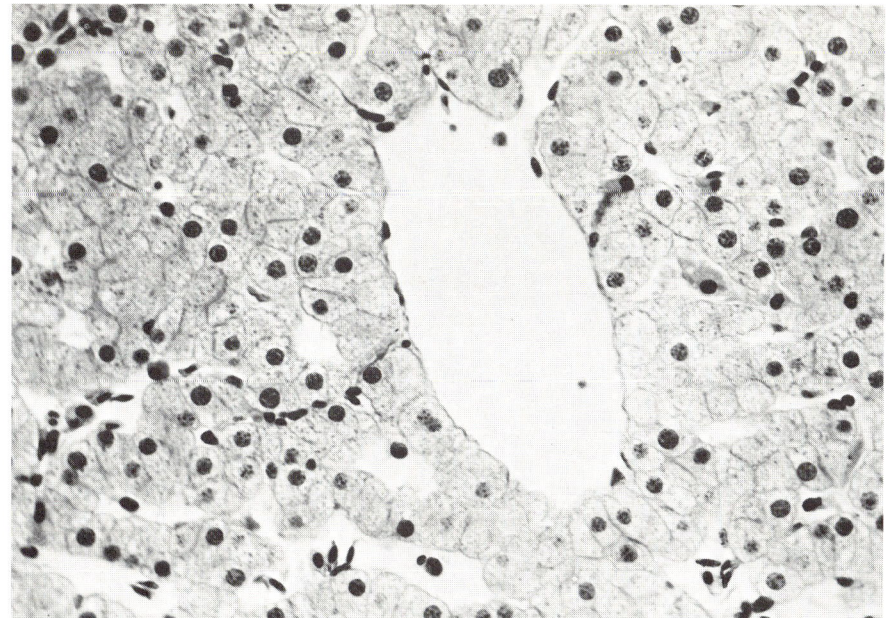

Fig. 7-3. Vein in the liver of a 7-month-old female *Ambystoma mexicanum*. The hepatic parenchyma shows no pigment cells nor hemopoietic (lymphoid) tissue. Masson's trichrome. ×125.

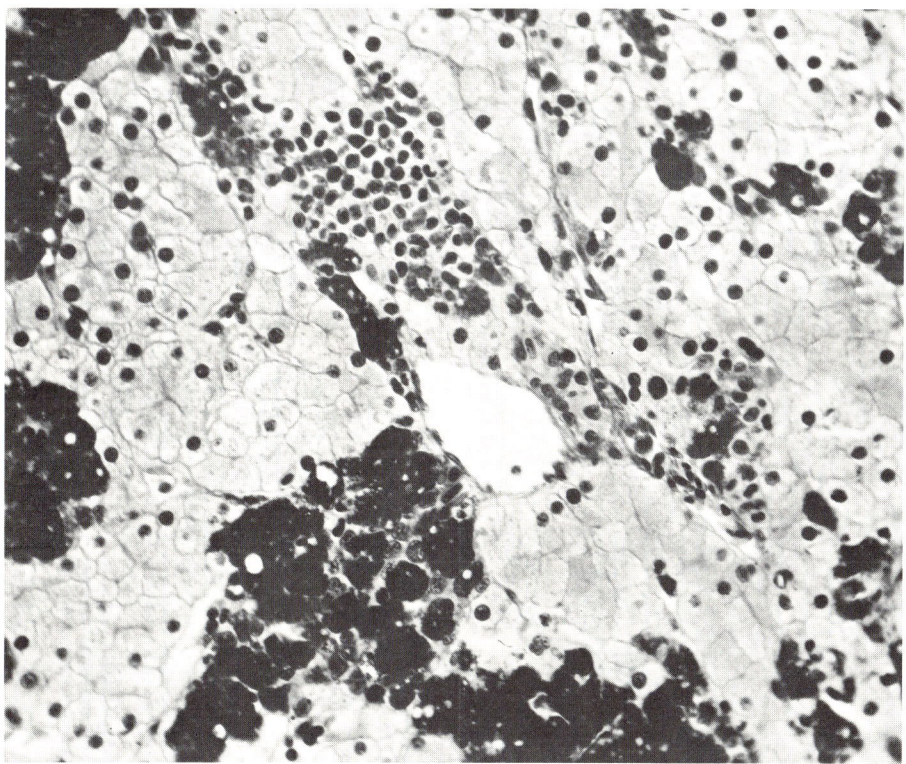

Fig. 7-4. Vein in the liver of female *Ambystoma mexicanum* of 8 years, 4 months. Clusters of cells of lymphoid and hemopoietic nature occur near the blood vessels and large pigment cells are abundant. Masson's trichrome. ×125.

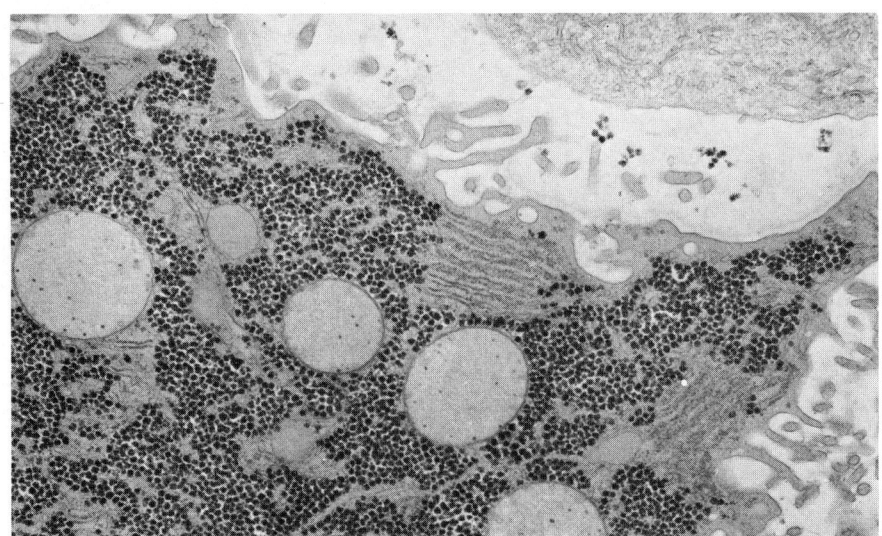

Fig. 7–5. Hepatic parenchymal cell of a very old specimen of *Ambystoma mexicanum*. Such cells often have abundant glycogen. They appear essentially similar to parenchymal cells in young livers, except for a considerable reduction in prominence of the mitochondrial cristae. Several mitochondria are seen as spheroid bodies in the figure. ×12,000.

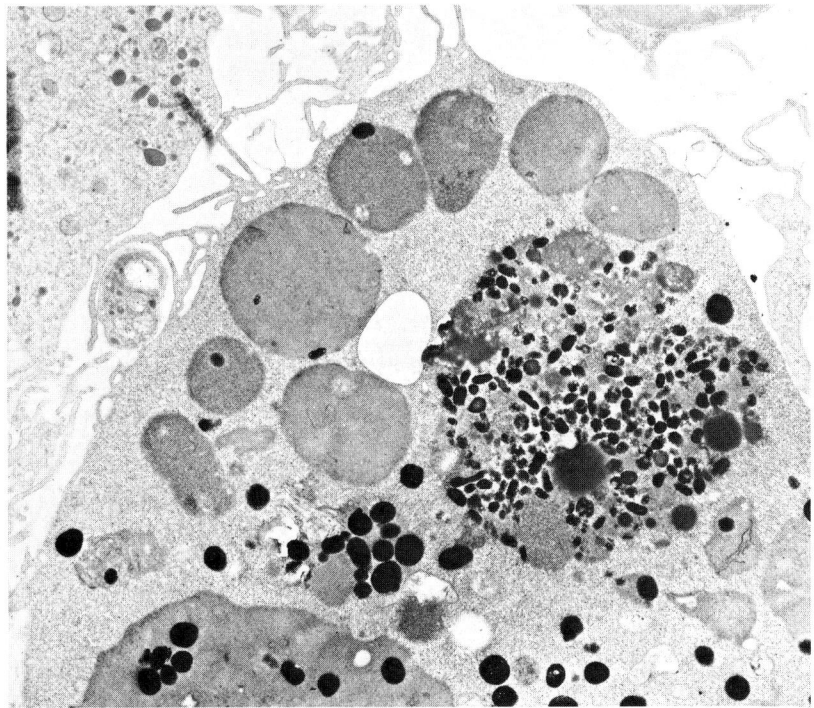

Fig. 7–6. Portion of a pigment cell in liver of senile *Ambystoma mexicanum*. Such cells are very numerous in old animals. They do not seem to be altered parenchymal cells. Some evidence for breakdown of dense granules is shown here, occurring in a mass of some less electron-dense substance. ×12,000.

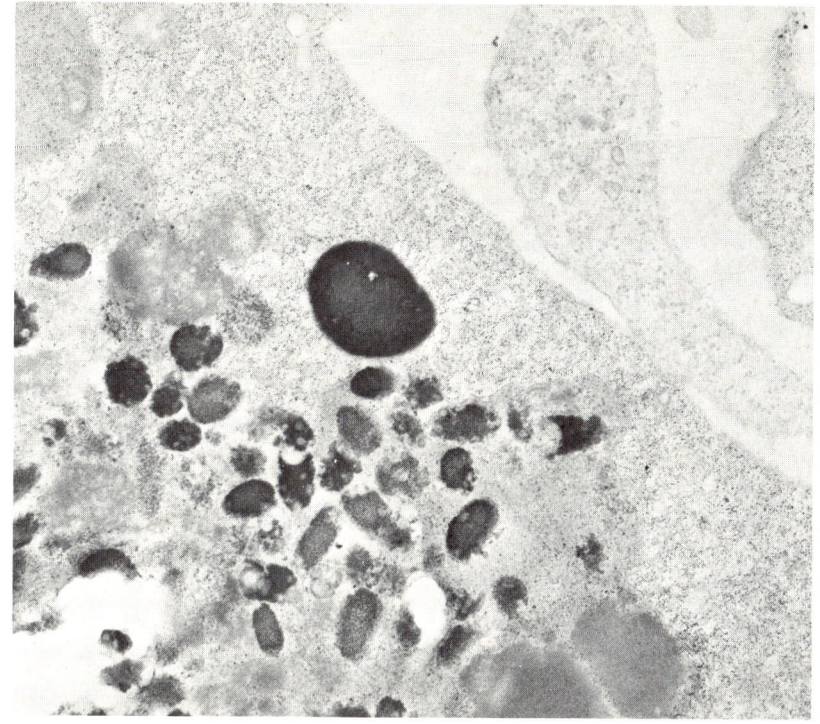

Fig. 7-7. Detail of dense pigment in senile Ambystoma liver. Disintegration of some granules seems to be occurring. ×41,600.

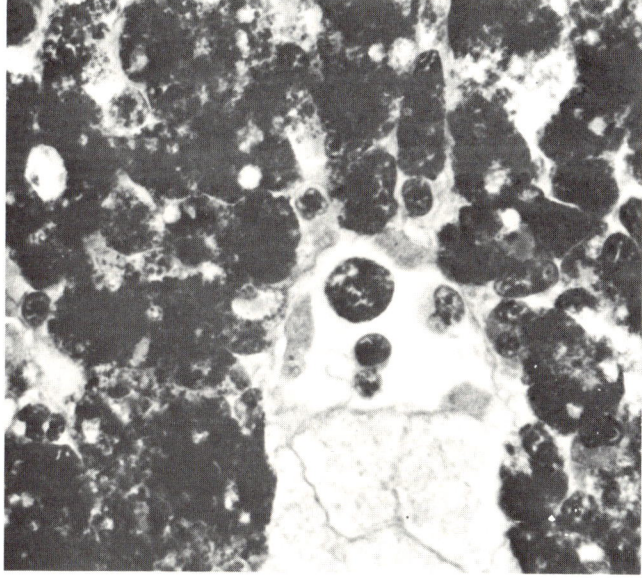

Fig. 7-8. A cluster of pigment cells in senile Ambystoma liver. The shape of these cells appears somewhat rounded in the tissue and clearly so when they are free in a vessel. This is another evidence that they are a type of cell different from those of the hepatic parenchyma. ×400.

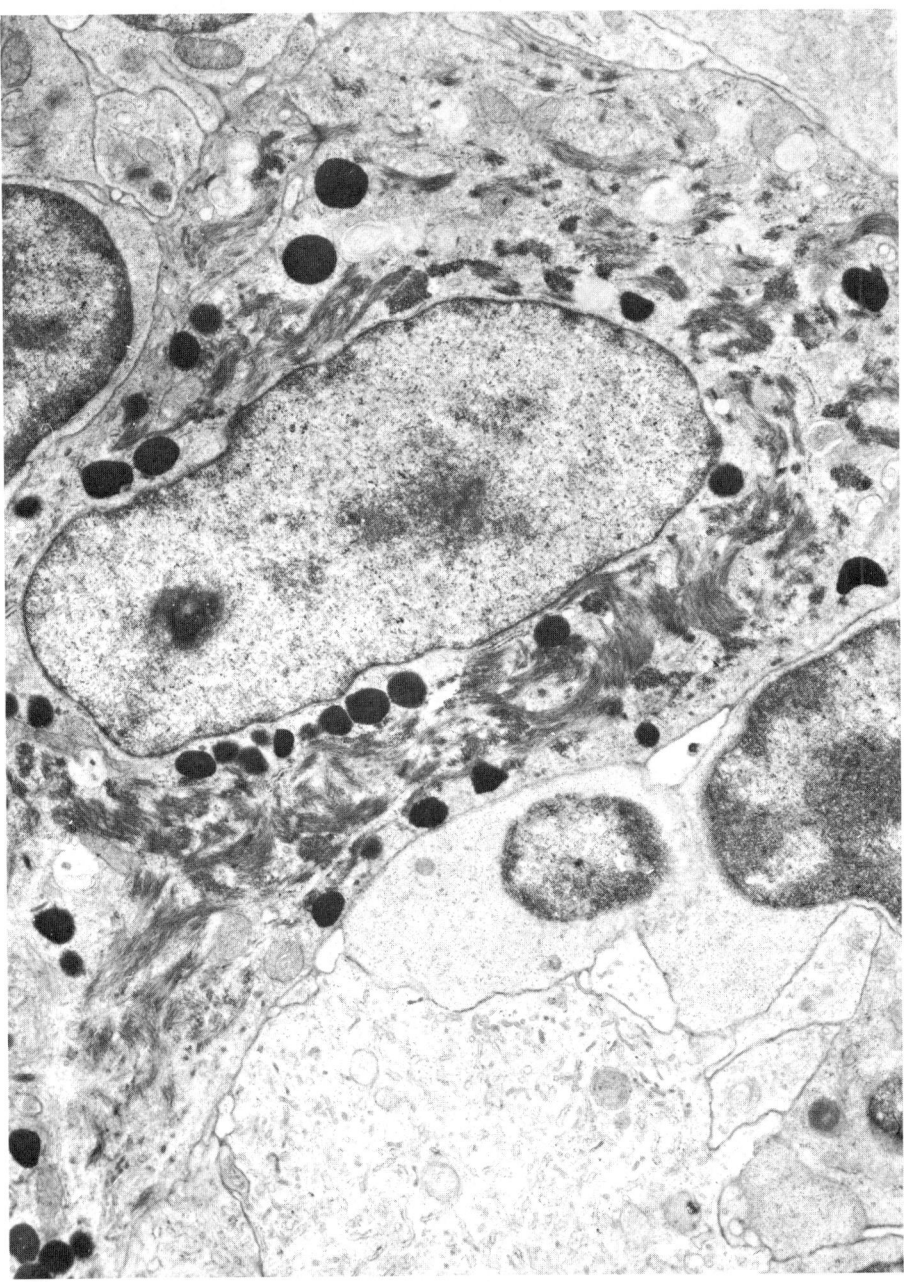

Fig. 7–9. A reticular epithelial cell in the thymus of an adult *Rana pipiens*. These cells can be identified by the presence of tonofilaments and they often show desmosomes and inclusions of various kinds. Karnovsky's Fixative, post fixed in Osmium tetroxide. Stain: Uranyl acetate and lead citrate. ×13,800. (Courtesy of S. D. Canaday.)

electron microscopy and, of course, before a large number of the investigations that have been carried out on lipofuscin. Further studies obviously are needed concerning the chemical and physical nature of this pigment, its source, and its effect on the economy of the liver. It is our opinion that these cells actually are macrophages, derived from the cells of von Kupffer (Fig. 7-8).

It seems probable that the pigmentary changes with age in the liver of the Axolotl is a fruitful field for work on the general metabolic changes in senescence. Correlation of these changes with possible changes in other parts of the amphibian body also should be of significance.

Thymus in an Adult Amphibian

While there is little information on the status of the endocrine glands with advancing age in the amphibians, it is of interest to find that the thymus is still a well-developed organ in the adult frog, *Rana pipiens* (Canaday, 1968). It is clearly divided into cortical and medullary zones as in mammals. The cortex consists chiefly of thymocytes which apparently are identical with small lymphocytes; while the medulla contains epithelial and other cells, together with an occasional Hassall's corpuscle.

Electron microscope study shows the epithelial reticular cells with desmosomes and tonofilaments, a well-developed Golgi apparatus, rough endoplasmic reticulum, and inclusions of various sizes (Fig. 7-9). Both lymphoid and epithelial elements in this adult amphibian would seem to be functional elements. It is true that the epithelial cells are in some cases associated with cysts of either intracellular or extracellular type.

Cooper and Mandell (1969) have studied interactions between thymus and bone marrow in adult *Rana pipiens* to see whether the thymus possesses the ability to regenerate immunological competence. As a criterion of such competence the workers took the ability to reject skin allografts within normal time limits. The dosage of irradiation and the amount of bone marrow used was varied in different experimental groups. Undamaged *thymic* tissue was provided by means of excision before the frog was irradiated, with re-implantation afterwards. Histological study showed considerably less cellular depletion in the spleens of frogs with replanted thymic glands than in those thymectomized with no re-implanatation. Influence of bone marrow on the rejection of the allografts was shown by shielding half of a limb in some frogs, not in others. The former had significantly shorter MSTs (median survival time *of the graft*) than the latter.

Of four pairs of experimental groups in which the effects of the non-irradiated thymus was studied, only one group had a significantly shorter MST than the thymectomized group.

REFERENCES

Andrew, W., 1961. An electron microscope study of age changes in nerve cells, with particular reference to chromophilia and to pigment accumulation in man and laboratory animals. J. Geront., vol. 16, no. 4, p. 388 (Abstract).

Brocas, J. and F. Verzár, 1961a. Measurement of isometric tension during thermic contraction as criterion of the biological age of collagen fibres. Gerontologia, vol. 5, pp. 223–227.

——— 1961b. The aging of Xenopus laevis, a South African frog. Gerontologia, vol. 5, pp. 228–240.

Canaday, S. D., 1968. Light and electron microscopy of the thymus in adult *Rana pipiens*. Anat. Rec., vol. 160, no. 2, p. 326 (abstract).

Comfort, A., 1964. Ageing: The Biology of Senescence. New York: Holt, Rinehart and Winston, 365 pp.

Cooper, E. L., and M. L. Mandell, 1969. Bone-marrow thymus interactions in amphibian allograft immunity. Anat. Rec., vol. 163, no. 2, p. 172 (Abstract).

Flower, S. S., 1925. Contributions to our knowledge of the duration of life in vertebrate animals. II. Batrachians. Proc. Zool. Soc., London, vol. 1, 269–289.

Jordan, H. E., 1931. The pigment content of the liver cells of urodeles. Anat. Rec., vol. 48, no. 2, pp. 351–366.

Noble, G. K., 1931. The Biology of the Amphibia. New York: McGraw-Hill.

Okamoto, H., 1925. Über die Leber und Milzpigmente der Kröte. Frankf. 2. Path., vol. 31, pp. 16–53.

Schneider, K. M., 1932. Zum Tode des Leipzigen Riesensalamanders. Zool. Gart., Lpz., vol. 5, p. 142.

Senning, W. C., 1940. A study of age determination and growth of Necturus maculosus based on the parasphenoid bone. Amer. J. Anat., vol. 66, p. 483–492.

Walterstoff, W., 1928. Triton (Cynops) pyrrhogaster 25 Jahre. Blatt. Aquar. Terrar. Kde., vol. 39, p. 183.

8
Reptiles and Birds

Reptiles

It is among the reptiles that we find the most remarkable longevity. Flower (1943) gives the growth curves for two turtles that were treated in captivity. The record from the first year to the fortieth shows a surprisingly protracted growth period.

There are at least five species of turtles that have lived over 100 years in captivity. Of these three have lived respectively 102, 123, and 152 years. There is, however, no method of age determination for wild turtles such as that furnished by the study of scales for fishes. While members of the other suborders of reptiles do not show as great a longevity as turtles, there are species of snakes, lizards, and crocodiles which have life spans ranging from 20 to at least 50 years. These reptiles generally live far longer than warm-blooded vertebrates of comparable size. Bourlière (1946) summarized much of the earlier data on longevity of reptiles.

Whether the larger species of tortoises (Fig. 8-1) have a longer life because of their greater determinate size or whether they are larger simply because they have a greater life span in which to grow is one of the interesting questions of zoological gerontology. It seems to be true that species with smaller individuals may also have an impressive longevity. The Loggerhead, *Caretta caretta*, for example, is listed as 33 years and the Snapping Turtle, *Macroclemmys temminickii*, as "over 58 years" (Flower, 1937).

According to Townsend (1931) the widely accepted idea that large Galápagos tortoises must be of very great age is not well based. Under favorable conditions in captivity, they attain a large size in a few years, and then there is a slowing of the growth rate. However, he states that there *are* records of both the Galápagos (Fig. 8-1) and Aldabra giant tortoises which lived over 150 years while under observation in tropical climates, when their lives ended on removal to cooler countries. Certain museum specimens weighing over 500 pounds must have been very old.

Observations on rattlesnakes in the wild condition (Fitch, 1949) indicate that growth is rapid in the young, slowing down but generally *continuing* in adults and seeming not to reach a stable level.

Snakes in the field have been marked for individual identification by clipping the ventral plates in various combinations.

Bryuzgin (1939) drew attention to an interesting phenomenon in some of the skull bones of reptiles, consisting in the appearance of "growth rings." These are

Fig. 8–1. The Galápagos turtle, *Testudo elephantopus*, a reptile which has a great longevity and may attain a weight of 500 pounds. (Courtesy of Dr. R. S. MacKay, reprinted with permission from Biomedical Telemetry: Sensing and Transmitting Biological Information from *Animals and Man*, second edition, John Wiley, Inc.).

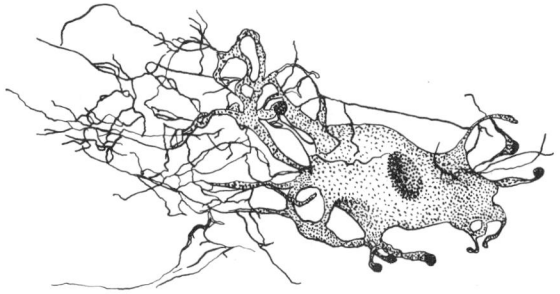

Fig. 8–2. Nerve cell in a sympathetic ganglion of a turtle, *Testudo nigrita*, of advanced age. The cell is an example of a type of overgrowth, being of giant size and showing luxuriant development of its processes to form a "fenestrated" structure. Method of Cajal. (Redrawn, from Bastai and Dogliotti, 1937, after Levi, 1926.)

particularly conspicuous in the anterior part of the ectopterygoid bone. He concluded that narrow, transparant bands correspond to the period of winter cessation of growth while the broad, less transparent bands correspond to the spring and summer period of active feeding and of growth. Bourlière followed up the work of Bryuzgin, studying the ectopterygoid of two species of snakes, *Natrix natrix* and *Vipera aspis*. Bourlière compared the body length of snakes from the field with that of snakes raised from eggs in a terrarium and the ages of which were accurately known. From this comparison and the examination of the ectopterygoid bones, he concluded that the method of Bryuzgin is a generally accurate means of age determination of snakes of the temperate zone, at least, and that snakes taken in the field can now be divided into a number of age classes.

In these snakes, Bourlière finds, growth during the first 5 years is very rapid and that it remains high even up to 15 years. Females have a much more rapid rate of growth than males, so that old females of *Natrix natrix* may be twice as long as males of a corresponding age.

There is an increase with age in the number of eggs in a "clutch" in the species which have been studied. Thus it appears that the fertility of snakes, like that of fishes, increases with age, along with the increase in size. Senescent infertility such as is found in warm-blooded vertebrates and, as we have seen, in some of the invertebrates, has seldon been found in these cold-blooded forms as a clear-cut phenomenon. An exception is the short-lived mosquito fish (*Gambusia affinis*) where Krumholz (1948) describes a decrease in size of broods with age and finally a period of senile sterility.

There is also some evidence that in the very oldest specimens of rattlesnakes (Klauber, 1936) and of haddock (Raitt, 1932) a slight decline of fertility has set in. We have described the findings of Rasquin and Hafter (1951) on the anatomical changes with age in testes of Astyanax. In relation to snakes, Volsoe (1944) found that in his oldest male specimens of *Vipera berus* growth of the testes had not kept up with growth of the animal as a whole. Changes with age in nerve cells of turtles have been described (Fig. 8–2.)

Among reptiles two general patterns of growth are seen: (1) a continuous growth at a declining rate, and (2) growth to a specific size reached relatively early in life, followed by cessation of growth. Both kinds of growth appear to occur in different species of chelonians (Sergeev, 1937).

Many reptiles show what appears to be a complete cessation of growth when a certain size has been attained.

Peabody (1961) was able to describe growth history by the study of the skeletons of crocodiles and lizards. He studied both recent and fossil specimens.

Birds

Ornithologists in recent times have devised a number of marking or labeling procedures which have made it possible to obtain data on life span of wild birds to add to the information which we have on domestic ones. By these means a general concept of life span of the majority of families living in the temperate zones has been gained.

In general, the small and fast-breeding species of birds are shorter lived than the large and slow-breeding ones (Bourlière, 1959). In the common house sparrow (*Passer domesticus*) there is a mortality rate of about 87 per cent for birds 1 to 4 months old and about 40 per cent for adult birds, (Summers-Smith, 1959) (these figures refer to urban populations). The average annual mortality for great tits (*Parus major*) in Holland was found to be 49 per cent (Kluijver, 1951).

The lowest mortality figures among birds are found in certain sub-Antarctic species. The yellow-eyed penguin (*Megadyptes antipodes*) showed a mortality of 12.9

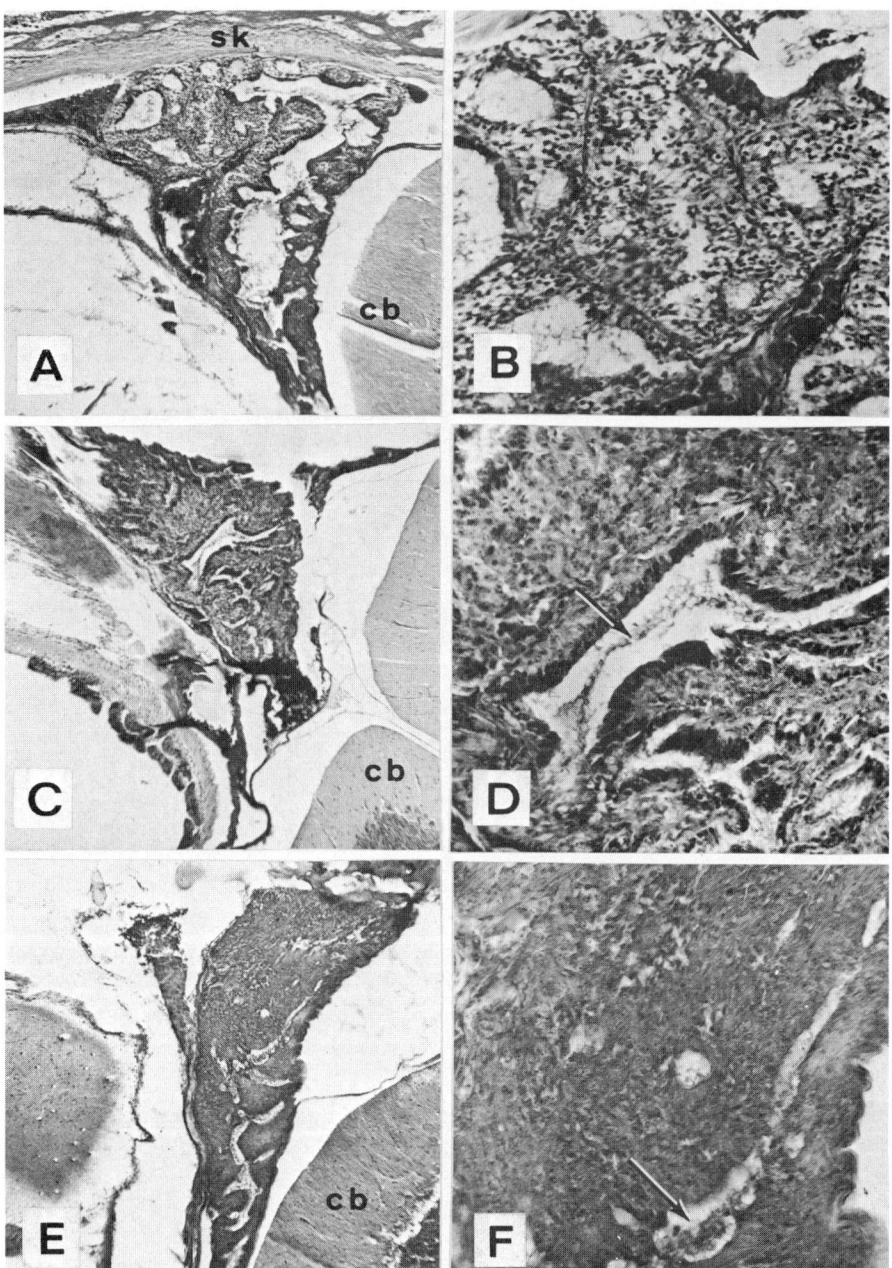

Fig. 8–3. Age changes in the pineal gland of the house sparrow, *Passer domesticus*. The figures are from median sagittal sections through the gland. (A) 15–30 days of age. ×90. (B) Higher magnification of same gland. Numerous intercellular spaces are seen. ×250. (C) 75–90 days of age. ×90. (D) Higher magnification of same gland. Intercellular spaces are few and there is a well-defined follicle wall. ×250. (E) 165–185 days of age. ×90. (F) Same gland at higher magnification. A solid cell mass and a dense interfollicular material are seen. ×250. sk = skull, cb = cerebellum. Arrow in B, D, and F points to follicle. (Courtesy of Dr. Charles L. Ralph.)

for adult birds (Richdale, 1957). The lowest figure is that for the royal albatross (*Diomedea epomorphora*) and is only 3 per cent! This species reproduces only every second year and may reach an age of 25 years.

According to Bourlière, then, the same general relationship between size and longevity is seen in birds as in mammals. Comfort (1964), however, does not agree with this statement.

Ralph and Lane (1968) have studied the pineal gland of the house sparrow (*Passer domesticus*). One hundred birds captured over a period of 2 years were studied. Ages of the birds, if they were 180 days or less, were determined by the skull ossification method of Nero (Wilson Bull., 63: 84, 1951). The older ones, unfortunately, cannot give an accurate age determination.

Two trends were seen with increasing age: the pineal gland decreased in volume and its character changed from a follicular to a more solid type. The youngest sparrows examined (20–35 days) showed wide lumina with several diverticula and many lacunae among the cell clusters (Fig. 8–3 A, B). In birds 120 days of age and more, the lumina are much reduced in size and number, while the cell mass is very solid in aspect (Fig. 8–3 E, F).

REFERENCES

Bourliere, F., 1959. Lifespans of mammalian and bird populations in nature, pp. 90–105 *in* Ciba Foundation Colloquia on Ageing, vol. 5, The Lifespan of Animals. Boston: Little, Brown and Co.

Bryuzgin, V. L., 1939. A procedure for investigating age and growth in Reptilia. Compt. rend. Acad. d. sc. URSS., vol. 23, pp. 403–405.

Clark, T. B., 1940. The relation of production and egg weight to age in White Leghorn fowls. Poultry Sci., vol. 14, p. 54.

Comfort, A., 1964. Ageing: The Biology of Senescence. New York: Holt, Rinehart and Winston, 365 pp.

Dominic, C. J., 1962. The ovary of the domestic pigeon, Columba livia, with special reference to follicular atresia. Proc. Ind. Sci. Congr., vol. 49, p. 405.

Fitch, H. S., 1949. Study of snake populations in Central California. American Midland Naturalist, vol. 41, pp. 513–579.

Flower, S. S., 1925. Contributions to our knowledge of the duration of life in vertebrate animals. III. Reptiles. Proc. Zool. Soc., London, vol. 2, pp. 911–981.

——— 1937. Further notes on the duration of life in animals. III. Reptiles. Proc. Zool. Soc., London, vol. 107, pp. 1–39.

——— 1945. Persistent growth in the tortoise Testudo graeca for thirty-nine years with other notes concerning the species. Proc. Zool. Soc., London, vol. 114, pp. 451–455.

Klauber, L. M., 1936. A statistical study of the rattlesnakes. Occ. Papers, San Diego Soc. Nat. Hist., vol. 1, pp. 2–24.

Kluijver, H. N., 1951. The population ecology of the Great Tit, *Parus m. major* L. Ardea, vol. 39, pp. 1–135.

Mackay, R. S., 1970. Bio-Medical Telemetry: Sensing and Transmitting Biological Information from Animals and Man, Second Edition, New York: John Wiley.

Payne, F., 1943. The cytology of the anterior pituitary of broody fowls. Anat. Rec., vol. 86, No. 1, pp. 1–13.

——— 1946. The cellular picture in the anterior pituitary of normal fowls from embryo to old age. Anat. Rec., vol. 96, pp. 77–91.

——— 1949. Changes in the endocrine glands of the fowl with age. J. Geront., vol. 4, pp. 193–199.

———— 1952. Cytological changes in the cells of the pituitary, thyroids, adrenals and sex glands of the aging fowl. Baltimore: Williams & Wilkins. Chapter 16 *in* Lansing, Problems of Aging, pp. 381–402.

Peabody, F. E., 1961. Annual growth zones in living and in fossil vertebrates. J. Morph., vol. 108, pp. 11–62.

Ralph, C. L. and K. B. Lane, 1968. Morphological changes in the pineal bodies of house sparrows (*Passer domesticus*) as a function of age. Anat. Rec., vol. 160, no. 2, pp. 412–413 (abstract).

Richdale, L. E., 1957. A Population Study of Penguins. Oxford: Clarendon Press, pp. 195.

Sergeev, A., 1937. Some materials to the problem of the reptile post-embryonic growth. Zool. Zh., vol. 16, p. 723.

Summers-Smith, D., 1959. The house sparrow, Passer domesticus: population problems. Ibis, vol. 101, pp. 449–454.

Townsend, C. H., 1931. Growth and age in the giant tortoise of the Galápagos. Zoologica, vol. 19, No. 13.

part III
Man and Mammals

INTRODUCTION TO PART III

Longevity of Mammals

Flower (1931) gave many interesting facts concerning the duration of life in the various orders and species of mammals. He pointed out that much of the "evidence" for great longevity for some mammals, such as the India elephant, is not well based, since the records for an animal said to be in its tenth decade may not go back more than half of that. He himself brought a number of elephants out of Africa and does not believe that these animals either have the longevity assigned to them by tradition.

Flower says (*ibid.*, p. 230): "The longest lived mammal is man, living to, and over, 100 years. The Primates, monkeys and lemurs, appear to be the longest-lived order of Mammals." The only mammals, other than man, which exceed 50 years are the Asiatic elephant and rarely, the horse.

In the order Chiroptera the fruit-eating bats are very long-lived.

There seems to be some difference of opinion concerning the effects of domestication on longevity. Flower says (p. 145): ". . . the life-spans of domestic animals such as the Cat, Dog and Horse have not been increased or decreased by the untold thousands of years that they have been under human control."

Bourliére (1946) says that captivity augments longevity and lets a species attain its maximum life span. Perhaps this refers more, however, to captivity for one or a few generations rather than to any long-range effects such as those which Flower was discussing.

Mammals in Relation to Lower Vertebrates

In the various orders of mammals there are considerable differences in the capacity for regulation of the temperature of the body. There are some rather striking correlations between relative lack of this capacity (i.e., relative "cold-bloodedness") and great longevity. Thus the bats, or Chiroptera, show little ability to regulate temperature, and instances of unusual longevity, *for size of the body*, are known in a number of species.

Flower (1931) cites various authors who have mentioned the long span of life of frugivorous bats in captivity. Small, free-living insectivorous bats also are long-lived (Bourliére, 1947). According to Trapido (1946) the vampire bat in captivity has attained an age of 12 years, while an age of $13\frac{1}{2}$ years has been recorded for *Rhinolophus ferrum-equissom*, and of 13 years for *Myotis myotis* (Bels, 1952).

Oxygen consumption is low in the bats during a large part of the day, and they produce only one or two young annually. In contrast, the shrews have a high metabolic rate, a high fecundity, and a short life span.

An important difference between animals which are able to regulate body temperature (homeotherms) and animals of similar size not able to do so (poikilotherms) is a greater basal heat production, i.e., a higher metabolic rate in the former (Prosser and Brown, 1961). Even at the same body temperature the difference in metabolic rate is found.

The great majority of mammals show a high degree of temperature-regulating capacity, and to this majority the human species belongs.

The Lifespan of Man

Whether the present life span of man is a "natural" one or one which, as many scientists have speculated, is shortened abnormally by factors other than a true senescence, we do not know. The age changes in systems, organs, and tissues which we describe in the present volume are, however, those which occur in the period of decline of the life which at present is allotted to the majority of members of our species.

A study of both the general functional and the clinical aspects of 197 men, aged 60 to 89, was made by Eiselt, Bosak, and Bojanovsky (1966).

They found that the number of cases of atherosclerosis, which was found in all men of over 70, and of emphysema and presbyopia, increased with age. Diagnosis of atherosclerosis did *not* correlate well with the lipid metabolism indices.

In relation to effects of physical culture, as given in the records of these men, the authors found that it tends to inhibit the development of emphysema and may have also helped to prevent right ventricular hypertrophy of the heart, but has no statistically significant influence on the development of atherosclerosis.

According to von Wüst (1965) the "decrease" in cancer mortality in persons over 80 years of age is only an apparent and a delusive one. At this age, he says, the covering of cancer cases is quite insufficient both because of incorrect diagnosis and a failure to report cases. The opinion that there is a reduced malignity of tumors in extreme senility also is incorrect, except for a decrease in incidence of metastases in the very old.

In man some of the general bodily changes with age are, of course, as much matters of common observation as of scientific scrutiny. Basilevich (1959), describing his studies on very old subjects ("longetarians," as he calls them) says: "An obvious external expression of the latter (atrophic changes) was their great leanness... their weight, with rare exceptions, was very slight. Only rarely did it go above 50 kg. for women and 60 kg. for men." He cites many earlier authors who found a similar decrease in weight and expresses his belief that any kind of overweight and especially a marked obesity in later decades should be regarded as a definitely pathologic feature.

It is of interest to note the numbers of longetarians studied by Basilevich. They are as follows:

Sex	Age in Years	Number of Subjects
Male	90–94	7
	95–99	7
	100–104	18
	105–110	5
	110–120	5
	Over 120	4
	TOTAL	46
Female	90–94	15
	95–99	9
	100–104	2
	TOTAL	26

These were inhabitants of Abkhasia in southern Russia. All were apparently healthy persons who were participating in society as respected members and who had not sought medical attention for many years preceding his study.

Other obvious changes of old age are in the skin (see Chapter 9). Basilevich (1959) says of his subjects (p. 122): "Their skin was almost always intensely damaged, under the destructive influence of time, especially on the exposed areas (face, neck, hands). It became slack, withered, attenuated, dry, wrinkled, and pigmented. Truly it looked like an old parchment." Further, he says (p. 122): "Such advanced involutionary skin changes as these longetarians exhibited we have rately seen on 60–80 year-old persons."

These very old subjects all showed some muscle atrophy, although it varied in degree between the robust farmers, where it was less, and the weaker city dwellers. Such atrophy can be seen not only on ordinary physical examination but also in more detail by study of x-rays.

The faces of the longetarians showed a high forehead with sparse hair of a gray, mixed-gray, or yellow-green color; eyes deep-set, mouth sunken with a greatly shrunken lower jaw, and prominent cheekbones. Deep wrinkles traversed the entire face. The nose and ears were prominent because of the atrophic changes of neighboring parts. Basilevich stresses the point that in the overwhelming majority of these very old persons the eyes "had remained young." For this reason, at least in part, these persons usually appeared younger than do the typical 70–80 year-olds.

These observations of Basilevich, and many others which he records, are important in that they probably represent an aspect of conditions as close to "normal" human aging as we can attain.

REFERENCES

Basilevich, I., 1959. The Medical Aspects of Natural Old Age. Review of Eastern Medical Sciences, Munich, Germany, 317 pp.

Bels, L., 1952. Fifteen Years of Bat Banding in the Netherlands. Natuurhistorisch Genootschap in Limburg, Maastricht, Holland.

Bourliére, F., 1947. La longévité des petis mammifères sauvages. Mammalia, vol. 2, pp. 111–115.

Eiselt, E., Vl. Bosák, and I. Bojanovsky, 1966. Biologische Aspekte alter Männer. Ztsch. f. Alternsforsch., vol. 19, pp. 279–292.

Flower, S. S., 1931. Contributions to our knowledge of the duration of life in vertebrate animals. V. Mammals. Proc. Zool. Soc., London, vol. 1, pp. 145–234.

Pearson, O. P., 1947. The rate of metabolism of some small mammals. Ecology, vol. 28, pp. 127–145.

Trapido, H., 1946. Observations on the vampire bat with special reference to longevity in captivity. J. Mammalogy, vol. 27, pp. 217–219.

Wüst, von G., 1965. Geshwulsthäufigkeit und Lebensalter. Ztsch. f. Alternsforsch., vol. 18, pp. 318–328.

9
The Skin and Fascia

The Skin

The skin is one of the most conspicuous features of the body of any person or animal. We frequently characterize individuals by the appearance of the skin. Age, of course, is one of the major factors in this appearance. The concept of the firm and elastic skin of a young person as compared with the pale or yellowed, somewhat flaccid and inelastic skin of the old person is widely held and often well-based.

The skin may be considered as an organ with a number of different functions. It forms a protective covering to prevent loss of water from the underlying tissues and to serve as a barrier against the entrance of harmful micro-organisms. Its myriads of sweat glands and its rich network of blood vessels make it an excellent temperature-regulating mechanism. It serves as a great receptor through which sensations of pain, temperature, and touch are communicated to the nervous system.

The skin consists of two major portions: (1) a deeper-lying connective tissue layer, and (2) a more superficial epithelial layer. The connective tissue portion of the skin is known as the dermis or corium, which may be described as a vast network of fibrous elements. Between and among the fibers a homogenous substance—the matrix—is found, which is "jelly-like" but varying in its consistency. In this environment then, we find the "inhabitants" of the dermis, the millions upon millions of cells.

A number of investigators have concentrated their studies on the fibrous elements of connective tissue of the dermis and on the changes with age in these structures.

Gross (1950) found an increase in the size of the fibrils in the skin as animals grew older. In the 60-day-old rat the average diameter was over twice that in the newborn one. The delicate fibrils of the very young rat are of the silver-staining or reticular type. Some authors believe that the delicate fibrils still are present in the mature animals but now are obscured by the coarser, collagenous, or white fibers which are laid down in the matrix and which become increasingly abundant (Nageotte and Guyon, 1930). In any event, the appearance of fibers in very young and in "grown-up" or mature animals is strikingly different. It seems probable that this represents a general tendency in connective tissue, as Schwarz

(1957) describes an increase in diameter of collagenous fibers in the cornea and sclera of the eye, in tendons, and in the meninges.

Apparently, the diameter of the fibrils becomes stabilized in mature individuals and in later life seems actually to decrease (Linken, 1955); but more work in needed on senile animal and human connective tissue to verify this last assertion.

A very important series of studies on some physical characteristics of the white fibers has been made by Verzár (1957). He found that fibers from tail tendons of senile rats are more capable of contraction against a heavier load, at the same shrinkage temperature, than those from young rats.

Thus far we have been concerned only with the white or collagenous fibers and with the delicate silver-staining fibers, the reticular type, which appear similar in many ways to the white ones. Distinct from both of these types of fibers are the elastic ones. These are yellow and show a high degree of elasticity. A peculiar alteration in a number of regions, as in skeletal and heart muscle, is an increase of elastic tissue with advancing age. In the skin we find an apparent increase in the amount of material taking the "elastin" stain, particularly in the skin of the neck. Collagen and elastin have nearly equivalent amounts of the amino acids, glycine, and proline. It has been suggested that the increase in amount of elastin in some places is due to actual transformation of collagen, the material of white fibers, into elastin, the material of the yellow fibers (Burton et al., 1955).

In human skin there are complications of individual variation in such factors as degree of exposure. Our study of the skin of laboratory rats showed the elastic network to be retained in its regular form in older animals and the elastic tissue seemed somewhat more abundant in them than in middle-aged rats (Andrew, 1951).

The wrinkles and lines of the aged human skin show interesting histological changes. The collagenic connective tissue tends to lose its woven, felt-like structure. Individual fibers are strongly basophilic, having lost the normal acidophilic character. Such changes occur in both the superficial and deep layers of the dermis (Himmel, 1903; Volarelli, 1920). The nuclei of fibroblasts often are elongated. It seems that these are the changes, combined often with prolonged or repeated muscular action as in frowning or squinting, which cause the wrinkles. Old skin is far less elastic than young and has an increasing difficulty in smoothing out after wrinkling.

There are some interesting age-related features of the matrix of ground substance. When dermal tissue is broken down in a Waring blender, the products from very young and from old individuals are surprisingly different. From the young, masses of highly refractile, jelly-like material are obtained, with only very few fibrils. From the old, on the other hand, abundant fibrils are obtained and almost none of the amorphous matrix. Thus, while it is difficult to present exact quantitative data, the matrix is definitely more abundant in the very young, and there is good evidence that it undergoes qualitative changes of chemical and physical character.

In the literature on the skin, including both the descriptions of its histological structure in textbooks and more detailed treatises, there is relatively little concerning the cell population of the dermis. Microscopic studies at low power show the cellular inhabitants lying among the fibers and presenting a fairly uniform appearance. When special stains and a high magnification are used, one can readily see that there are several distinct kinds of cells. These cells are of the same general type whether one is looking at the dermis of a human being or of a laboratory animal such as the mouse, rat, or guinea-pig. In a study of the human skin, Andrew and Sato (1964) attempted: (1) to define types of cells present, and (2) to determine the total population and differences in the proportions of the various kinds of cells.

The most common type of cell in the dermis is the fibroblast (Fig. 9-1). It is usually a spindle-shaped cell with tapering processes at either end, containing a moderate amount of cytoplasm, and a nucleus which generally stains lightly and shows one or more nucleoli. Fibroblasts usually form 50 to 60 per cent or more of the total cell population of human dermis. Some fibroblasts show a dark, rather homogeneous nucleus, a scanty cytoplasm, and are smaller than the other fibroblasts.

The next most abundant type of cell in the dermis is the macrophage. These are large cells with abundant cytoplasm and an ovoid or rounded form, usually not showing processes. They are motile cells which wander about among the fibers. They are phagocytic and will devour micro-organisms as well as debris from degenerating cells or other sources. Macrophages constitute, according to Andrew and Sato, from 18 to 35 per cent of the total number of cells in human dermis.

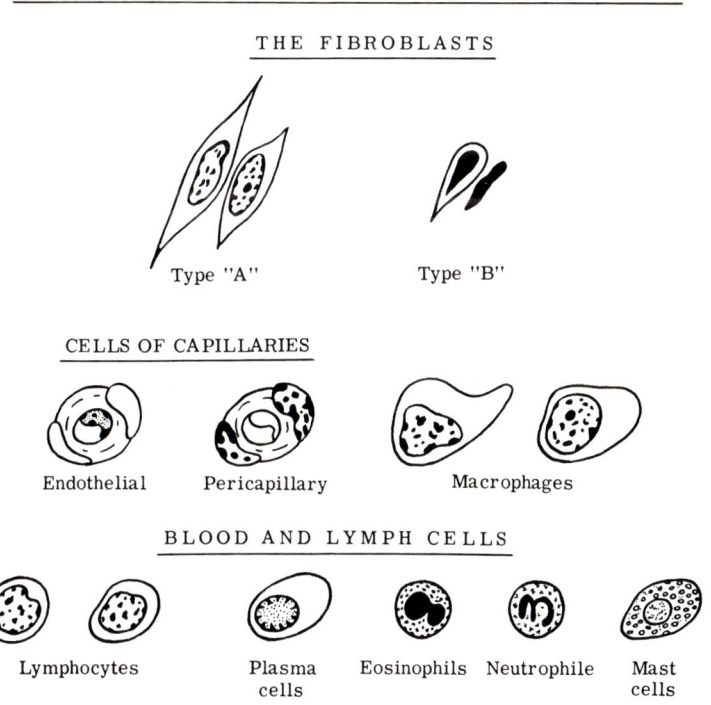

Fig. 9-1. The various kinds of cells found in the dermis. The relation to age of the general density of cell population and the proportions of the various cell types are discussed in the text.

The third type of cell is the endothelial cell of the capillaries. It is advantageous to include these cells in the investigation to provide quantitative information on the vascularization of the tissue at different ages. Endothelial cells range from 7 to 9 per cent of the total number. They are flattened cells with the nuclei causing a bulging of the central portion. They resemble fibroblasts in general structure, but with the electron microscope they show some distinctive features.

Closely related spatially to the endothelial cells is the fourth cell type, the pericapillary cell, a rather inconspicuous element constituting from 2 to 4 per cent of the total number of cells.

There are five other types of cells in the dermis; the lymphocytes and plasma cells are usually associated with the lymphatic circulation and with lymphoid tissue, although lymphocytes constitute an important part of the cells of circulating blood. In the dermis the lymphocytes constitute from 1 to 5 per cent of the cells. They are widely scattered and their source is obscure, but it seems probable that they are migrants from the vessels. The plasma cells are considered as cells of the tissue rather than from the vessels although they are weakly motile. They constitute only a fraction of 1 per cent of the cells. It will be recalled that they, as well as the lymphocytes, are considered very important in the production of antibodies. The lymphocytes of the dermis are small cells with relatively scanty, basophilic cytoplasm, and deeply staining nuclei. The plasma cells, supposed by many to be derived from lymphocytes, have somewhat more cytoplasm and an eccentric nucleus with its chromatin usually in a characteristic "cart-wheel" pattern.

The neutrophils and eosinophils are similar to the cells of the same name which occur in the circulating blood. They each constitute only a fraction of 1 per cent of the total population, with the eosinophils less numerous than the neutrophils. They are motile cells, and apparently are migrants from the blood vessels. In conditions of acute inflammation the neutrophils are greatly increased in number, while in allergic conditions of the skin eosinophils become very numerous.

The last type of cell, the mast cell, varies from less than 1 per cent up to about 3 per cent. These cells are tissue cells with coarse granules crowding the cytoplasm. The granules usually take what is called a metachromatic stain, i.e., when treated with a blue dye such as toluidine blue, they present not a blue, but a different, somewhat lavender color. Mast cells are found in a number of places in the body in addition to the dermis, as in the heart, uterus, liver, and submucosa of the alimentary tract. They are considered important in the manufacture of the anticoagulant substance, heparin. Several research reports on skin, heart, and alimentary canal have described a decrease in these cells with advancing age (Hellstrom and Holmgren, 1950; Constantinides and Rutherdale, 1957; Lindholm, 1959) yet Marx et al. (1960) found an increase with age in mast cells in rat liver, lung, kidney, and thymus; and Simpson and Hayashi (1960) found mast cells more numerous in dermis of old male mice of the C57 Black strain and less numerous in old female mice than in young ones of the same sex. In addition to the nine types of cells which have been described, there are some cells which did not seem to fit well into any classification and this small group has been designated simply as "unclassified cells."

Andrew and Sato (1964) studied 419 specimens of human abdominal skin. These were persons who had died of various causes but in whom there were no known dermatological conditions. There were three groups of subjects: 144 Caucasian males, 172 Japanese males, and 103 Japanese females. The biracial nature of the material seemed advantageous in determining whether age differences override racial differences.

The cells in a given area were counted, then the slide moved to the adjacent area and an additional count made, continuing in this manner until a total area of 34,810 sq. μ (0.034810 square mm), had been studied for total cell population in each human subject. For studies of the types of cells, the number of any given type was taken as a per cent of the total.

To evaluate vascularity a greater area was used because of the relative paucity of the endothelial cells. The area covered, then, for endothelial cell population (E.C.P.) for each subject was 203,918 sq. μ.

Tables indicate the results of the counts on the total number of the cells in the dermis and the results are summarized in a stereogram (Fig. 9-2). All of the Cau-

Table 9-1. Total Number of Cells in Dermis of Human Abdominal Skin in Autopsy Cases (Caucasian Male)

Age in years (cases)	Mean value ± standard error (c.v.)									
under 1 (11)	90.90 ±4.86 17.7%	under 1 (90.90)								
1–10 (10)	83.50 ±4.01 15.19%	-	1–10 (83.50)							
11–20 (10)	76.70 ±2.42 9.93%	-	-	11–20 (76.70)						
21–30 (11)	69.18 ±2.39 11.47%	**	**	-	21–30 (69.18)					
31–40 (12)	58.66 ±4.39 25.94%	**	**	**	-c	31–40 (58.66)				
41–50 (11)	50.90 ±4.48 29.13%	**	**	**	**	-	41–50 (50.90)			
51–60 (10)	54.20 ±4.08 23.81%	**	**	**	-	-	-	51–60 (54.20)		
61–70 (24)	45.50 ±1.79 19.34%	**	**	**	**	**	-	-	61–70 (45.50)	
71–80 (25)	50.72 ±1.86 18.09%	**	**	**	**	-	-	-	-	71–80 (50.72)
81– (20)	46.15 ±1.60 15.53%	**	**	**	**	**	-	-	-	-

** = The difference between two groups in highly significant at a probability level of 0.01.
* = The difference between two groups is significant at a probability level of 0.05.
- = There is no significant difference between two groups.
c.v. = coefficient of variation = $\dfrac{\text{standard deviation}}{\text{mean}} \times 100$.

Tukey studentized range test for comparing each group mean with each other group mean, 9-1 to 9-7.

From Andrew et al., 1964, Gerontologia, vol. 10, pp. 1–19. Reprinted with permission.

casian specimens are from males, having come from the Veterans' Hospital at Indianapolis. Table 9-1 for Caucasian males indicates that the highest concentration of cells occurs in the youngest individuals and a considerably higher population in the first 4 decades of life than in the later decades. In the 9 columns to the right is a comparison of each group to each of the other groups. Here a - sign indicates a lack of statistically significant difference, a single asterisk * a significant difference, and a double asterisk **, a highly significant difference. For example, the difference between the mean value of the 11-20 and each of the older age groups (31-40 and over) is highly significant.

In the Japanese male series (Table 9-2) a rather strikingly similar trend is observed, beginning with a mean value of over 109 cells, remaining well above 60 cells through the 41-50 year group, and dropping into the 42 and 45 cell figures in the persons over 70 years of age.

In the Japanese female series (Table 9-3) the figures vary somewhat more but again the trend is evident. The rather pronounced plateau of the 11-20 through the 51-60 decades may mean a real difference between the sexes in the rate of cell population decrease. A similar study of a series of female subjects of a different race would be of considerable value to correlate the results.

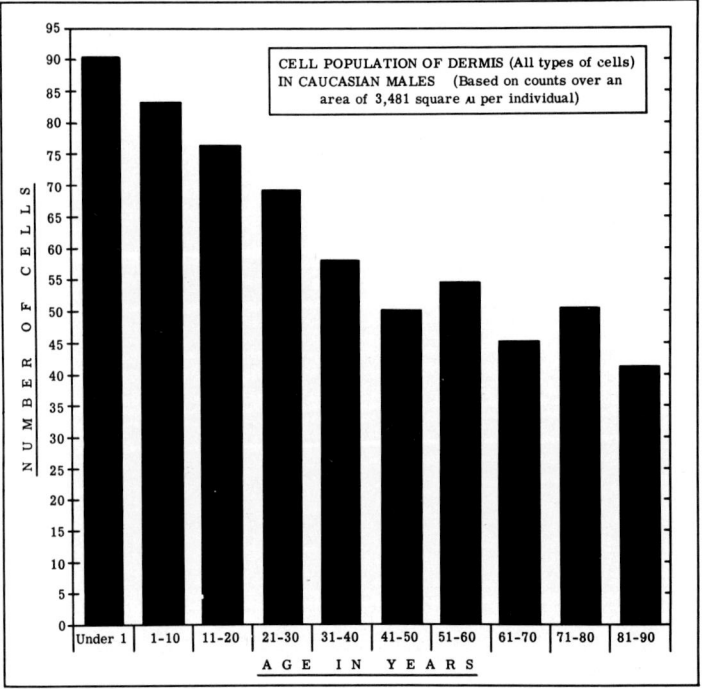

Fig. 9-2. Cell population of the dermis of abdominal skin in a series of 144 Caucasian males. The high level in the first two decades descends to somewhat of a plateau in the middle decades but the oldest individuals show the least dense population. (After Andrew *et al.*, 1964.)

The Skin and Fascia 77

Stereograms present the change in total cell population in graphic manner. A stereogram (Fig. 9-2) shows the trend in the Caucasian male; Stereogram 9-3 presents the picture for the Japanese, both males and females. Thus graphically shown, the aging factor certainly seems more striking than do the differences between the sexes at any given age. The data obtained for the total cell population presents a clear picture and, on the whole, a consistent one.

The decrease in numbers of endothelial cells in later life is not a very consistent one in these three series, and it would not seem justified to assign a decrease in number of capillaries as a cause for decrease in total cell population.

The classification of cells has revealed two small but probably significant age differences in human subjects. These concern, first, the fibroblasts, in particular the subdivisions of this very large group of cells, and second, the mast cells. In individuals under 1 year of age the fibroblasts with small cell bodies and dark nuclei, which have been called Type B, form a relatively small per cent of the total and those with larger bodies and lightly staining nuclei, Type A, a relatively large

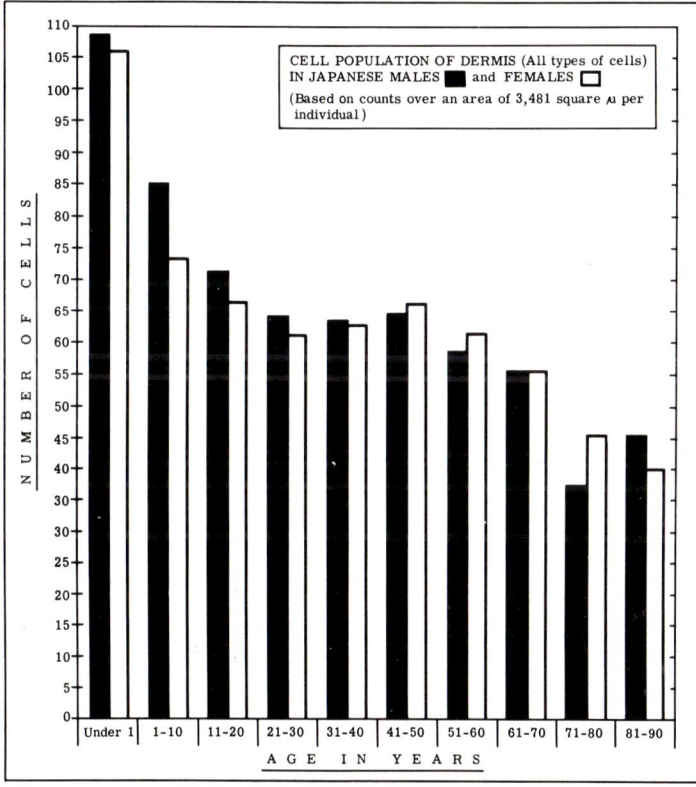

Fig. 9-3. Cell population of the dermis in two series of human subjects, Japanese males (172 individuals) and Japanese females (103 individuals). A trend similar to that for Caucasian males is seen for both sexes, but the decrease with advancing age is somewhat more marked. (After Andrew et al., 1964.)

Table 9-2. Total Numbers of Cells in Dermis of Human Abdominal Skin in Autopsy Cases (Japanese Male) in 34,810 Sq. μ

Age in Years (cases)	Mean Value ± Standard Error (c.v.)									
under 1 (7)	109.14 ± 7.09 17.16%	under 1 (109.14)								
1–10 (7)	85.85 ± 5.10 15.70%	**	1–10 (85.85)							
11–20 (15)	72.40 ± 3.33 17.83%	**	-	11–20 (72.40)						
21–30 (17)	64.23 ± 2.68 17.23%	**	**	-	21–30 (64.23)					
31–40 (23)	63.08 ± 2.57 19.53%	**	**	-	-	31–40 (63.08)				
41–50 (20)	64.30 ± 2.44 17.01%	**	**	-	-	-	41–50 (64.30)			
51–60 (35)	58.17 ± 1.69 17.20%	**	**	**	-	-	-	51–60 (58.17)		
61–70 (29)	55.48 ± 1.64 15.08%	**	**	**	-	-	-	-	61–70 (55.48)	
71–80 (16)	42.18 ± 1.20 11.40%	**	**	**	**	**	**	**	**	71–80 (42.18)
81– (3)	45.33 ± 1.35 5.16%	**	**	**	-	-	-	-	-	-

For explanation of symbols see Table 9-1.
From Andrew *et al.*, 1964, Gerontologia, vol. 10, pp. 1–19. Reprinted with permission.

per cent. In the Japanese male group the percentage of Type B goes from 4.42 in the "under one year" group up to 15.18 in the 71–80 year group.

In the Japanese female series, however, the "under-one year" group has 10.5 per cent of the total cell number of the Type B fibroblasts and there is a rather indeterminate set of figures in the later decades.

The percentage of Type B cells is fairly high in the Caucasian male series through the 11–20 years decade, then very low until the oldest group, the 81–90, in which it rises remarkably.

The mast cells, the general nature of which has been described above, show a tendency to rise in the last several decades of life in the Japanese males and in the last decade in the Japanese females and the Caucasian males as well.

Of the other "minor" cell types, with reference to their numerical frequency, the lymphocytes are perhaps most significant. The plasma cells, eosinophils, and neutrophils are present in the dermis at all ages but no evidence of age change in their numbers has been found. In the lymphocytes, however, the Japanese series, both male and female, show a relatively very small number in the individuals under 1 year. This corresponds in an interesting manner with the relatively low count of

Table 9-3. Total Numbers of Cells in Dermis of Human Abdominal Skin in Autopsy Cases (Japanese Female) in 34,810 Sq. μ

Age in Years (cases)	Mean Value ± Standard Erroe (c.v.)									
under 1 (8)	106.59 ± 4.95 13.11%	under 1 (106.59)								
1–10 (10)	73		1–10							
11–20 (7)	66.42 ± 4.53 18.22%	**		11–20 (66.42)						
21–30 (1)	61.10 ± 2.44 12.90%	**	-		21–30 (61.10)					
31–40 (16)	62.93 ± 2.55 16.25%	**	-	-		31–40 (62.93)				
41–50 (14)	66.71 ± 3.12 17.52%	**	-	-	-		41–50 (61.71)			
51–60 (12)	61.00 ± 2.22 12.60%	**	-	-	-	-		51–60 (61.00)		
61–70 (16)	55.68 ± 2.16 15.51%	**	-	-	-	-	-		61–70 (55.68)	
71–80 (13)	50.07 ± 2.30 16.55%	**	*	-	*	**	*	-		71–80 (50.07)
81– (6)	39.33 ± 1.41 8.79%	**	**	**	**	**	**	*	-	

For explanation, see Table 9-1.
From Andrew et al., 1964, Gerontologia, vol. 10, pp. 1–19. Reprinted with permission.

lymphocytes in the epidermis which was found in the abdominal skin of young laboratory animals in an earlier study (Andrew and Andrew, 1956). The number increased in middle age and still more so in old age in the rat.

In the Caucasian male series, the earlier decades show fewer lymphocytes in general, but the "under-one-year" group does not have a very low percentage and variations are considerable among the decades.

In the Caucasian males (the only Caucasians available) and in both sexes of the Japanese the number of dermal papillae protruding into the epidermis in a given area of the very youngest groups is relatively smaller. In the 11–20 decade and young adult groups the counts showed the highest value. Beginning with the middle-aged group the number of papillae decreases continuously with increasing age.

The decrease with age is highly significant statistically (Tables 9-4 through 9-6). The number of papillae in the oldest group is approximately half of that found in the youngest group. In older groups, not only does the number of papillae decrease, but there is also a change in their shape. The papillae become flat, with a loss of the rete pegs.

Table 9-4. The Number of Papillae in Given Area (203,918 Sq. μ) of Abdominal Skin of Caucasian Male

Age in Years (cases)	Mean Value ± Standard Error (c.v.)	under 1	1–10	11–20	21–30	31–40	41–50	51–60	61–70	71–80
under 1 (11)	20.8 ±1.71 27.3%	under 1 (20.8)								
1–10 (10)	21.3 ±1.12 16.6%	–	1–10 (21.3)							
11–20 (10)	22.8 ±1.27 17.6%	–	–	11–20 (22.8)						
21–30 (11)	24.2 ±1.26 17.3%	–	–	–	21–30 (24.2)					
31–40 (12)	19.0 ±1.30 22.6%	–	–	–	*	31–40 (19.0)				
41–50 (11)	18.7 ±0.84 14.8%	–	–	–	*	–	41–50 (18.7)			
51–60 (10)	17.2 ±1.14 20.9%	–	–	–	**	–	–	51–60 (17.2)		
61–70 (24)	14.9 ±0.88 29.0%	**	**	**	**	–	–	–	61–70 (14.9)	
71–80 (25)	15.4 ±0.29 15.8%	**	**	**	**	–	–	–	–	71–80 (15.4)
81– (20)	10.8 ±0.64 26.6%	**	**	**	**	**	**	**	*	**

For explanation, see Table 9-1.
From Andrew et al., 1964, Gerontologia, vol. 10, pp. 1–19. Reprinted with permission.

The difference in number of papillae in a given area between Caucasian males and Japanese males is highly significant. The number in Japanese males is higher than in Caucasian males in all age groups.

Several investigators have studied the thickness of the epidermis at different ages. Evans, Cowdry, and Nielson (1943) found a change in thickness in the antecubital fossa from 33.8μ in persons 19–30 years of age to 27.3μ in persons 80–94 years of age. Thuringer and Cooper (1950) spoke of atrophy of epidermis although the number of dividing cells was somewhat greater in old persons.

In the large series of skin specimens in which quantitative studies were made on the dermis, the epidermis also was studied. Figures 9-4 and 9-5 show skin sections from a 34-year-old female and an 83-year-old female subject respectively. They are prepared with a connective tissue stain known as Masson's stain. Again, the loss of rete pegs may be seen. Figure 9-6A shows the epidermis of the abdominal skin of a 3-year-old female child. The appearance of the epidermis usually is described in relation to configuration of its basement membrane, the negative image of dermal papillae and other protrusions. The so-called "rete pegs" which appear like down-growths of the epithelium, are conspicuous. The epidermis be-

Table 9-5. *The Numbers of Papillae in Given Area (203,918 Sq. μ) of Abdominal Skin of Japanese Male*

Age in Years (cases)	Mean Value ± Standard Error (c.v.)								
under 1 (7)	14.1 ±1.01 18.95%	under 1 (14.1)							
1–10 (7)	15.5 ±1.76 29.86%	-	1–10 (15.5)						
11–20 (15)	17.8 ±0.72 15.70%	-	-	11–20 (17.8)					
21–30 (17)	18.1 ±0.69 15.90%	-	-	-	21–30 (18.1)				
31–40 (23)	17.4 ±0.55 15.31%	-	-	-	-	31–40 (17.4)			
41–50 (20)	14.3 ±0.55 17.48%	-	-	**	**	**	41–50 (14.3)		
51–60 (35)	11.4 ±0.45 23.49%	-	*	**	**	**	**	51–60 (11.4)	
61–70 (29)	10.1 ±0.59 31.26%	*	**	**	**	**	**	-	61–70 (10.1)
71–80 (16)	9.5 ±0.50 21.00%	**	**	**	**	**	**	-	-	71–80 (9.5)
81– (3)	7.8 ±1.73 38.31%	*	**	**	**	**	**	-	-	-

For explanation, see Table 9-1.
From Andrew *et al.*, 1964, Gerontologia, vol. 10, pp. 1–19. Reprinted with permission.

tween pegs is relatively thick. Figure 9–6B shows the skin of a 67-year-old female subject. Here the rete pegs have been "erased" but the general thickness is not much altered. Figure 9–6C shows the skin of an 83-year-old male, in which not only are the rete pegs gone but this particular area shows a remarkable decrease in thickness.

While the structure of individual sebaceous glands often is well preserved, there is some evidence of atrophic changes in such glands in old age (Epstein, 1946).

It must be remarked that Evans, Cowdry, and Nielson (1943) felt that a good deal of the difference in epidermis thickness of skin of old and young human subjects is due to the much greater tendency of the more elastic young skin to "shrink" or pull together on removal. They find that epidermis *removed* from the dermis by their special method shows less of this "age" difference. The process of dermo-epidermal separation itself may, however, introduce other artefacts.

Weidman (1942) said: "As to the epiderm, there is no doubt that it undergoes atrophy with age." Yet, in speaking of keratotic lesions, he cites the process of senility as predisposing to hyperplasia of the epidermis. Cowdry and Andrew

82   Man and Mammals

Table 9-6. The Numbers of Papillae in Given Area (203,918 Sq. μ) of Abdominal Skin of Japanese Female

Age in Years (cases)	Mean Value ± Standard Error (c.v.)	under 1 (13.9)	1–10	11–20 (16.2)	21–30 (18.5)	31–40 (17.0)	41–50 (14.6)	51–60 (11.2)	61–70 (8.9)	71–80 (10.6)
under 1 (8)	13.9±0.95 19.3%	under 1 (13.9)								
1–10 (1)	14.00		1–10							
11–20 (7)	16.2±1.19 19.4%	-		11–20 (16.2)						
21–30 (10)	18.5±0.78 13.4%	*		-	21–30 (18.5)					
31–40 (16)	17.0±0.96 22.0%	-		-	-	31–40 (17.0)				
41–50 (14)	14.6±0.57 14.6%	-		-	*	-	41–50 (14.6)			
51–60 (12)	11.2±0.91 28.1%	-		**	**	**	-	51–60 (11.2)		
61–70 (16)	8.9±0.68 30.6%	**		**	**	**	**	-	61–70 (8.9)	
71–80 (13)	10.6±0.52 17.8%	-		**	**	**	**	-	-	71–80 (10.6)
81– (6)	8.5±0.70 20.1%	*		**	**	**	-	-	-	-

For explanation, see Table 9-1.
From Andrew et al., 1964, Gerontologia, vol. 10, pp. 1–19. Reprinted with permission.

(1950) point out that in senile keratoses there often are areas of marked atrophy of epidermis accompanying those of hypertrophy.

Epidermis of Laboratory Animals

Andrew (1951), studying epidermis of abdomen, back, and ear in Wistar Institute rats, found no significant alteration in thickness with age in any of these regions when animals of middle age (300 days) were compared with senile ones (900–1000 days). The degree of variation in thickness in a given region is, however, greater in senile rats.

Some conspicuous differences are seen when the individual layers of the epidermis are compared. A stratum granulosum is lacking in epidermis of young and middle-aged rats in the skin of abdomen, back, and ear, and is only slightly developed on the soles of their feet. Senile rats (900–1000 days), on the other hand, show a well-differentiated stratum granulosum, with coarse basophilic granules in the cells, in all of these regions (Fig. 9–7C).

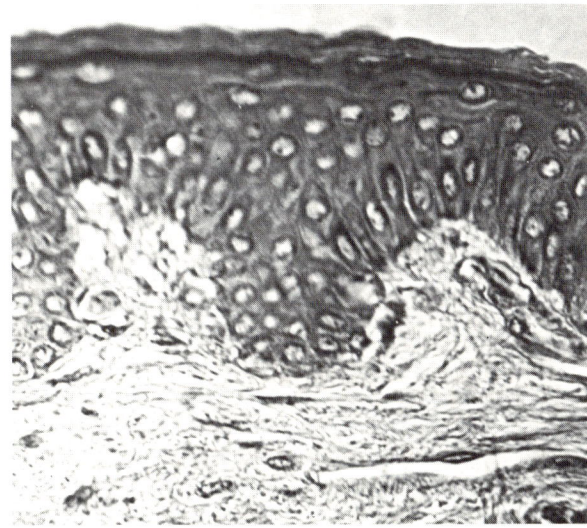

Fig. 9-4. Abdominal skin of a 34-year-old Caucasian female subject. Surgical. Masson's connective tissue stain. ×250.

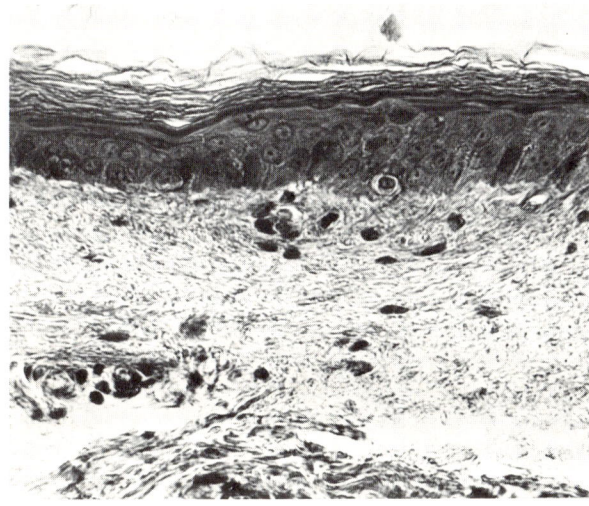

Fig. 9-5. Abdominal skin of an 83-year-old Caucasian female subject. Surgical. Masson's connective tissue stain. ×250.

The stratum corneum in the senile animals consists of lamellae which seem more tightly adherent and which are more eosinophilic than those in young and middle-aged ones (Fig. 9-7B).

The stratum germinativum of the rat also shows an interesting qualitative change. Nuclei of the cells in younger animals are generally spheroidal and clear; while in old ones they are more often ovoid, orientated with the long axis at right angles to the surface of the skin, and have a more granular aspect. While quantitative investigation was not undertaken on this feature, the nuclei in younger animals seem to be larger and present an appearance of crowding, the senile rats showing apparently wider areas of cytoplasm with more prominent "intercellular bridges" in the spinous layer.

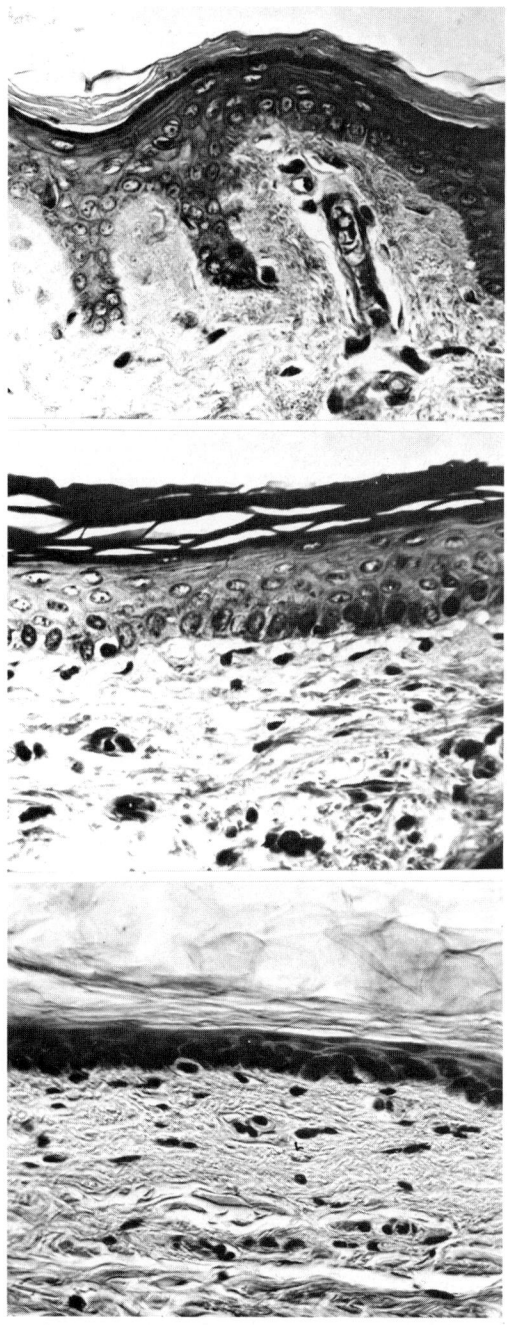

Fig. 9–6. Abdominal skin of Caucasian subjects of different ages. (A) Three-year-old female. The downward projections of the epidermis are conspicuous. Surgical. Iron hematoxylin and Orange G. ×250. (B) Sixty-seven-year-old female. The epidermo-dermal junction is smooth in many places, as in this field. Iron hematoxylin and Orange G. Surgical. ×250. (C) Eighty-three-year-old Caucasian male. The epidermo-dermal junction is smooth and the epidermis in many areas, as here, is thin. Surgical. Iron hematoxylin and Orange G. ×250.

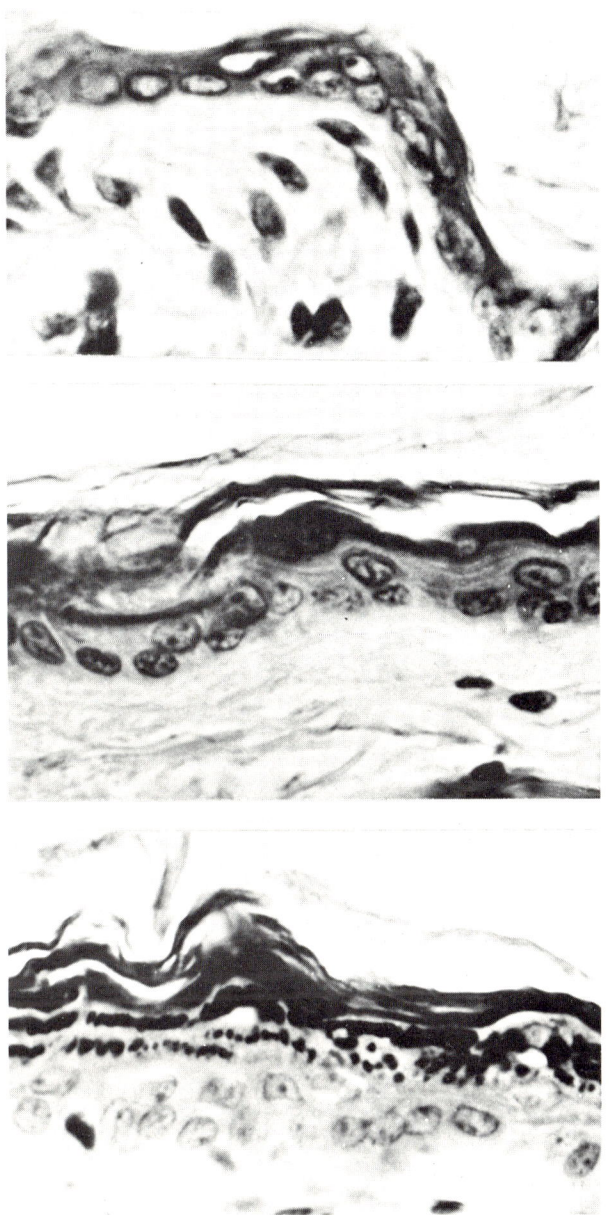

Fig. 9–7. Age changes in the skin of Wistar Institute rats. (A) Skin from the back of a 21-day-old male Albino rat. The nuclei are generally clear and present a crowded appearance. Harris' hematoxylin and eosin. ×810. (B) Skin from the back of a 900-day male Albino rat. A relative prominence of the cytoplasm is seen and the stratum corneum is more closely adherent. Harris' hematoxylin and eosin. ×810. (C) Skin from the back of a 1000-day male Gray Norway rat. A great degree of development of the stratum granulosum, common in old rats, is seen here. Harris' hematoxylin and eosin. ×810.

In several reports (Andrew and Andrew, 1949; 1956; Andrew, 1968) we have described the presence of lymphocytes in the epidermis which apparently have migrated from the dermis. The possible relationship of these cells to clear cells and to cells of Langhans lies somewhat outside of the study of age changes as such but may be of importance in relation to the renewal of the epidermis or to its defensive reactions.

The occurrence of areas where small groups of cells or individual cells are of an aberrant type is frequent in senile man and rat, according to our findings. Some cells may show the binucleate condition or separation of the single nucleus into lobes. Others may show an internal partition of the nucleus, or there may be a great hypertrophy of the nucleolus.

In the rat we have found a tendency in some areas for the cells to round up and separate from adjoining cells, to become more eosinophilic and even to form keratin deep within the stratum germinativum (Fig. 9-8).

A quantitative analysis of the cell population of the dermis, such as that by Andrew and Sato (1964) for the human subject, has not been undertaken for the rat nor apparently for other laboratory animals. General observation, however, shows a very dense cell population in dermis of 21-day-old rats, a much less dense one in animals of 300 days, and in 900–1000 day animals a generally sparse but rather widely varying population. The dermis of 21-day rats showed very delicate fibers compared with that of middle-aged animals. The "coarseness" in the latter

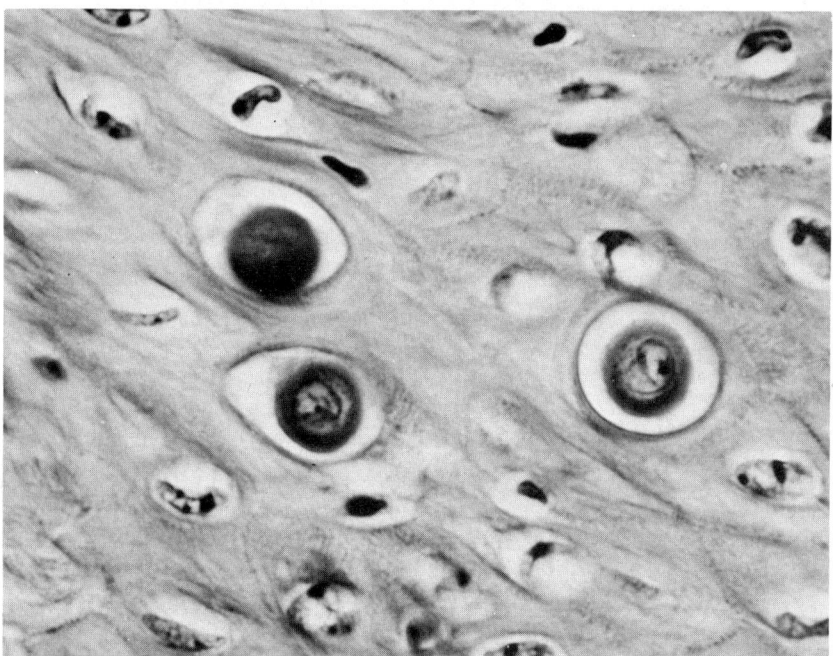

Fig. 9-8. Aberrant cells in the abdominal epidermis of a 900-day male Albino rat. In both rat and man aberrant cells, either singly or in small groups, as here, become increasingly frequent with advancing age. Heidenhain's iron hematoxylin and eosin. ×900.

may represent an aggregation into bundles. The dermis in our senile rats (Andrew, 1951) showed no degenerative changes in elastic fibers. One old animal, however, a 1000-day male Albino, presented many areas of its abdominal skin with deeply basophilic fibers in the process of disintegration into short threads and granules. The degenerating fibers appeared to be collagenous. Epidermis over these areas of fiber degeneration was highly vacuolated and contained many basophilic granules.

Localized Changes in Human Skin

The changes in the epidermis of the rat which we have mentioned, both nuclear and cytoplasmic, are only detectable by use of the microscope. There is in senile human skin, however, a variety of pathological lesions, many of which are not of great significance for the organism as a whole but some of which may prove fatal.

Lentigo senilis, consisting of brown, irregular areas, usually on the dorsal surfaces of the hands and forearms and on the face, seldom are seen before the fourth decade and tend to increase in size and number with age. There seems to be no relation between them and exposure to sunlight. No connection between general condition of health and these splotchy pigmented areas has been traced. According to Cawley and Curtis (1950) these areas show numerous clear cells and a heavily pigmented basal layer. There is no tendency to malignancy in them.

Pigmented naevi are generally considered as congenital malformations of the skin. They do not seem to increase in number with age but do undergo further differentiation, with the development in them of fusiform cells containing fibrils (Lund and Strobbe, 1949).

Senile keratoses are lesions occurring most often on exposed areas, again on the backs of the hands and forearms and on the face. The keratosis is a slightly raised area, irregular in shape and presenting a thin, adherent covering of keratin. These lesions are classed as "precancerous." The epidermal cells show a variety of abnormalities: discrepancies in cell size, hypertrophy of nucleoli, tendency to internal separation of nuclei, and amitotic nuclear division (Figs. 9–9 and 9–10). Aggregations of lymphocytes often occur in the underlying dermis. Microincineration studies yield less mineral ash from the lesions than from normal epidermis.

Another lesion common in senile skin is the seborrheic keratosis (Verruca senilis). It occurs as either single or multiple lesions and is most frequent on the trunk, although limbs, face and other areas may be sites. The seborrheic keratosis is rounded or ovoid, raised, yellow-brown to brownish-black and presents to the touch an oily feeling. Actually, this lesion has no relation to the sebaceous glands and is named from the oily nature of its surface. In the epidermis of this type of keratosis are many cysts, some of which contain laminated keratin, reminding us of our incidental findings in some areas of skin of senile rats. The seborrheic keratosis is not considered as a precancerous lesion.

The brilliant red, slightly raised areas of considerable frequency in older persons (although found also at earlier ages) are known as telangiectases and commonly called "blood-spots." The lesion consists of a group of wide capillaries in

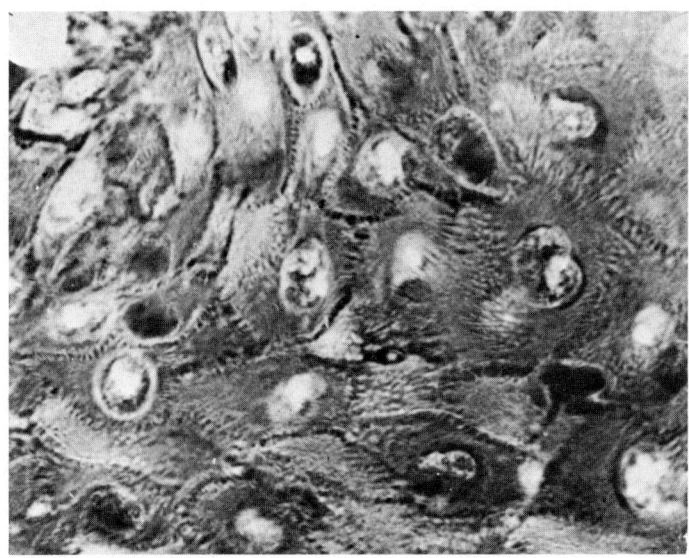

Fig. 9-9. Spinous cell layer in a senile keratotic lesion. Variation of size and appearance of cells is seen. Perinuclear "vacuoles" are common. ×600.

the dermis of the area involved. It is most common on the trunk. Weidman (1942) considered it as a purely hyperplastic vascular change.

In senile women another local change of common occurrence is the cutaneous tag or Acrochordon, often found in numbers on thorax and neck. They appear to be papillomas, showing a hyperplastic dermal connective tissue with loose, spongy structure and with wide blood and lymph channels. They probably are associated with hormonal deficiency.

The nails are a portion of the integument which provide a more or less permanent record, in chronological order, of the events of synthesis which lead to their formation. Lavelle (1968) made measurements on nail growth in rats, hamsters, mice, rabbits, and guinea pigs at different ages and over a total period of nine weeks for each age group.

The results showed no consistent change in the rate of nail growth with advancing age for rats, hamsters, or guinea pigs. In mice and rabbits there appeared to be a slight *increase* in the rate of nail growth with advancing age. This points up perhaps, the danger of having preconceived notions about what *should* happen in aging.

Age Changes in Fine Structure of Skin

There has been relatively little work in this field. Detailed studies of the fine structure of the "normal" fibroblast have been made and were reviewed by Movat and Neil (1962). Our own studies on human skin show the fibroblasts in the dermis of young and middle-aged persons to have the typical structure as described in the literature. The mitochondria have well-developed cristae and a clear matrix. The

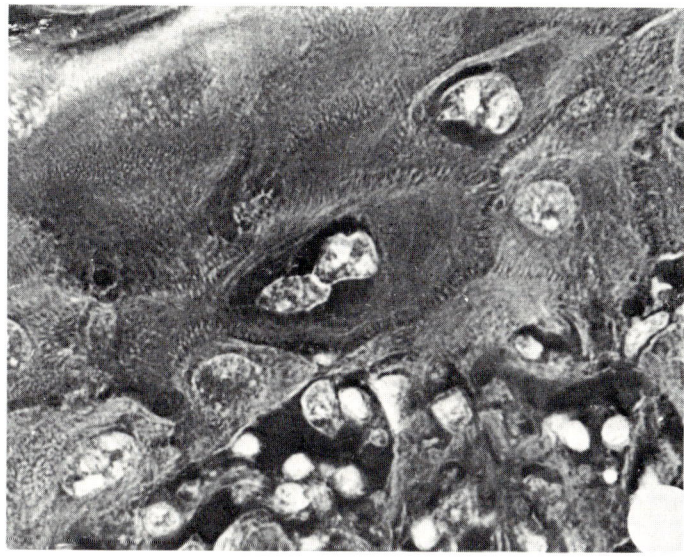

Fig. 9-10. Basal portion of epidermis in a senile keratotic lesion. The cell near to center of field shows a double nucleus in which separation of daughter nuclei has not yet occurred. Evidence indicates that binucleate cells in these lesions often arise by amitotic division. ×600.

endoplasmic reticulum is canalicular in profile and the channels are studied with ribosomes (Fig. 9-11). In the senile skin which we have studied with the electron microscope, which unfortunately represents relatively few individuals, the dermal fibroblasts show mitochondria in which the cristae are not regularly arranged and in which small, dense particles occur in the matrix. The endoplasmic reticulum is often of a vesicular nature (Fig. 9-12).

Examination of the fine structure of fibroblasts and the areas about them, in both young and old dermis, does serve to emphasize the probable relationship between these cells and the extracellular environment. These cells frequently contains masses of amorphous material which seem to be cast out to become matrix already present. While collagen fibrils with the striking 640 Å periodicity are seen close to many fibroblasts, there usually are, closest to them, some very fine fibrils of smooth type, completely lacking in visible periodic structure; and such smooth fibers occur also in the cytoplasm.

It seems, then, from the standpoint of changes with age, that changes in matrix and in fibers, and hence in the gross structure and appearance of the skin, may be dependent upon the more subtle changes within the cells of the dermis.

Subcutaneous Tissue and Fascia

The subject of aging of the subcutaneous tissue and of the various fasciae of the body is linked very closely with that of aging of connective tissue as such, for this tissue largely constitutes these structures, as well as being found in many other

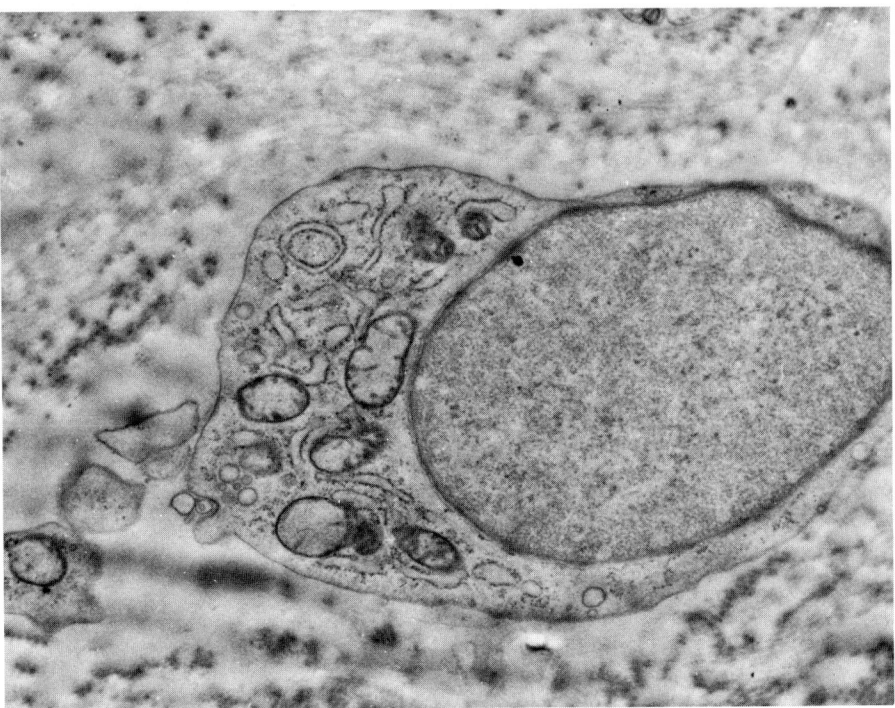

Fig. 9-11. Fibroblast in the dermis of abdominal skin of a 52-year-old male subject. The mitochondria show the classical structure. The endoplasmic reticulum is largely canalicular. Surgical. Palade's fixative. Millonig stain. ×23,000.

places throughout the body, including the stromal portions of glands, the submucosal layers, serosal or adventitial layers of many organs, and the outer layers of blood and lymph vessels.

Some of the constituents of connective tissue, compared with the cells, are relatively inert metabolically and thus should provide important models for the physical and chemical changes that occur in the body with the passage of time (Bensusan, 1958).

A thorough consideration of age changes in the non-cellular components of connective tissue would necessitate a fairly comprehensive preliminary discussion of what is known of their physicochemical makeup. This would be somewhat beyond the range of an *anatomical* discussion of age changes. Only the more general features of these constituents will be described here.

Collagen

There appears to be a decrease and in some cases even a disappearance of extractable collagen from connective tissues with increasing age (Banfield, 1952; Orekhovitch *et al.*, 1948, 1958). Such change probably is due to a cessation of synthesis of new collagen, together with a precipitation of recently formed extrac-

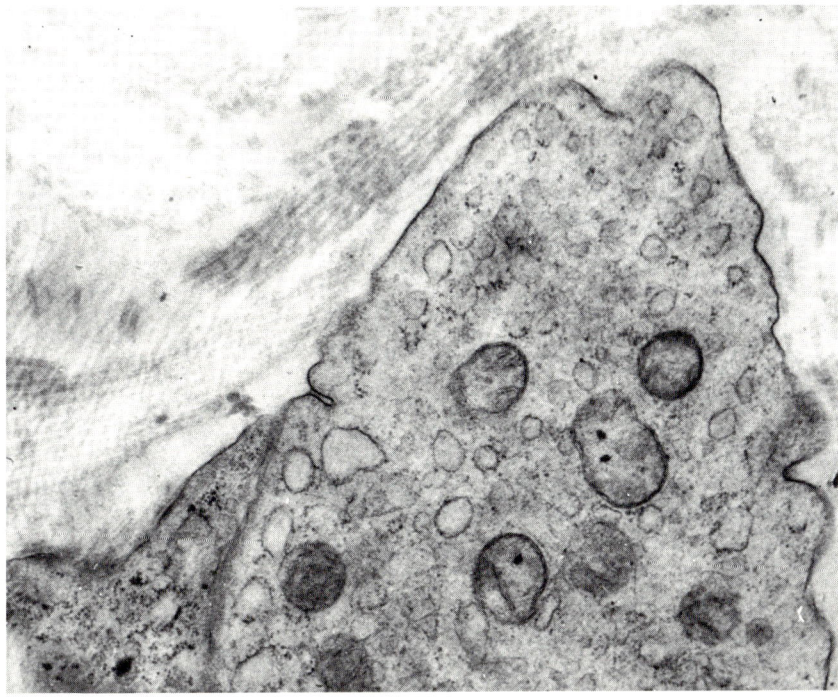

Fig. 9-12. Fibroblast of abdominal dermis of an 82-year-old male subject. Some mitochondria are of aberrant shape and appearance. The endoplasmic reticulum is chiefly vesicular. Surgical. Palade's fixative. Millonig stain. ×32,600.

table collagen in the form of fibrils which are not soluble. Formation of the fibrils with their typical periodicity appears to be the result of collagen molecules seeking and finding their most stable state of association. The process results in a decrease of the distance between the molecules and a more perfect crystalline organization.

Gross (1958a, b) has done extensive experimental work on the precipitation of collagen from solutions, using *in vitro* studies. According to him, the rate of synthesis of the extractable fraction of collagen (in skin) is proportional to the rate at which the animal is growing. On cessation of growth, whether this comes about due to the natural limitation of age, or from starvation, the synthesis of collagen also ceases. The collagen fraction extractable with neutral salt, present partly in molecular dispersion in the ground substance, partly as loose aggregates, and partly as freshly formed fibrils, becomes insoluble within a few days. The actual aging effect in a population of collagen molecules is, both *in vivo* and *in vitro* a relatively short process. The longer time for change *in vivo* is due to the continued synthesis of new collagen.

Evidence for a progressive increase in degree of *order* within the collagen fibrils has been obtained in wide-angle X-ray diffraction studies on various connective tissue structures (Chordae tendineae: Feitelberg and Kaunitz, 1949; tendo caleaneus: Hartmann, Gottam and Fricke, 1957).

Elastin

Chemical changes have been found in elastin with advancing age, at least in the elastin of the media of the aorta. Lansing, Rosenthal, and Alex (1950) showed an increase in calcium content. Purified finely ground elastin showed an age-related increase in histidine, arginine, threonine, and in the dicarboxylic acids (Lansing et al., 1951). In contrast, elastin from the pulmonary artery did not show such changes. In regard to physical features of elastin, several authors, including Krafka (1940) and King (1957) have shown a decrease in extensibility of the elastin derived from the aorta.

The Ground Substance or Matrix

Either simple teasing or fragmentation in a Waring blender shows a striking difference between connective tissue of the dermis of newborn and aged human subjects. The newborn yields few or no fibers but small pieces of a jelly-like ground substance which are highly refractile. The aged subject yields many fibers and very little ground substance. While this striking qualitative difference can be demonstrated, it is not yet possible to make accurate quantitative statements about the relative amounts of ground substance and fibrous material.

Determination of one component of the matrix such as hexosamine, mucopolysaccharide, or hyaluronic acid can be carried out but apparently does not indicate the true amount nor the condition of that matrix. Different portions of the same tissue may vary, indeed, in their mucopolysaccharide content (Meyer et al., 1957).

The relative proportion of hexosamine to collagen has been stated (Sobel et al., 1953, 1956, 1958a, 1958b) to be an indicator of senescence in bone and skin in man and in a number of animals. They consider the hexosamine to be an index of the amount of matrix or "gel," and the collagen to be an index of fibrous material. It is interesting that chronically ill patients have a much lower hexosamine-collagen ratio than normal subjects.

REFERENCES

Andrew, W., 1951. Age changes in the skin of Wistar Institute rats with particular reference to the epidermis. Amer. J. Anat., vol. 89, pp. 283–320.

——— and N. V. Andrew, 1956. An age difference in proportions of cell types in the epidermis of abdominal skin of the rat. J. Geront., vol. 11, pp. 18–27.

———, R. H. Behnke, and T. Sato, 1964. Changes with advancing age in the cell population of human dermis. Gerontologia, vol. 10, pp. 1–19.

Burton, D., D. A. Hall, M. K. Keech, R. Reed, H. Saxl, O. B. Tunbridge, and M. J. Wood, 1955. Apparent transformation of collagen fibrils into 'Elastin.' Nature, London, vol. 176, pp. 966–969.

Cawley, E. P., and A. C. Curtis, 1905. Lentigo senilis. Arch. Dermat. Syph., Chicago, vol. 62, pp. 635–641.

Constantinides, P. and J. Rutherdale, 1957. Effects of age and endocrines on the mast cell counts of the rat myocardium. J. Geront., vol. 12, pp. 264–269.

Cowdry, E. V. and W. Andrew, 1950. Some cytochemical and cytologic features of senile keratosis. J. Geront., vol. 5, pp. 97–111.

Epstein, S., 1946. Dermatitis in the aged. Geriatrics, vol. 1, pp. 369–383.

Evans, R., E. V. Cowdry, and P. E. Nielson, 1943. Aging of human skin. Anat. Rec., vol. 86, pp. 545–566.
Gross, J., 1950. Connective tissue fine structure and some methods for its analysis. J. Geront., vol. 5, pp. 343–360.
Hall, D. A., 1966. Connective Tissue—its role in the aging process. *In* "Perspectives in Gerontology," Springfield, Ill.: Charles C Thomas, pp. 125–133.
Hellstrom, B. and H. J. Holmgren, 1950. Numerical distribution of mast cells in the human skin and heart. Acta Anat., vol. 10, pp. 81–107.
Himmel, J. M., 1903. Zur Kenntnis der senilen Degeneration der Haut. Arch. f. Dermat. u. Syph., Wien u. Leipz., vol. 64, pp. 47–60.
Holečková, E. and Chvapil, M., 1965. The effect of intermittent feeding and fasting and of domestication on biological age of the rat. Gerontologia, vol. 11, pp. 96–119.
Linke, K. W., 1955. Elektronmikroskopische Untersuchung uber die differungzierung der interzellularsubstanz der menschlichen lederhaut. Z. Zellforsch., vol. 42, pp. 331–343.
Lavelle, C. 1968. Effect of age on the rate of nail growth. J. Geront., vol. 23, pp. 557–559.
Lund, H. Z. and G. D. Stobbe, 1949. The natural history of the pigmented nevus; factors of age and anatomic location. Amer. J. Path., vol. 25, pp. 1117–1155.
Ma, C. K. and E. V. Cowdry, 1950. Aging of elastic tissue in human skin. J. Geront., vol. 5, pp. 203–210.
Marx, L., M. Hirota, C. A. Printup, M. A. Warnick, and W. Marx, 1960. Effects of exposure of 5°C and advancing age on tissue mast cells and heparin in male rats. Amer. J. Physiol., vol. 198, pp. 180–182.
Milch, R. A., 1966. Aging of Connective Tissues. *In* "Perspectives in Experimental Gerontology," pp. 109–124.
Nageotte, J. and L. Guyon, 1930. Reticulin. Amer. J. Path., vol. 6, pp. 631–654.
Schwarz, W., 1957. "Morphology and differentiation of the connective tissue fibers." *In* Connective Tissue, R. E. Tunbridge, ed. Springfield, Ill.: Charles C Thomas, pp. 144–156.
Simpson, W. L. and Y. Hayashi, 1960. Distribution of mast cells in the skin and mesentery of BALB/c and C57BL mice. Anat. Rec., vol. 138, pp. 193–201.
Sobel, H., H. A. Zutraven, and J. Marmorsten, 1953. The collagen and hexosamine content of the skin of normal and experimentally treated rats. Arch. Biochem. Biophys., vol. 46, pp. 221–231.
——— and J. Marmorsten, 1956. The possible role of the gel-fiber ratio of connective tissue in the aging process. J. Geront., vol. 11, pp. 1–7.
———, S. Gabay, E. T. Wright, I. Lichtenstein, and N. H. Nelson, 1958. The influence of age on the hexosamine-collagen ratio of dermal biopsies from men. J. Geront., vol. 13, pp. 128–131.
———, E. T. Wright, and S. Gabay, 1958. Urinary steroids and the hexosamine-collagen ratio of dermal punches. Gerontologia, vol. 2, pp. 59–63.
Thuringer, J. M. and Z. K. Cooper, 1950. The mitotic index of the human epidermis, the site of maximum cell proliferation, and the development of the epidermal pattern. Anat. Rec., vol. 106, p. 255.
Verzar, F., 1957. "The aging of collagen." *In* Connective Tissue, R. E. Tunbridge, ed. Springfield, Ill.: Charles C Thomas, pp. 208–221.
Volarelli, F. F., 1920. Sulla Istologia E Sulla Patogenesi Delle Rughe. R. Accademia Dei Fisiocritici in Siena Atti Seri 8 V. 12, pp. 649–658.
Weidman, D., 1942. "Aging of the Skin." *In* Cowdry's Problems of Aging, 2d ed., Baltimore: Williams & Wilkins.

10

The Skeleton

Cartilage

Evidences of age change in the skeleton are more conspicuous and more readily preserved after death than are those in the great majority of the softer tissues of the body. Calcification of cartilage as a localized process may begin in relatively young adults but becomes pronounced only with advancing age. In very old persons actual formation of bone within the cartilage is of frequent occurrence. Noback (1949) has shown this change in the cartilages of the larynx. Cartilages of the thoracic basket and of other regions of the body also show calcification and, not infrequently, bone formation. Physiological effects of this tissue change include a lessened capacity for respiratory movements and a decreased elasticity of the skeleton in general.

The type of change leading to bone formation involves cellular proliferation and probably differentiation of chondroblasts or chondrocytes into osteoblasts.

Amprino and Bairati (1933) describe the changes in human cartilage which occur with growth and in the aging process.

Frankly degenerative changes, however, also occur in the aging process in all three main kinds of cartilage; hyaline, elastic, and fibrous.

In the *albuminoid* type of regressive change there is a deposition of proteid granules in the matrix. As degeneration proceeds, death of cartilage cells occurs, the cartilage substance in some areas becomes fluid, and cavities of varying size are formed. In the asbestiform type of degeneration, masses of rigid-appearing fibers are seen in parallel array, seeming almost to "crowd out" the matrix. This type of degeneration usually is seen only at very advanced ages. In the costal cartilages, in the cartilaginous structures of the respiratory tree, and in other such "permanent" cartilage, the various kinds of degenerative processes usually begin in the central portions and spread centrifugally.

In the normal differentiation of cartilage, there is what may be called an "aging" of individual cells. As precartilage changes into cartilage in an 8-day chick embryo, for example (Silberberg and Silberberg, 1961) the matrix becomes firmer while the nuclei shrink, cytoplasm disintegrates, and eventually some cells break down. If this process is to be succeeded by bone formation, the matrix, at first increased in density, undergoes a dissolution and often calcium is precipitated in it prior to its replacement by bone matrix.

The cycle of growth, maturation, hypertrophy, and degeneration of cartilage cells in the cartilage columns at the epiphyseal lines during ossification has been followed by Scott and Pease (1956) and by Policard (1958).

The cell capsules of the hypertrophied cells, although calcified, are opened by capillaries advancing from the bone marrow. The fate of the individual "senescent" cells appears to vary: some die, some survive as elements of the bone marrow, their "cartilaginous" character now lost, and others remain protected in their capsules. Rather surprisingly, a peak of glycogen content is reached before death. Whether this is due to a decreased expenditure of energy with a piling up of the stored food material or to some other cause is not known.

Eventually the growth and ossification zones themselves undergo a decline. Proliferative and hypertrophic activity slow down and stop, while the cell columns become separated from each other by large quantities of matrix. The columnar architecture becomes difficult to recognize. In the epiphyseal plates areas of partially calcified and degenerate cartilage are seen. Eventually the bone marrow is more or less walled off from the cartilage by a mass of bone. Ossification processes may invade the cartilage.

In the articular cartilages the reproduction of cells is said to occur by amitosis in older individuals (Elliot, 1936). Growth may be seen in such cartilage at rather advanced ages and *degenerative* processes seem to occur simultaneously with growth in many instances. As degenerative change begins to predominate, the cartilage matrix becomes increasingly acidophilic. Wagoner, Rosenthal, and Bowie (1941) found that respiratory activity of individual cartilage cells decreases with age but that glycolytic activity does not. With the appearance of actual degenerative change of the albuminoid or asbestiform type and the formation of cysts, erosions of the joint surfaces occur. A combination of growth and of regressive changes is often seen in degenerative joint disease of the aging.

Glycogen begins to appear in cartilage cells as soon as intercellular matrix begins to be laid down (Guizetto, 1910; Sundberg, 1924).

Fat appears in the cartilage cells at various times according to type and location of the cartilage. It is questionable whether either glycogen or fat, unless present in a very great quantity, is indicative of regressive changes in the cytoplasm.

The nucleus pulposus in the center of the fibrocartilaginous disc between the vertebrae shows its nature as a notochordal "rest" with large, fluid-filled cells, in the human subject until about 10 years of age. It is then replaced by fibrocartilage. About the third decade, the cartilage shows a decrease in number of cells and a loss of mucoid substance. Large bundles of coarse fibers appear and pores develop among them. After 30 years of age, pigment, in the form of auto-fluorescent Sudanophilic granules makes its appearance. Hansen (1956) believes that it may be partly lipofuscin ("age pigment") and partly hemosiderin (from a breakdown of hemoglobin). Calcification is often seen beginning at about this age as other degenerative processes begin. A number of clinical conditions of the back occurring beyond middle age are thought to be associated with those changes.

Bone

The earlier concept of bone as a somewhat static tissue has changed gradually to one of a tissue undergoing various dynamic changes through almost the whole of life: changes including appositional growth, resorption, and modeling. Even the gross aspect of bone is very different in old age from that found in youth and middle age (Figs. 10–1 and 10–2).

Hevelke and Falk (1965) investigated the magnesium content of the dry substance of human intervertebral discs in 92 subjects. The content is highest in early youth, of the order of 115 mg per cent. It then shows a progressive decrease to 65 mg per cent in the seventieth year but in later old age, and in the case of marked degenerate processes, the content again rises to reach 116 mg per cent in the 81–90 year group. The authors consider the increase to be an expression of excessive slagging.

It has long been recognized that in old age the substance of the bone becomes more "brittle," more fragile, and less resilient. The greater vulnerability of bone fracture in persons of advanced years, and the slowness and difficulty of healing,

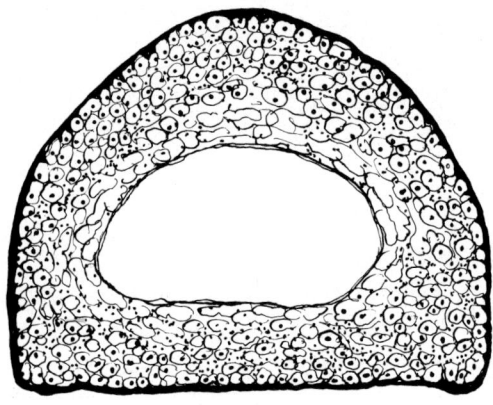

Fig. 10–1. Cross section of the diaphysis of a phalanx of a 39-year-old female subject. The bone is of normal thickness. (Redrawn from Amprino and Bairati, 1936.)

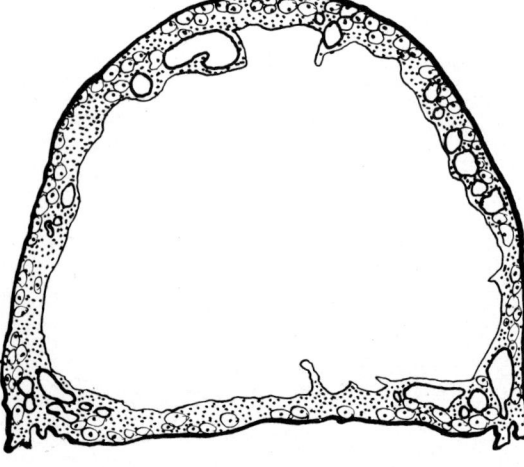

Fig. 10–2. Cross section of the diaphysis of a phalanx of a 90-year-old female subject. The bone is very thin and the medullary cavity much enlarged. (Redrawn from Amprino and Bairati, 1936.)

make this a very serious problem in this period of life. Where senile osteoporosis is in an advanced stage, the central medullary canals of many bones may be greatly enlarged while the compact bone around it is far narrower than in young people. In such cases there has been a long-sustained predominance of the resorptive over the osteogenetic processes. According to Amprino and Bairati (1936) the phenomena of appositional growth and of functional adaptation continue even into senility. The Haversian canals, however, often increase in diameter with age, the additional space being filled with fibrous or adipose tissue. Osteoclasts may increase in number. It is not at all clear which of various factors are of most importance in bringing about the changes seen in aging bone: changes in the osteogenetic cells *per se*, endocrine deficiencies, or sclerosing of the blood vessels supplying the bone.

Tonna (1966) has described a striking difference in the number of empty lacunae in femoral bone when 5 week (almost none), 54 week (3.7 per cent), and 106 week (31 per cent) mice are studied.

The sequence of age changes in the intercellular components of bone has been reviewed by Bourne (1956a, 1956b). The matrix in young bone shows metachromasia, while the delicate fibers are argentophilic and tend to form networks, being pre-collagenous in nature. Later, collagen fibers develop and in the matrix large amounts of PAS-positive mucopolysaccharides and phosphatase are seen. In the bone of the very aged, metachromatic staining and the PAS reaction become weaker, while phosphatase activity disappears.

Younger bone shows finer crystals of hydroxypatite. With advancing age there are increasingly coarser crystals and larger fibers (Robinson and Watson, 1955). According to Weidenreich (1930) the ash content of bone in infancy is one-half, in middle age four-fifths, and in senescence seven-eighths of the dry weight of bone.

In older individuals Bohatirchuk (1954) has found that new bone may be formed as a simple age change, with no association with disease of the joints. Males show a preponderance of this process over females (Bohatirchuk, 1955).

Bugyi (1966) has made a roentgenological study of men who had been active in sports earlier in life, following them from 60 years up to over 75 years. The somatotypes were determined according to the method devised by Sheldon (1947). The roentgenological measurements showed a rather sudden reduction of musculature occurring between the fiftieth and sixtieth year of life, with a reduction of the subcutaneous fatty tissue between the sixtieth and seventieth years. No change in the width of the bones of upper arm, thigh, and leg was found, but there was a substantial narrowing of the cortex of the shafts of the bones.

There are many factors which may modify the genetic pattern of aging of the skeletal parts of the body. The major ones, however, are nutritional and endocrine. Often it is impossible to separate the effects of these two factors because of their close interaction. If prolongation of the span of life occurs, the final degree of senescence of skeletal parts will be the same as that seen in the more rapid aging (Silberberg and Silberberg, 1950, 1955).

The incidence of radiographically demonstrable osteoporosis is surprisingly high in older persons, according to the findings of Gitman and Kamholtz (1965). They carried out routine X-ray studies of the dorso-lumbar spine on all new admissions to a large geriatric facility. Readings of the films were carried out without knowledge of age or sex of the patient. Factors evaluated were:

1. "demineralization" as judged by degree of "whiteness," depending on density of the bone;
2. atrophy, as shown by thickness of the vertebral cortices and disappearance of trabeculae in the spongy bone; and
3. softening as shown through (a) invasion, by pressure, of the disc into the bodies of adjacent vertebrae, (b) collapse, and (c) exaggerated curves of the spine.

The incidence was determined in 1,518 persons over the age of 65 years. Over-all incidence for females was 65.8 per cent, and for males, 21.5 per cent.

The question whether osteoporosis is related to osteoarthrosis has been investigated in an extensive series of old mice. Silberberg and Silberberg (1962) studied the legs of 321 animals ranging in age from 24 to 34 months. Osteoarthrosis is always associated with surface changes and is not to be confused with other joint changes of age which include atrophy, hyalinization and calcification, hyperplasia, hypertrophy and degeneration of cartilage. Osteoarthrosis itself is of hypertrophic or ulcerative type or may be a combination of the two. In this study osteoporosis as a descriptive term was restricted to those instances in which vascular spaces were enlarged. Considerable bone atrophy, evidenced by narrowing of the bone cortex, may occur without osteoporosis, thus defined.

It was found that osteoarthrosis was significantly more common in male than in female mice and the lesions were also distinctly more severe in the male. Interestingly, however, the lesions generally did not increase with advancing age beyond the degree observed in mice 18 to 24 months of age.

Osteoporosis, on the other hand, was significantly more frequent in female than in male mice. There was neither a positive nor a negative correlation between osteoarthrosis and osteoporosis.

The views concerning the morphology and the mechanisms of calcium deficiency in bone have been controversial in a number of respects. One hypothesis, which may be called the osteoclast hypothesis, says, in effect, that any resorptive process in bone, including osteoporosis, must pass first through a stage in which *both* organic and inorganic constituents are destroyed. It is assumed that the multinucleate cells known as osteoclasts are primarily responsible for such destruction. There are clear-cut examples, however, in which bone resorption occurs without osteoclasts. These include cases of constant pressure on bone, as by an aneurysm or a tumor, and of resorption caused by a rapidly growing bone tumor.

A second mechanism proposed to explain bone resorption is that of halisteresis, in which calcium *only* is supposed to leave the bone.

Bohatirchuk (1960) has employed microradiography to test for halistereis. He found conspicuous increase in X-ray transparency, evidenced by an increase in blackness of the bone in radiographs in the eighth and ninth decades (Figs. 10–3 and 10–4). Some increase was also found in about 75 per cent of skeletons of all normal persons in the seventh decade.

In relation to the micromorphological pattern of aging bone, he finds irregularities in calcium impregnation, with large areas of the calcified fibers denuded of their calcium and with remnants of the calcium appearing as small grains or dots.

The Skeleton 99

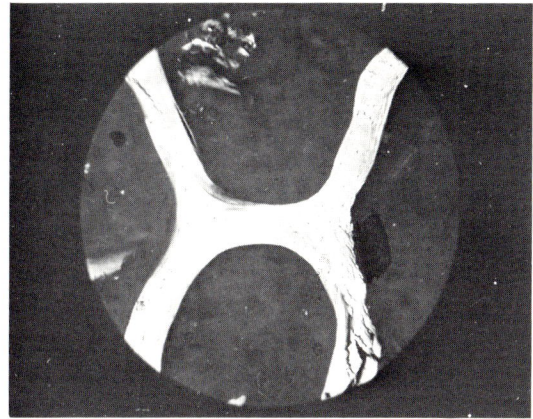

Fig. 10-3. Microradiograph of bone of a subject 22 years of age. The image is an intense white, characteristic of an abundance of calcium. (Courtesy of Dr. F. P. Bohatirchuk.)

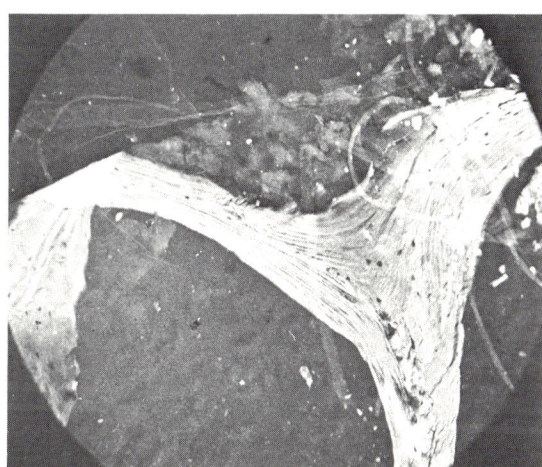

Fig. 10-4. Microradiograph of bone of a subject 91 years of age. The image is relatively pale, indicating a much smaller concentration of calcium. (Courtesy of Dr. F. P. Bohatirchuk.)

REFERENCES

Amprino, R. and A. Bairati, 1933. Studi sulle trasformazioni delle cartilagini dell'uomo nell'accrescimento e nella senescenza. Ztschr. f. Zellforsch. u. mikr. Anat., vol. 20, pp. 143–205.

——— 1936. Processi di ricostruzione e di riassorbimento nella sostanza compatta delle ossa dell'uomo. Ztschr. f. Zellforsch. u. mikr. Anat., vol. 24, pp. 439–511.

Bastai, P. and G. C. Dogliotti, 1957. Fisiopatologia e Patologia Speciale della Vecchiaia. Luigi Pozzi, Editore, Rome, 228 pp.

Bohatirchuk, F., 1954. Some microradiographical data on bone aging. Brit. J. Radiol., vol. 27, pp. 177–182.

——— 1955. Aging vertebral column (macro- and historadiographical study). Brit. J. Radiol., vol. 28, pp. 389–404.

——— 1960. Micro-morphological data on aging bone atrophy as seen in microradiographs and colored specimens. J. Geront., vol. 15, no. 2, pp. 142–148.

Bourne, G. H., 1956a. Phosphatase and Bone. *In* The Physiology and Biochemistry of Bone, G. H. Bourne, ed. New York: Academic Press, pp. 251–286.

———— 1956b. Vitamin C and Bone. *In* The Physiology and Biochemistry of Bone, G. H. Bourne, ed. New York: Academic Press, pp. 539–580.

Elliott, H. C., 1936. Studies on articular cartilage: growth mechanisms. Amer. J. Anat., vol. 58, pp. 127–145.

Gitman, L. and T. Kamholtz, 1965. Incidence of radiographic osteoporosis in a large series of aged individuals. J. Geront., vol. 20, no. 1, pp. 32–33.

Guizetto, P. 1910. Das Glykogen im menschlichen Knorpelgewebe. Centralbl. f. allg. Path. u. path. Anat., Jena, vol. 21, pp. 481–489.

Hansen, H. J., 1956. Studies on the pathology of the lumbosacral disc in female cattle. Acta orthop. scand., vol. 25, pp. 161–182.

Noback, G. J., 1949. Correlation of stages of ossification of the laryngeal cartilages and morphologic age changes in other tissues and organs (abstract). J. Geront., vol. 4, no. 4, p. 329.

Policard, A. and C. A. Baud, 1958. Les structures inframicroscopiques normales et pathologiques des cellules et des tissues. Paris: Masson et Cie.

Robinson, R. A. and M. L. Watson, 1955. Crystal-Collagen Relationships in Bone as Observed in the Electron Microscope. III. Crystal and collagen morphology as a function of age. Ann. N.Y. Acad. Sci., vol. 60, pp. 596–628.

Scott, B. L., 1956. Electron microscopy of the epiphyseal apparatus. Anat. Rec., vol. 126, pp. 465–495.

Silberberg, R. and M. Silberberg, 1950. Growth and articular changes in slowly and rapidly developing mice fed a high-fat diet. Growth, vol. 14, pp. 213–230.

Silberberg, M. and R. Silberberg, 1955. Diet and life span. Physiol. Rev., vol. 35, pp. 347–362.

———— 1962. Osteoarthrosis and osteoporosis in senile mice. Gerontologia, vol. 6, no. 2, pp. 91–101.

Sundberg, C., 1924. Das Glykogen in menschlichen embryonen von 15, 27 und 40 mm. Z. ges. Anat., vol. 73, pp. 168–246.

Tonna, E. A., 1957. The histochemistry of aging periosteum. J. Bone Joint Surg., vol. 39A, p. 674. (Proceedings).

———— 1958. Histologic and histochemical studies on the periosteum of male and female rats at different ages. J. Geront., vol. 13, pp. 14–19.

———— 1959. The histochemical nature and possible significance of the subperiosteal reversal lines of aging rate femora. J. Geront., vol. 14, pp. 425–429.

———— 1959. Posttraumatic variations in phosphatase and respiratory enzyme activities of the periosteum of aging rats. J. Geront., vol. 14, pp. 159–163.

———— 1966. A study of osteocyte formation and distribution in aging mice implemented with H^3-Proline Autoradiography. J. Geront., vol. 21, pp. 124–130.

———— 1959. Mitochondrial changes associated with aging of periosteal osteoblasts. Anat. Rec., vol. 34, pp. 739–759.

Tonna, E. A. and N. Pillsbury, 1959. Changes in the osteoblastic and mitochondrial population of aging periosteum. Nature, vol. 183, pp. 337–338.

Wagoner, G., O. Rosenthal, and M. A. Bowie, 1941. Studies of the cell in normal and arthritic bovine cartilage. Amer. J. Med. Sci., vol. 201, pp. 489–495.

Weidenreich, F., 1903. Das Knochengewebe. *In* Handbuch mikr. Anat. Menschen, W. v. Mollendorff, ed. Berlin: J. Springer, vol. II, Teil 2, pp. 391–486.

11
The Muscular System

Smooth Muscle

Smooth muscle is a tissue which is very widespread in the body and of great importance in almost all places where automatic, involuntary motion is required. By "automatic" we would wish to convey the idea that a given stimulus always will bring about a given response, either contraction or relaxation of the muscle fibers, without the intervention of the will. Smooth muscle is found in the walls of the blood vessels, as conspicuous layers along most of the length of the alimentary tract, in much of the respiratory tree, in the urogenital system, and in a number of other places. Usually it forms sheets made up of very many cells in close relationship to one another, occasionally showing actual cytoplasmic anastomoses (Thaemert, 1959) between cells. In some places it is present in strands, made up of relatively few cells, as the scrotum and in the muscularis mucosae of the small intestine, where groups of smooth muscle cells penetrate the core of each villus.

The evidence that changes with age occur in smooth muscle is derived more from the apparent results of such changes than from descriptions based on microscopic study. In the alimentary tract, for example, particularly in the colon, there is an increased incidence of diverticula, sac-like protrusions of the wall. Such diverticula are often sites of inflammation, and diverticulitis is a far more common finding in older individuals, generally from 60 years on. Long ago, Hausemann (1896) showed that diverticula can be produced by distending, at autopsy, the colons of aged persons with water under pressure, while this could not be done with the colons of young persons.

Although visible changes in the smooth muscle of the arterial wall are not conspicuous in material which we have studied, we find individual muscle cells in the media which show an accumulation of some type of material, probably lipid, to form a large intracellular vacuole (Fig. 11-1) (Andrew et al., 1959). The relationship between such changes in smooth muscle cells and the rare atherosclerotic lesions of rats which affect primarily the intimal layer of the vessel, is not clear; but Strong and Geer (1966) have shown that in the vicinity of intimal streaks of the aorta in man, smooth muscle cells may show large accumulations of lipid.

The smooth muscle in the wall of the uterus shows a great hypertrophy of the individual cells during prgenancy. The studies by Csapo (1948, 1949, 1950) showed important relations between the sex hormones and the contractile pro-

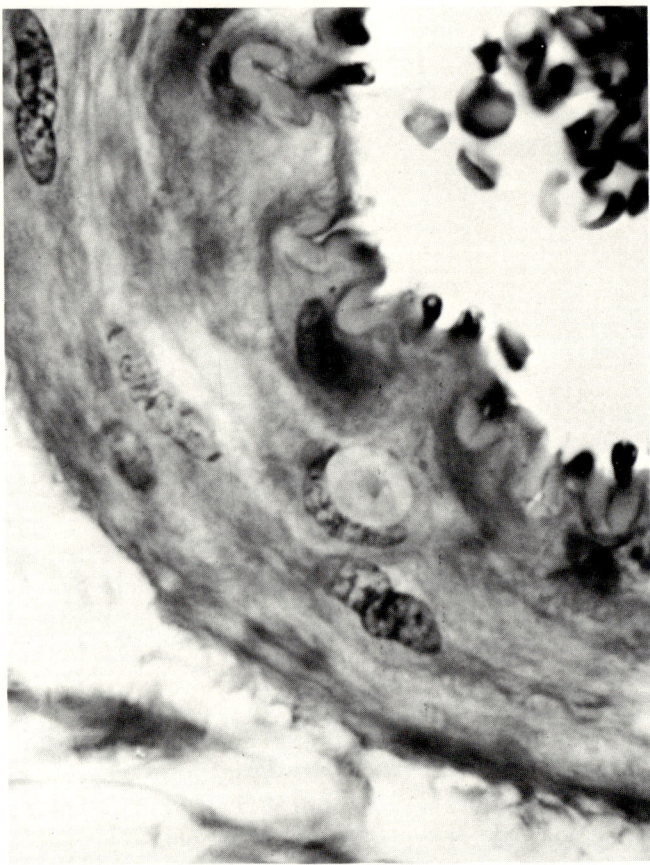

Fig. 11-1. Smooth muscle in the media of an artery in hind limb muscle from a 25-month-old female McCollum rat. A large vacuole has formed in one of the smooth muscle cells and the nucleus is deformed. Vacuoles such as this are conspicuous features in arteries of the muscles of old animals. Ehrlich's hematoxylin and eosin. ×1352.

teins, i.e., the actomyosin complex of smooth muscle. Removal of the ovaries led to a great decrease in concentration of actomyosin. Estrogen treatment restored the normal content. In pregnancy the actomyosin content increases.

Csapo and Gergely (1950) found that ovariectomy caused ATP to decrease by 50 per cent and creatine phosphate to disappear completely from the uterus.

Earlier work (Loeb et al., 1948; Wolfe et al., 1942) had found age changes in the wall of the uterus, with an increase in amount of collagenous connective tissue. Rolle and Charipper (1949) found an increase of connective tissue between the longitudinal and circular layers of smooth muscle in the uterus of the golden hamster. Here the circular smooth muscle was decreased in amount, while the longitudinal remained largely unchanged. There is a tendency for reticular connective tissue in the wall of the uterus to change to collagenous with advancing age. Suntzeff *et al.* (1940) found evidence that the connective tissue itself is influenced by estrogens or their lack. Thus the changes in the uterine wall seem to be due to hormonal influences. Whether such influences affect smooth muscle in other parts of the body has not been determined.

One form of arterial sclerosis may have its primary change in the smooth muscle of the media. This is the condition described by Monckeberg in 1903 and named after him Monckeberg's sclerosis. It occurs chiefly in the medium-sized or "muscular" arteries. There is, however, still a question as to whether, even here, the primary changes may be in the connective tissue with degeneration of smooth muscle as a secondary phenomenon.

Skeletal Muscle

The units of skeletal muscle are very long fibers each of which contains many nuclei, and hence may be said to be syncytial in character. These elements appear to be long-lived and few evidences of their multiplication or of that of their nuclei are seen in adults. Resembling thus the neurons, they might be expected to show changes during senescence of the organism.

There is one group of muscles that is subject throughout life to activity which is apt to be similar in amount and in kind in most individuals. These are the extrinsic ocular muscles: superior and inferior, medial and later recti, and superior and inferior oblique. Bucciante and Luria (1934) drew attention to changes with age in these muscles. First to appear are alterations in arrangement of the myofibrils. Closely following or accompanying this change is an increase in the amount of sarcoplasm. Disintegration of some myofibrils occurs and individual fibers come to vary considerably in size. The early change usually shows the myofibrils running obliquely, intertwining, and giving a plaited aspect. Later, a central retraction of fibrils occurs, the fibrils retaining their striations but often forming irregular plexuses, condensing into tightly packed balls or forming other and varied patterns. As these changes are occurring, areas of fibril-free sarcoplasm are seen, at first at the periphery of the fiber, then coming to occupy more and more of its volume.

The nuclei, often enlarged, show conspicuous centrally located nucleoli. As the myofibrils undergo final degeneration, large isolated clumps of sarcoplasm are seen along the fiber and these islands of sarcoplasm often contain lipofuscin, the "age pigment."

A peculiar formation is the so-called *Ringbinde*, a striated ring about a fiber, of myofibrillary origin. It was described first by M. Heidenhain (1918) and since then by Wohlfart (1938) and Bergstrand (1938).

Elastic connective tissue fibers are more abundant in the extrinsic ocular muscles than in other skeletal muscles; and in them, at least, an increase with age is seen.

An increase of collagenous connective tissue fibers with age appears to be a progressive process in the superior rectus muscle.

The extrinsic ocular muscles are not the only ones in which the *Ringbinde* and other kinds of alterations of myofibrils have been described. According to Malan (1934) such changes are seen in tensor tympani muscle of old cats and dogs and according to Bach, Lederer, and Dinolt (1941) similar changes are seen in some muscles of the aging human larynx. The *Ringbinde* formation was identified also in a supposedly normal temporal muscle (Greenfield et al., 1957) and in a section of diaphragm (Bergstrand, 1938).

Görnig (1967) has re-investigated the question of the connective tissue in the medial and lateral rectus muscles and in the inferior oblique, studing material from 21 persons ranging from premature newborn to senile. Very few elastic fibers are present at birth, they increase consistently during growth, and remain generally constant in the adults. This author finds an increase in old age but believes that the elastic material is functionally inferior. Collagenous tissue shows a remarkable increase in old age. In some areas very few cells are seen in this tissue and it is almost homogeneous.

As for the muscle itself, there is a decrease in population of nuclei by as much as a third of the number found in the fibers of premature babies.

Other skeletal muscles are more difficult to study in relation to the aging process, partly because of the variation in their functional history in different individuals, and partly because of the doubt concerning the identification of age changes on the one hand and pathological changes on the other.

In a study in which an attempt was made to corerlate histological and chemical characteristics of some tissues of rats of different ages, Andrew et al. (1959) found an increase in the proportion of extracellular to total water in skeletal muscle of hind limbs and a decrease in the potassium (cellular) component of fat-free material.

Muscle from the calves of the same animals was studied histologically. There was a degenerative change in the muscle following a rather constant pattern. Early change consisted in an increase in number of nuclei which were arranged in characteristic long rows. Mitotic figures were not seen.

In the cytoplasm a frankly destructive process occurs. The myofibrils appear to lose their striations and either disappear or become obscured in the sarcoplasm which, in many places, is gathered up into rounded masses of varying size (Fig. 11-2). A delicate connective tissue replaces the degenerating fibers. In some places groups of adipose cells develop. Although the degeneration is marked, no leucocytic reaction occurs. Lymphocytes in the replacing connective tissue do not seem more abundant than in other connective tissue nearby. Nor is there evidence of phagocytosis of cellular debris by macrophages or other cells. One type of cell does appear, however, to have some relation to the degenerative process. This is the *mast cell*. Such cells are definitely more abundant in the muscles of older rats. They have a special affinity to areas of degenerating fibers, often lying closely pressed to them either singly or in groups.

The degenerative process as described here is very similar to that found by Berg (1956). In a series of Sprague-Dawley rats numbering 568 animals he found muscle deterioration with advancing age, especially marked in the hindlimbs. The distribution was patchy. Fragmentation and disintegration of fibers was seen and adipose tissue infiltration occurred. In earlier stages muscle nuclei showed evidence of multiplication and often occurred in long rows, as in the animals which we studied. Berg does not mention an association of mast cells with the degeneration of muscle tissue.

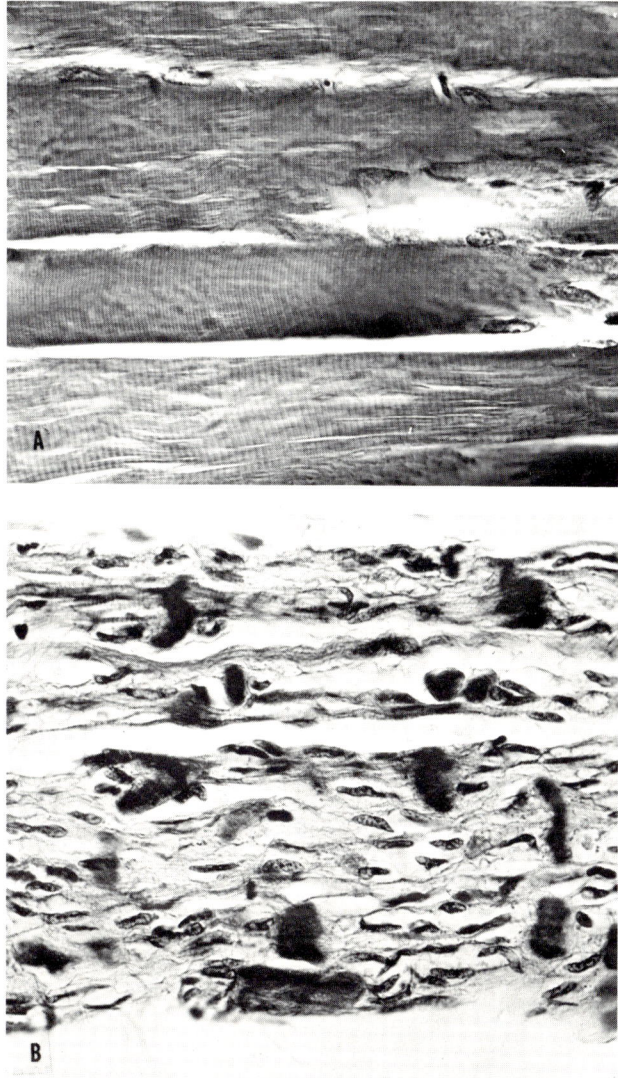

Fig. 11-2. (A) Normal muscle fibers in the gastrocnemius of an adult McCollum rat. Striations are prominent and there is little connective tissue among the fibers. Harris' hematoxylin and eosin. ×570. (B) Degenerate area in the gastrocnemius muscle of an old McCollum rat, 33 months of age. Large fragments are seen as remnants of muscle among the replacing connective tissue. Harris' hematoxylin and eosin. ×570.

Lowry *et al.* (1942) compared muscle tissue, both skeletal and cardiac, from rats of middle age (603 days) with that from senile rats (988 days). They found the extracellular fluid almost doubled in the senile rats. This is the reverse of the alteration occurring from birth to maturity and is a change in the same direction and of about the same magnitude as in our aging animals.

Human muscles have been studied by Frantzell and Ingelmark (1951). They correlated strictly morphological with radiological and chemical studies. Gastrocnemii from autopsies on 175 subjects ranging from birth to 80 years of age were studied microscopically, and the biceps brachii was studied in about one third of these subjects. The sections were both longitudinal and transverse. They were thick (30–50μ) and stained with Sudan III for fat. There was a progressive increase of fat just beneath the epimysium of both muscles and also in the perivascular and interstitial regions of the muscles. By the fourth decade the perimysial fat in many cases was abundant enough to separate the primary muscle bundles and bring about a honeycomb pattern in the sections. In subjects over 50 fatty replacement often was so extensive as to present central areas of degeneration where fat could be observed surrounded by muscle tissue.

Sex differences were inconsiderable but there was a general impression that the process was more marked in the females [unlike the findings on the rat by Berg (1956) and Andrew (1959)]. There was, however, a definitely greater degree of fatty change in the gastrocnemius than in the biceps brachii.

Chemical studies carried out parallel to the microscopic ones showed that fat generally could not be seen histologically until it exceeded 7 per cent of the dry weight of the muscle. It would seem, then, that this percentage represents the amount of "intracellular" fat within the fibers.

Radiological studies, by a special technique, were made on 222 healthy subjects and confirmed the autopsy studies. Frantzell and Ingelmark conclude that skeletal muscle tissue is decreased in volume during aging and is replaced by adipose connective tissue. They admit that many factors, including amount of exercise, nutritional state, and genetic constitution are of importance. Yet, the overall picture of type of change seems a consistent one.

Myocardium

As might be expected, a large number of authors have noted the appearance of the myocardium in aging persons. An increase in size of fibers until the 6th decade is followed by a decrease (Fig. 11-3). In old age, hypertrophy of individual nuclei is seen (Fig. 11-4). It is interesting to learn that the myocardium of the atria and of the ventricle show differences not only in their histological structure but also in the type of age change which they undergo. The myocardium of the atria in the adult heart is more loosely arranged and normally has a larger content of connective tissue than that of the ventricles.

In the fetus near term in the myocardium of the left atrium the collagenous and elastic connective tissue are inconspicuous. In the right atrium the connective tissue is slightly more prominent. In both ventricles connective tissue is minimal at this time.

During the first decade of post-natal life elastic fibers increase in the atrial myocardium and collagenous and elastic fibers increase in the ventricles, particularly about the blood vessels.

By the beginning of the third decade adipose tissue is generally seen among the muscle fibers in the atria. It is found particularly around the blood vessels.

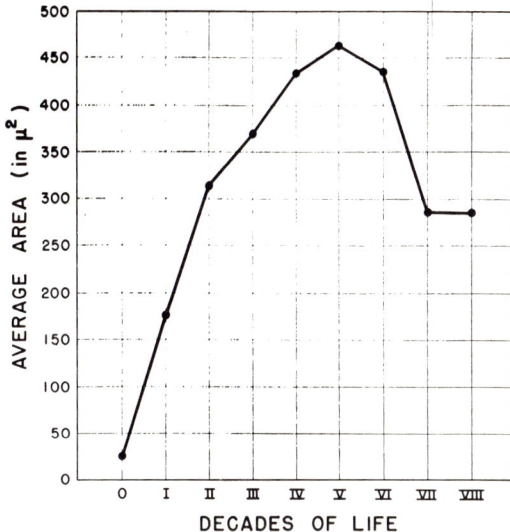

Fig. 11-3. Graph to show the change in average size (cross-sectional diameter in microns) of cardiac muscle fibers with advancing age. (Redrawn, with modifications, from Dogliotti, 1931.)

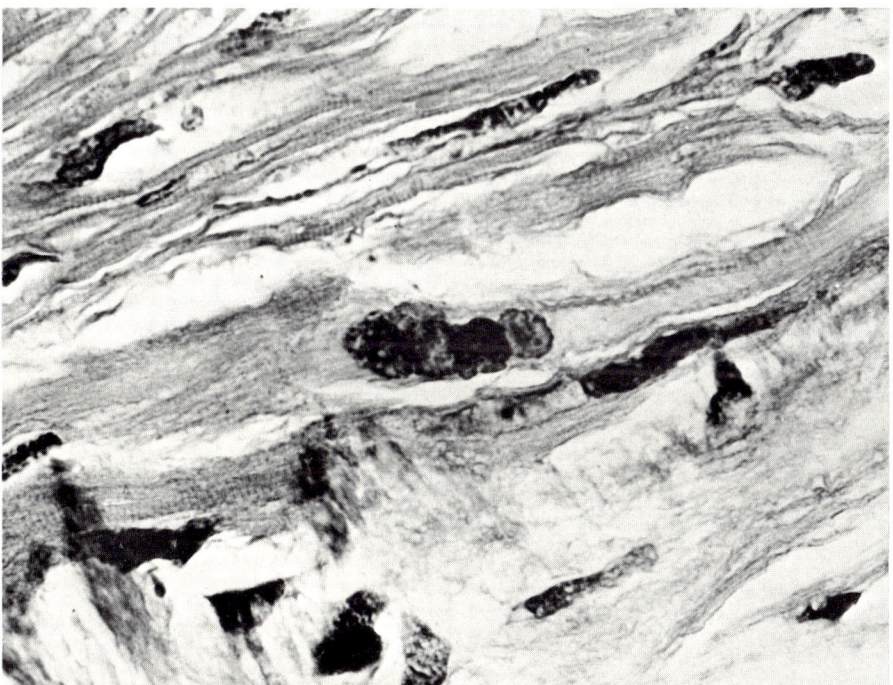

Fig. 11-4. Large, aberrant nucleus in cardiac muscle of a very old specimen of *Ambystoma mexicanum*. Such aberrant nuclei have been described in senility in cardiac muscle of the baboon (*Papio anubis*) by Katzberg (1970). Masson's trichrome. ×400.

The right atrium usually shows a greater accumulation than the left. Elastic fibers now are numerous in the atrial myocardium. The ventricles at this time also show some adipose tissue accumulations chiefly about the blood vessels, but these are less conspicuous than those in the atria.

By the time of middle age the penetration of adipose tissue into the right atrium often is pronounced. It appears to be derived largely from the fatty pads of the epicardium. As middle age moves on into senescence, there is a tendency for the fatty tissue to be replaced by fibrous tissue, both collagenous and elastic. The whole process is much more pronounced in the atria than in the ventricles. The latter do show, however, an increase in amount of elastic connective tissue among the myocardial fibers (Dogliotti, 1931) and a much increased prominence of the reticular network surrounding the myocardial fibers.

In senility these changes continue in a progressive way, while fatty infiltration of the ventricles often becomes conspicuous, being now definitely more prominent in the right than in the left ventricle (Lev and McMillan, 1961).

In regard to the individual myocardial fibers, a gradual increase in size occurs from birth to maturity (Fig. 11-3). In old age there seems to be a slow diminution in size of muscle fibers with, however, increasing variability in diameter.

The amount of yellow-green pigment—lipofucsin—in the human myocardium has been shown to be proportional to age and to increase at a rate of approximately 0.6 per cent of the total intracellular volume per decade (Strehler et al., 1959). This pigment and related substances have been the object of investigations for over a third of a century (see Hamperl, 1934). Their chemical and physical properties have been reviewed by Strehler (1964).

In cardiac muscle the pigment granules accumulate at the poles of the nuclei. From this location it often extends in long bead-like arrangements. In many cases the granules seem to have a 1-1 relation to the striations of the muscle.

With advancing age lipofuscin accumulates in the myocardium of all four chambers of the heart. Great accumulation of pigment with decrease in size gives rise to the condition of "brown atrophy" of the heart.

Munnell and Getty (1968) found the rate of accumulation of pigment in the cardiac muscle of the dog to be approximately 5.5 times faster than that seen in man when one goes by the calendar year.

Since the physiological aging of the dog is approximately five times that of man, the rate of accumulation with age is very similar in dog and human subject.

Katzberg (1970) has recently described hypertrophy of some nuclei in cardiac muscle of the East African baboon, *Papio anubis*, as a phenomenon of aging. As the animal approaches senility, such nuclei become very prominent and numerous. They occur side by side with nuclei of normal size.

Katzberg conjectures that they may be the expression of some pathological process due to infection or to the ingestion of toxic organic or inorganic substances. It is not clear whether these nuclei are polyploid.

REFERENCES

Andrew, W., N. W. Shock, C. H. Barrows, Jr., and M. J. Yiengst, 1959. Correlation of age changes in histological and chemical characteristics in some tissues of the rat. J. Geront., vol. 14, no. 4, pp. 405–414.

Bach, A. C., F. L. Lederer, and R. Dinolt, 1941. Senile changes in the laryngeal musculature. Arch. Otolaryng., Chicato, vol. 34, pp. 47–56.

Berg, B. N., 1956. Muscular dystrophy in aging rats, J. Geront., vol. 11, pp. 134–139.

Bergstrand, C. G., 1938. Zur morphologie der quergestreiften Ringbinden. Z. mikr.-anat. Forsch., vol. 44, pp. 45–55.
Bick, E. M., 1955. The physiology of the aging process in the musculoskeletal apparatus. Geriatrics, vol. 10, pp. 274–277.
Bucciante, L. and S. Luria, 1934. Trasformazioni nella struttura dei muscoli volûntari dell'uomo nella senescenza. Arch. Itâl. Anat. Embriol., vol. 33, pp. 110–187.
Csapo, A., 1948. Actomyosin content of uterus. Nature, London, vol. 162, pp. 218–219.
——— 1949. Studies on adenosine-triphosphatase activity. Acta Physiol. Scand., vol. 19, pp. 100–114.
——— 1950. Actomyosin formation of estrogen action. Amer. J. Physiol., vol. 162, pp. 406–410.
Frantzell, A. and B. E. Ingelmark, 1951. Occurrence and distribution of fat in human muscles at various age levels; a morphologic and roentgenologic investigation. Acta Soc. Med. Upsal., vol. 56, pp. 59–87.
Greenfield, J. G., G. M. Shy, E. C. Alvord, and L. Berg, 1957. Livingstone, Edinburgh: An Atlas of Muscle Pathology in Neuromuscular Diseases.
Hamperl, H., 1934. Arch. path. Anat. Physiol., vol. 292, pp. 1–51 (cited from Strehler, 1964).
Heidenhain, N., 1918. Uber progressive Veränderungen der Muskulatur bei Myotonia Atrophica. Beiträge Path. Anat., vol. 64, pp. 198–225.
Katzberg, A. A., 1970. Relationship of age to incidence of macronuclei in cardiac muscle fibers of the East African Baboon, Papio anubis. Proc.IX Int.Congr.Anats., Leningrad, p. 62 (abstract).
Loeb, L., V. Suntzeff, and E. L. Burns, 1938. The effects of age and estrogen on the stroma of vagina, cervix and uterus in the mouse. Science, vol. 88, pp. 432–433.
Lowry, O. H., A. B. Hastings, T. Z. Hull, and A. N. Brown, 1942. Histochemical Changes Associated with Aging. II. Skeletal and Cardiac Muscle in the Rat. J. Biol. Chem., vol. 143, pp. 271–280.
Malan, E., 1934. Étude d'histologie comparée sur quelques modifications particulieres des fibres du tensor tympani dues a la senescence. Arch. Biol., Paris, vol. 45, pp. 355–375.
Mönckeberg, J. G., 1903. Über die Reine Mediaverkalkung der Extremitätenarterien und ihr Verhalten zur Arteriosklerose. Virchows Arch., vol. 171, pp. 141–167.
Munnell, J. F. and R. Getty, 1968. Rate of accumulation of cardiac lipofuscin in the aging canine. J. Geront, vol. 23, pp. 154–158.
———, ——— 1968. Nuclear lobulation and amitotic division associated with increasing cell size in the aging canine myocardium.
Rolle, G. K. and H. A. Charipper, 1949. The effects of advancing age upon the histology of the ovary, uterus and vagina of the female golden hamster Cricetus auratus. Anat. Rec., vol. 105, pp. 281–297.
Strehler, B. L., 1964. On the histo-chemistry and ultrastructure of age pigment. Advances in Geront. Res., vol. 1, pp. 343–384. New York: Academic Press.
Suntzeff, V., R. S. Babcock, and L. Loeb, 1940. Reversibility of hyalinization in mouse uterus produced by injections of estrogen, and changes in mammary gland and ovaries after cessation of injections. Amer. J. Cancer, vol. 38, pp. 217–233.
Swigart, R. H., H. S. Riley, J. Withers, and J. B. Rogers, 1961. A comparison of the distribution of cardiac glycogen in young and old guinea pigs. J. Geront., vol. 16, no. 3, pp. 239–242.
Wohlfart, G., 1938. Zur Kenntnis der Altersveränderungen der Augenmuskeln. Z. mikr.-anat. Forsch., vol. 44, pp. 33–44.
Wolfe, J. M., E. Burack, W. Lansing, and A. W. Wright, 1942. The effects of advancing age on the connective tissue of the uterus, cervix, and vagina of the rat. Amer. J. Anat., vol. 70. pp. 135–165.
Yiengst, M. J., C. H. Barrows, and N. W. Shock, 1959. Age Changes in the Chemical Composition of Muscle and Liver in the Rat. J. Geront., vol. 14, pp. 400–404.

12
The Blood Vascular System

It seems probable that many of the changes with age which have been observed in arteries, veins, capillaries, and the heart itself are an intrinsic part of fundamental tissue changes. Nevertheless, the circulatory system has such a special role in the economy of the body and there is such a number of seemingly special features related to its aging, that discussions of the process in this system often have occupied a unique place in the more general consideration of senescence.

That some changes occur in the smallest vessels, or in the regulatory mechanism related to them is indicated by the change in response to stimuli which they show at more advanced ages. This change was shown by Di Palma and Foster (1942) for the extremities. Knobloch (1954) demonstrated that the intensity of the erythema brought about by topical application of mustard oil is greatest in the third and fourth decades and least in the seventh decade.

Medium-sized and Large Arteries

In larger vessels, it is possible to isolate and test the tissue elements to find the probable causes of altered ability to react. Burton (1951) found a greater rigidity of the fibrous tissue in the walls of the largest arteries. Wilens (1937) pointed out a progressive loss of elasticity of the wall and Krafka (1940) described decreased elasticity of the connective tissue. The elastic tissue appears to show not only a loss of capacity to recoil after stretching but also a loss of tensile strength and stretchability (Haas, 1943).

A seemingly paradoxical fact is the increase in total amount of elastic tissue in the aorta (Schwarz, 1937). Such increase occurs in arteries where pressure is high and not in others, such as the pulmonary artery, where there may be an actual *decrease* in old age. Age changes are more pronounced in the ascending aorta and in its arch than in the remaining portions; and the arteries of the lower extremities show greater change than those of the upper. The progressive thickening of the wall of the aorta, due chiefly to changes in the media, is much more evident in the first two portions than in the descending thoracic and abdominal segments (Buerger and Hevelke, 1956).

It has been shown that the matrix of the *media* of the human aorta exhibits a decrease in metachromatic property in later years. Some calcification and degeneration of elastic tissue, and to a lesser extent, of smooth muscle, is occurring. Deposition of new collagenous fibers then takes place (Bunting and Bunting, 1953).

The *intima* of the human aorta in subjects up to 20 years of age shows an endothelial layer over an intact internal elastic lamella. Between 21 and 40 years, a layer of collagenous tissue appears between endothelium and internal elastic lamella and gradually increases in thickness (Figs. 12-1, 12-2). A number of medium-sized arteries show this type of change but it is not seen in the pulmonary artery under normal conditions (it may occur in pulmonary hypertension). In later stages of intimal thickening the processes may be irregular and even lead to development of an asymmetrical lumen in the vessel (Handler, Blache, and Blumenthal, 1952).

In many arteries the internal elastic lamella shows changes, including increased density, straightening, splitting, fraying, or thickening. Calcification often follows the pattern of fragmented fibrous tissue. The coronary arteries show an invasion of musculoelastic tissue into the inner half of the media (Lansing, Blumenthal, and Gray, 1948).

The cerebral arteries show very prominent age changes of the internal elastic lamella, which may begin as early as the latter part of the third decade. By the end of the fourth decade, or somewhat later, a complete replication of the lamella is found (Betetto, 1953). Much splitting, and a loss of normal staining properties of the elastic tissue occur in old age (Baker, 1937).

Age changes in the intima occur in the arteries of mammals other than man but they are by no means always strictly comparable.

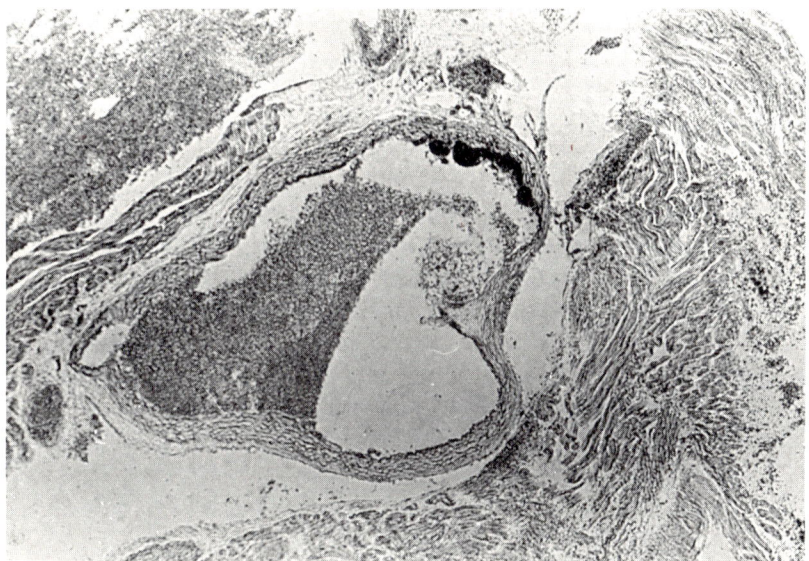

Fig. 12-1. Transverse section of aorta of a (C57BL/6J) mouse fed on a high cholesterol-high fat diet for 10 weeks. The lesion is confined to a relatively restricted area. Frozen section stained with oil red O. ×80. (Courtesy of J. S. Thompson.) (Reprinted with permission from J. Ather. Res., vol. 10, 1969.)

The media of human arteries shows an increase of collagenous tissue with age. This increase tends to separate the original common mass of smooth muscle and probably causes some atrophic change in that tissue. It is thought that these microscopic changes bring out increasing tortuosity of the vessels and a widening of the lumen, features often seen in medium-sized arteries in advanced age.

Both large and medium-sized arteries in man may show concentrations of new collagenous tissue at the junction of media and adventitia. According to Troitzkaja-Andreewa (1931) these areas probably correspond to the course of vasa vasorum.

The increase of collagen in the arteries appears to have no clear relation to the atherosclerotic process in the intima. Thus, whether the latter is or is not to be considered as a manifestation of normal aging, the same question concerning new collagen formation does not depend upon that answer. The relative constancy of increase of collagen and its lack of association with any inflammatory process seem to favor considering it as a phenomenon of physiological aging.

For the elastic fibers, Schwartz (1954) has shown that the ultrastructural aspect in the young is one of abundant, tortuous fibrils in a finely granular matrix, while that for the old is one of straight and stiff-appearing fibrils in a matrix which is more coarsely granular. Visualization of calcium shows it in the young as fine, needle-like deposits, while in the old it appears as larger, irregular needles or ovoid granules.

The changes that take place in the elastic tissue in the media which lead to fraying and fragmentation of the fibers in that layer and to calcium deposition, occur in a number of other kinds of mammals—the rabbit, cow, and horse in advanced age (Fox, 1933). On the other hand, they appear not to occur in the dog (Fox, 1933), nor in the mouse (Blumenthal *et al.*, 1950). Some species of birds show marked atherosclerotic change and *Columba* has proved an excellent experimental subject for studies on the process (Clarkson, *et al.*, 1959; Lofland *et al.*, 1966).

A number of investigators (Lansing, Alex, and Rosenthal, 1950; Handler *et al.*, 1952; Gray, 1953; Schwarz, 1954) have described the progressive increase with age in mineral content, primarily calcium, in the media of the arteries and either within or on the surface of the elastic fibers. This increase, unlike that in the intima, is not grossly detectable. Direct chemical anlaysis, microincineration, and other microscopic and ultamicroscopic methods, however, indicate the constancy of its occurrence and of its association with other changes in the elastic fibers.

There are topographical differences in elastocalcinosis in different arteries. Thus the more common site in the renal artery is near the external elastic lamella, while in the coronary artery it is near the internal elastic lamella (Lansing *et al.*, 1948), while the iliac arteries show a concentration about equal in these two locations (Blumenthal *et al.*, 1950). In the splenic artery there may occur a calcification of the outer portion of the media following degeneration of smooth muscle, rather than in relation to elastic tissue (Handler *et al.*, 1952).

Thompson (1969) has made experimental studies on the arteries of mice by feeding of cholesterol-high diets over a long period of time. Frozen sections were stained with oil red O. Lipoid deposits are clearly demonstrated. All members of the C57BL/6J strain developed atheromatous lesions of the wall of the aorta in the region of the aortic valves after about 20 weeks on the diet. By 30 weeks the lesions were "full blown" and contained many fat laden foam cells. The endothelium may be lifted or even become discontinuous in the area of the lesions.

From the genetic aspect, it is of interest that mice homozygous for the gene ob (obesity) became very obese but did not seem to be more susceptible to atheromatous change than did those of the same strain without the ob gene.

Smaller Blood Vessels

There is evidence for a progressive reduction of capillaries in the myocardium and in other parts of the body (Bastai and Dogliotti, 1937).

Increase in the amount of reticular connective tissue of many of the capillaries can be demonstrated by silver impregnation methods and actual obliteration of the lumen can be seen (Fig. 12-2). A qualitative change is an increasing irregularity of the capillary network, in which angular tortuosities appear in older individuals, usually near the venous ends, although arterioles also may share in the irregularity. Intravascular aggregation of red blood corpuscles then becomes more common.

Studies of living venules in the bulbar conjunctival bed (Ditzel, 1956) show sacculations and irregularities of these vessels and increased intravascular aggregation of red blood corpuscles. There also is infiltration by a hyaline material.

The small vessels in the walls of the aorta—the vasa vasorum—tend to grow inward from the adventitia to the media with advancing age. Thus, capillaries

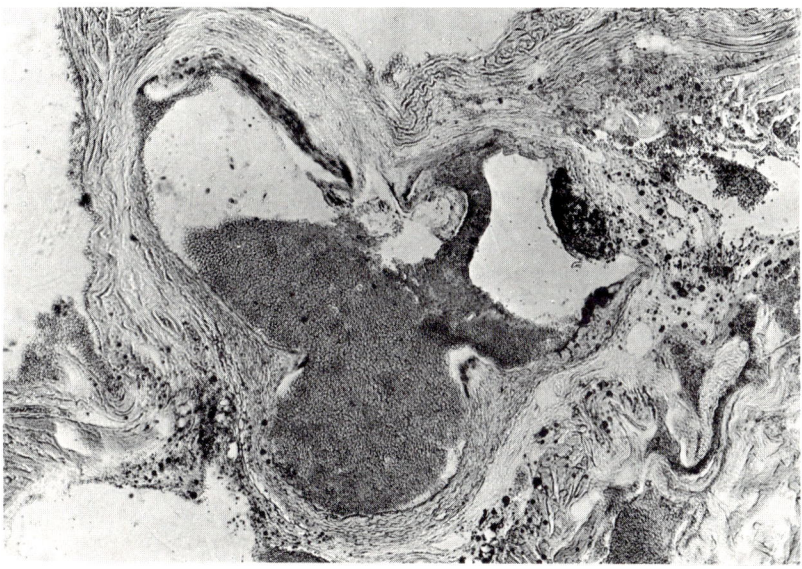

Fig. 12-2. Transverse section of aorta of a (C57BL/6J) mouse fed on a high cholesterol-high fat diet for 25 weeks. Several lesions are seen. Two of them are on either side of the origin of the coronary artery. Frozen section stained with oil red O. ×80. (Courtesy of J. S. Thompson.) (Reprinted with permission from J. Ather. Res., vol. 10, 1969.)

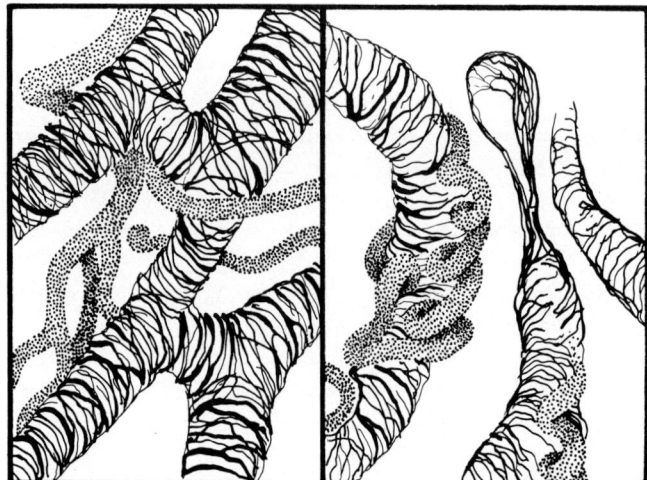

Fig. 12-3. Capillaries in the lung of a senile human subject. Thickening of the sheath of reticular connective tissue is seen on both left and right. On the right, narrowing and obliteration of the capillary passage have occurred. Method of Achucarro for reticular tissue, modified by Amprino. (Redrawn from Bastai and Dogliotti, 1937.)

often are seen in the outer third of the media in old subjects and occasionally in the middle third.

The Veins

Veins often show some increase in the thickness of the wall with age. A loss of muscle fibers in the media may be associated with an increase of connective tissue and with focal deposition of calcium in the intima, probably as in the arteries, in relation to elastic connective tissue.

Hypertrophy of smooth muscle, and of elastic and collagenous tissue, internal to the internal elastic lamella may produce irregular thickenings narrowing the lumen. Vacuoles and lacunae may appear in the ground substance (Lev and Saphir, 1951).

The valves of the veins show an increase of connective tissue, some new formation of thin elastic laminae, and fatty infiltration on the inner side of the valve (Saphir and Lev, 1952).

The Heart

The endocardium, or lining of the heart, shows different degrees of age change in the four chambers. The left atrium is the only one in which a relatively diffuse thickening of the endocardium occurs. In this chamber most of the lining shows some degree of opacity by the end of the first year. By the age of 20 there is plaque formation and pronounced corrugation of the posterior wall. Thereafter there is a general increase of thickness until the sixth decade. Fat spots may be seen in the fifth and sixth decades.

Lev and McMillan (1961) describing the histological changes in the left atrium, say that maximal hypertrophic changes are seen at 5 years of age. In the fourth decade, according to them, the entire endocardium of this chamber is the seat of hypertrophy, while collagen infiltration is common and sclerotic change is more intense than in the right atrium. According to them, the maximal changes are found in the sixth decade.

The right ventricle shows the least degree of endocardial thickening with age. By adult life small opacities are seen on papillary muscles and trabeculae but few other changes occur in these areas with advancing age.

Histologically, mild elastic hypertrophy of focal type is seen in this chamber by the fifth year. Sclerotic changes are more prominent, with focal deposits of collagenous tissue in the endocardium, while there is a loosening of the subendocardium. In the second and third decades endocardial sclerosis proceeds further in the absence of significant hypertrophy. In the sixth decade reduplication of layers is seen in areas of hypertrophy, and fatty tissue infiltrates portions of the endocardium and subendocardium. Lev and McMillan (1961), however, stress the fact that even in this decade there are areas of endocardium which appear histologically just as they did at birth.

In the left ventricle endocardial thickening is more marked and by middle age fat spots are seen in this chamber in some hearts, especially on the anterior wall and on the septum.

The endocardium of the left ventricle is similar to that of the right in showing patchy areas of hypertrophy and sclerosis and yet showing many regions in which no apparent histological change is seen throughout life. Both sclerosis and hypertrophy are greater here than in the right ventricle. By the fifth year, sclerosis and hypertrophy are considerable on the upper septum. These changes progress and in the sixth decade adipose tissue is seen in the subendocardium. The most advanced sclerosis is seen in the seventh and eighth decades. Again, however, even at these more advanced ages, many regions appear histologically as they did at birth.

In the right atrium, while thickening is not as prominent as in the left atrium, the fat spots appear in the fourth and fifth decades.

In the right atrium, endocardial hypertrophy is seen in the eighth month of fetal life; it is of the elastic (fibroelastic) or musculoelastic type. By the fifth year the hypertrophy is maximal in the anterior and posterior walls and in the septum, and has spread to other parts of the chamber. Fat infiltration is seen in the second decade. In the fourth decade, hypertrophic and sclerotic changes are more widespread. The fifth decade shows a beginning of atrophy of the smooth muscle in the endocardium; while the maximal sclerotic changes occur in the sixth decade.

Skrzypczak (1966) studied the age changes in the Pars membranacea of the interventricular septum in hearts of 14 persons without history of disease in this organ. The ages ranged from newborn to 87 years.

He found a decrease in the number of cells with increase in the amount of intercellular substance. Histochemically, an increase is seen in the acid mucopolysaccharide content. The elastic tissue, which generally increases with aging of the endocardium, does so here also, but does exhibit clear signs of deterioration at advanced ages.

Valves of the Heart

The valves are essentially specializations of the endocardial lining of the heart. They show differences in kind and degree of age change in both gross and microscopic aspect.

The left atrioventricular or mitral valve commonly shows plaques in its anterior leaflet in the third decade. Near the line of closure these plaques are fibrous beneath the endothelium, while further back from the edge, in the body of the leaflet, they are yellow atherosclerotic plaques. In the next two decades these changes are accentuated, while thickening of chordae tendineae and the atrioventricular ring occurs. Calcification is common in the mitral valve after 60 years of age.

The right atrioventricular or tricuspid valve shows relatively less change in the young adult, with only occasional thickening of its anterior and septal leaflets. In the sixth decade, however, there is considerable thickening and fibrous plaques are prominent beneath the endothelium. The smaller amount of mechanical stress may account for the more moderate "age" changes in the tricuspid as compared with the bicuspid valve.

In the pulmonary semilunar valve there is remarkably little change up to age 60. The cusps may show some thickening with commissural adhesions.

The aortic pulmonary valve shows considerable fibrous thickening at the base of the cusps even in the third decade, while adhesions of the cusps may be seen. Increased approximation and adhesion occur in the fourth decade. In the fifth decade the cusps along the line of closure, together with the noduli arantii, are thickened, while definite atherosclerotic alteration usually is seen in the posterior cusp. The sixth decade brings also a focal calcification of the cusps. By the eighth decade, prominent atherosclerosis and calcification, especially at the bases of the sinuses, has occurred, while adhesion of the cusps usually is marked. Thus this valve of the left heart, like the mitral valve, undergoes much greater change than its fellow of the right heart.

The valves of the heart in rats show a gradual increase with age (10 days to 3 years) in the number of collagen fibrils (Nakao *et al.*, 1965). These fibrils are more densely packed and arranged in a more parallel manner in old rats.

In the valves the elastic fibers also show age changes. The number of elastic fibers increases with age and the structure of individual fibers is altered. In a given fiber the central core becomes thickened. The central core is an amorphous, electron-light zone, the intermediate zone is one of moderate electron density, and the outermost zone is a sheath of fine fibrils. The fine structural changes may be related to the elongation, distortion, and gross thickness of valves often seen in old age.

Epicardium

The age changes in the human epicardium have been described by Nelson (1940) and by McMillan and Lev (1960). An increasing opacity occurs from birth to maturity. Peculiar patches known as "soldiers' plaques" make an appearance.

A gradual accumulation of adipose tissue beneath the epicardium takes place and follows a distinctive pattern, traced by McMillan and Lev (1960). However, the fat in the various regions may become confluent by the fifth decade.

The epicardium consists of a single layer of flattened cells—mesothelium—of some elastic and many collagenous fibers, and a small amount of smooth muscle. A distinct layering is seen by the time of maturity. The dense collagenous fibers of the deeper portion become separated from the loose, reticular, submesothelial tissue by an elastic lamina. In the fourth decade the so-called "solders' plaques" or pericardial milk spots make an appearance. These are patches of fibroblastic tissue in the submesothelial reticular tissue. They may show tags or protuberances and may contain groups of lymphoid cells. These plaques become increasingly frequent from middle age onwards.

Cardiac Nodes and Bundle of His

Distinctive changes occur in the conduction system of the heart with advancing age. In the sinoatrial node there is an increase of elastic and reticular fibers. After about age 40 there is an apparent loss of muscle fibers and hypertrophy of some of the remaining ones, and also a fatty infiltration. In the atrioventricular node and bundle of His the elastic and reticular fibers increase while infiltration by adipose tissue into the A-V node begins around age 30 and becomes marked by age 50. Fatty infiltration may extend into the bundle and its branches.

Erickson and Lev (1952) have described the changes in the atrioventricular node and bundle of His, while Lev (1954) has described those in the sinoatrial node. Rondolini (1937) made observations on the entire conduction system in both fetal and postnatal life.

Glycogen Content

The glycogen content of the heart of the guinea pig changes with age ($5\frac{1}{2}$ months to 50 months) (Swigart, et al., 1961). Each of the four chambers in older animals showed a higher mean concentration than that of the same chamber in young animals. The authors believe that the higher concentration may represent a case of specific organ adaptation as the total capacity of the animal for adaptation to a stressful (possibly hypoxic) situation is decreasing.

Ultrastructure of Vessel Wall

Ultrastructural aspects of connective tissue in the wall of the aorta in aging mice of several strains have been studied by Karrer (1961). He found in the tunica media an increasing amount of collagen, with interruptions and rarefactions of elastic laminae. There was also a fraying or fragmentation of the innermost elastic fibers and disconnections between elastic laminae and smooth muscle cells. In mice of one strain, the recessive-obese hypercholesteremic, hyperglycemic strain, he found subendothelial deposition of collagen. In one "normal" mouse, not of this strain and only 5 months old, he found a subendothelial proliferative "lesion" with apparently newly formed collagen and elastin and elongated cells. Such a lesion he interpreted as "possibly" parallel to the initial lesions seen in atherosclerosis.

The Blood-forming (Hemopoietic) Tissue

The red marrow (Fig. 12-4) gradually decreases in amount through childhood and is replaced by yellow marrow. The major difference between the two forms of marrow is, of course, the larger amount of adipose tissue and the lesser amount of truly hemopoietic tissue in the yellow marrow.

Whether a continuation of this process occurs in adults and during senescence is somewhat controversial. From study of aspirated sternal marrow from 100

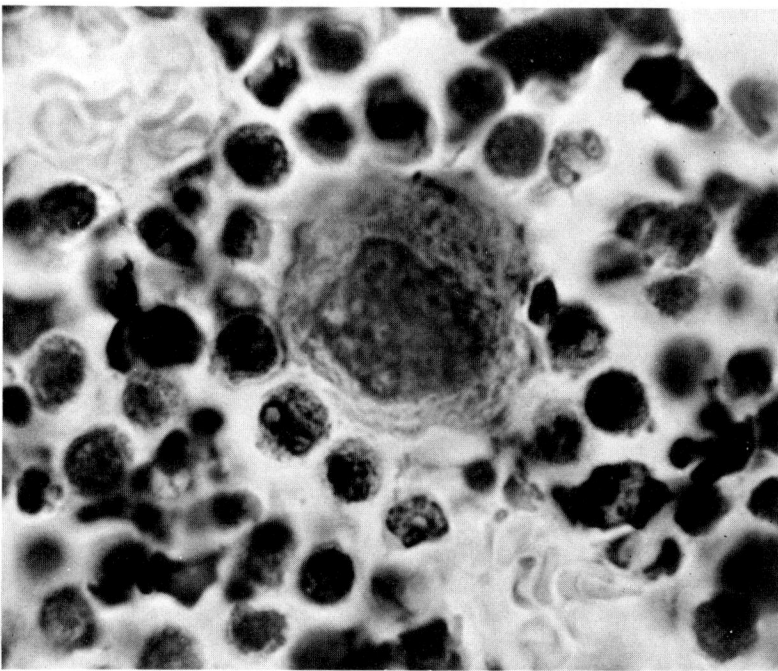

Fig. 12-4. Typical red bone marrow. Adult rat. With advancing age much of the red marrow is replaced by yellow marrow which contains large quantities of adipose tissue. The large cell at the center of the field is a megakaryocyte, a platelet forming cell. The cells just below it to the left, with lobated nuclei, are myelocytes, differentiating to granular leucocytes. Below the megakaryocyte to the right, the cell with relatively clear nucleus and conspicuous nucleolus is a hemocytoblast. Hematoxylin eosin-azure stain. ×1,000.

human subjects of 65–95 years Reich, Swirsky, and Smith (1944) concluded that the bone marrow of the aged is relatively unchanged and in some subjects may show an even increased cellularity.

Marrow specimens from 21 patients from 60 to 90 years old were studied by Cheli and Giordano (1947) and compared with the marrow of younger persons. They saw no definite decrease in cellularity. Segerdahl (1934) did report a probable decrease in amount of active marrow and ascribed it to increase in number of adipose tissue cells. However, Plum (1941) found no constant relationship between the cellularity of sternal marrow and age.

Although the activity of the hemopoietic tissue of the bone marrow in old persons seems to be sufficient to maintain blood cell counts much the same as in youth, a recent study (Nelius et al., 1968) has shown that the marrow in senility is less able to respond to the functional Pyrexal (R) test. A group of 107 healthy persons, 60–96 years of age, in old people's homes of Berlin, were compared with 11 students 20–25 years of age. Injection of the lipopolysaccharides, which stimulate the granulocytopoietic activity, showed no qualitative difference but the increase in leukocytes was significantly lower in old than in younger persons.

The Blood

The results of many studies on blood volume, on the numbers of erythrocytes, and on the amount of hemoglobin show little, if any, evidence of consistent change with age. One hundred subjects were studied by Shapleigh, Mayes, and Moore (1952)—50 males 60–94 years and 50 females 60–90 years of age. They found a slight but significant decrease in the erythrocyte, hemoglobin, and packed erythrocyte values from those found in healthy young adults. Since the decrease is somewhat more pronounced for men, the hematologic sex differences grow smaller with age. There seem to be minimal, if any, qualitative changes in the erythrocytes.

In the studies of Shapleigh, Mayes, and Moore, there was also no significant deviation with age from the general nor the differential leucocyte counts of healthy young adults, with the possible exception of a slight tendency toward an increase of eosinophils, which may be a manifestation "secondary to the atrophic skin changes commonly seen in the aged" (see Chapter 9 of this book), and a slight tendency to an increase in numbers of monocytes, which might have been artefactual, since supravital preparations were used to determine these.

In regard to possible qualitative differences in leucocytes of young and old persons, an earlier investigator (Dobrovici, 1904) concluded that the neutrophils are in general more "mature" in old persons, and Olbrich (1947) described increased lobulation of neutrophil nuclei. Shapleigh et al. found no qualitative changes in leucocytes. These authors found no consistent change in platelet count, and they say (p. 302): ". . . there is no evidence that hemorrhagic disorders, when seen in old persons, are related directly to the process of aging."

Data on sedimentation rate in relation to age are conflicting.

Ewing and Tauber (1963) have described changes in the blood of a common laboratory animal, the mouse.

Male C57 BL/6 Jax mice (2–24 months old) showed a significant linear regression of hemoglobin content, erythrocyte count, mean corpuscular hemoglobin, and mean corpuscular hemoglobin concentration with increased age. Hematocrit and mean corpuscular volume showed significant quadratic regression with age.

Fine structural and histochemical study of the blood and blood cells with advancing age would seem to be a promising field for future investigation.

Bertolini et al. (1965) say that there is a decrease in the enzyme content of red blood corpuscles in aged subjects, as compared with adults. Enzymes involved are acid phosphatase, true cholinesterase, glutamic-oxalecetic and glutamic-pyruvic traminases, and diaphorase. Catalase and glucose-6 phosphate dehydrogenase did not show a change.

An extensive study of the hemogram in beagles, making use of a total of 433 animals (Dougherty and Rosenblatt, 1965) showed no change in mean leucocyte values with age but a downward trend of the erythrocytess with aging.

In cattle the total numbers of leucocytes are decreased with advancing age after sexual maturity, while a relative increase of neutrophils and decrease of lymphocytes occur. (Riegle and Nellor, 1966). The total plasma protein and gamma globulin are increased and the albumins decreased.

REFERENCES

Blood Vessels

Baker, A. B., 1937. Structure of the small cerebral arteries and their changes with age. Amer. J. Path., vol. 13, pp. 453–461.

Bastai, P. and G. C. Dogliotti, 1937. Fisiopatologia e Patologia Speciale della Vecchiaia. Luigi Pozzi, Editore, Rome, 228 pp.

——— 1955. Die gerontologischen Studien in Italien. Z. Altersforsch., vol. 8, pp. 278–283.

Betteto, G., 1953. Modificazioni strutturali dell'arteria e della vena centrale della retina in rapporto all'accrescimento ed alla senescenza. Ann. Ottalm. clin. Ocul., vol. 79, pp. 79–92.

Blumenthal, H. T., A. I. Lansing, and S. H. Gray, 1950. The interrelation of elastic tissue and calcium in the genesis of arteriosclerosis. Amer. J. Path., vol. 26, pp. 989–1009.

Buerger, M. and G. Hevelke, 1956. Do human beings have the age of their blood vessels? Angiology, vol. 7, pp. 137–151.

Bunting, C. H. and H. Bunting, 1953. Acid mucopolysaccharides of aorta. A.M.A. Arch. Path., vol. 55, pp. 257–264.

Burton, A. C., 1951. On physical equilibrium of small blood vessels. Amer. J. Physiol., vol. 164, pp. 319–329.

Clarkson, T. B., R. W. Pritchard, M. G. Netsky, and H. B. Lofland, 1959. Atherosclerosis in pigeons: its occurrence and resemblance to human atherosclerosis. Arch. Path., vol. 68, pp. 143–147.

Di Palma, J. R. and F. I. Foster, 1941. The segmental and aging variations of reactive hyperemia in human skin. Amer. Heart J., vol. 24, pp. 332–344.

Ewing, K. L. and O. E. Tauber, 1964. Hematological changes in aging male C57 BL16 Jax mice. J. Geront., vol. 19, no. 2, pp. 165–167.

Fox, H., 1933. In E. V. Cowdry's Arteriosclerosis, Josiah Macy, Jr., Foundation, ed., New York: Macmillan, pp. 157, 163, 164, 179.

Handler, F. P., J. O. Blache, and H. T. Blumenthal, 1952. Comparison of aging processes in the renal and splenic arteries in the Negro and white races. Arch. Path., vol. 53, pp. 29–53.

Hass, G. M., 1943. Elastic tissue. III. Relations between the structure of the aging aorta and the properties of the isolated aortic elastic tissue. Arch. Path., vol. 35, pp. 29–45.

Karrer, H. E., 1961. An electron microscope study of the aorta in young and in aging mice. J. Ultrastruct. Res., vol. 5, no. 1, pp. 1–27.

Knobloch, H., 1954. Das Verhalten der Capillaren im Altersablauf. Verh. dtsch. Ges. inn. Med., vol. 60, pp. 895–896.

Krafka, J., Jr., 1940. Changes in the elasticity of the aorta with age. Arch. Path., vol. 29, pp. 303–309.

Lansing, A. I., H. T. Blumenthal, and S. H. Gray, 1948. Aging and calcification of the human coronary artery. J. Geront., vol. 3, pp. 87–97.

Lev, M. and O. Saphir, 1951. Endophylebohypertrophy and phlebosclerosis, popliteal vein. A.M.A. Arch. Path., vol. 51, pp. 154–178.

Lev, M., and J. B. McMillan, 1961. Ageing changes in the heart. Chapter 20, pp. 325–349 in Structural Aspects of Aging, edited by G. H. Bourne, London: Pitman Medical Publishing Co.

Lofland, H. B., T. B. Clarkson, L. Rhyne, and H. B. Goodman, 1966. Interrelated effects of dietary fats and proteins on atherosclerosis in the pigeon. J. Atheroscl. Res., vol. 6, pp. 395–403.

Saphir, O. and M. Lev, 1952. The venous valve in the aged. Amer. Heart J., vol. 44, pp. 843–850.

Thompson, J. S., 1969. Atheromata in an inbred strain of mice. J. Atheroscler. Res., vol. 10, pp. 113–122.

Troitzkaja-Andreewa, A. M., 1931. Zur Kenntnis der Altersveranderungen der Arterien. (Über die Altersfibrose der Arterienwand.) Frankfurt. Z. Path., vol. 41, pp. 120–135.

Wilens, S. L., 1937. The postmortem elasticity of the adult human aorta. Its relation to age and to the distribution of intimal atheromas. Amer. J. Path., vol. 13, pp. 811–834.

Hemopoietic Tissue and Blood

Bertolini, A. M., F. M. Quarto di Palo, and L. Gastaldi, 1965. Diaphorase I, catalase and glucose-6-phosphate dehydrogenase activity in the erythrocytes of aged subjects. Gerontologia, vol. 10, pp. 167–173.

Cheli, E. and A. Giordano, 1947. Contributo allo studio del midollo osseo e del sangue periferico nella senilita. Arch. "E Maragliano" pat. c. clin., vol. 2, p. 18.

Dobrovici, A., 1904. Les leucocytes du sang chez les viellards. Compt. rend., Soc. de biol., vol. 56, p. 970.

Doughty, J. H. and L. S. Rosenblatt, 1965. Changes in the hemogram of the beagle with age, J. Geront., vol. 20, no. 2, pp. 131–138.

Ewing, K. L. and O. E. Tauber, 1964. Hematological changes in aging male C57 B1 16 Jax mice. J. Geront., vol. 19, pp. 165–167.

Nellius, D., P. Neumann, and H. Strobbe, 1968. Beitrag zum qualifativen und quentitativen Verhalten der Granulozytopoese im Alter. Z. Alternsforsch, vol. 21, pp. 115–126.

Nelson, A. A., 1940. Pericardial Milk Spots. Arch. Path., vol. 29, pp. 256–262.

Olbrich, O., 1947. Blood changes in the aged. Edinburgh Med. J., vol. 54, pp. 306, 649.

Reich, C., Swirsky, M. and D. Smith, 1944. Sternal bone marrow in the aged. J. Lab. and Clin. Med., vol. 29, p. 508.

Riegle, Gail D. and John E. Nellor, 1966. Changes in blood cellular and protein components during aging. J. Gerontol., vol. 21, pp. 435–438.

Segerdahl, E., 1934. Über Knochenmarkspunktion. Acta Med. Scand. Suppl., vol. 59, p. 173.

Shapleigh, J. B., S. Mayes, and C. V. Moore, 1952. Hematologic valves in the aged. J. Geront., vol. 7, no. 2, pp. 207–219.

The Heart

Dogliotti, G. C., 1931. La struttura del miocardio dell'uomo nei vari individui e nelle varie eta. Ricerche di anatomia microscopica del miocardio su 200 cuori normali in rapporto alla senescenza ed alla costituzione individuale. Ztschr. Anat. u. Entwcklngsgesch., vol. 96, pp. 680–723.

Erickson, E. E. and M. Lev, 1952. Aging changes in the human atrioventricular node, bundle, and bundle branches. J. Geront., vol. 7, pp. 1–12.

Lev, M., 1954. Aging changes in the human sinoatrial node. J. Geront., vol. 9, pp. 1–9.

McMillan, J. B. and M. Lev, 1959. The Aging Heart, I. Endocardium. J. Geront., vol. 14, pp. 268–283.

——— 1958. The Aging Heart, II. Valves. Joint annual meeting of Amer. Soc. of Clin. Pathologists, Chicago, November.

——— 1960. The Aging Heart, III. Myocardium and Epicardium. Vth International Gerontological Congress, San Francisco, August.

Nakao, K., P. Mao, J. Ghidoniand, and A. Angrist, 1966. An electron microscope study of the aging process in the rat heart valve. J. Geront., vol. 21, no. 1, pp. 72–85.

Rondolini, G., 1937. Transformazioni nella struttura del sistema di conduzione del cuore nell'uomo durante il periode fetale e post-natale. Z. Anat., vol. 106, pp. 782–806.

Swigart, R. H., H. S. Riley, J. Withers, and J. B. Rogers, 1961. A comparison of the distribution of cardiac glycogen in young and old guinea pigs. Geront., vol. 16, no. 3, pp. 239–242.

13
The Lymphocyte and Lymphoid Tissue

The Lymphocyte

In the past 10 years a great deal of new information has been acquired concerning lymphoid tissue and, in particular, the lymphocyte. Among the most important facts is the remarkable transformation that these cells, cultured from peripheral blood, will undergo when treated with phytohemagglutinin (Nowell, 1960; MacKinney *et al.*, 1962). They transform into large cells of primitive or stem type and undergo mitotic division. Cooper and Inman (1963) have found that profound modifications of their fine structure and of their metabolism occur.

Andrew (1965) has described marked changes in lymphocytes in epithelial layers (Figs. 13–1 and 13–2).

A further description of the cells arising by transformation of lymphocytes is given by Tanaka, Epstein, Brecher, and Stohlman (1963) as follows: "Fortunately, we need no longer rely on the presence of transitional forms, since isotopic studies have clearly indicated the origin of the proliferating large cells from small lymphocytes. The products of transformation are described as cells with large and occasionally bizarre nuclei with prominent nucleoli. They have ample cytoplasm containing abundant ribosomes, multivesicular bodies, a well-developed Golgi apparatus, and variable fat vacuoles as well as peculiar granular inclusions."

Thus the most recent findings tend to vindicate the earlier investigators who looked upon the small lymphocyte as having a potential for transformation even in the adult body into cells of other types, although, of course, the isotope studies have not proved their ability to develop into macrophages, fibroblasts, or eosinphilic granulocytes, as concluded by research workers who studied this type of problem on prepared sections and in tissue culture (Maximow, 1928).

Another surprising feature of the lymphocyte which has been described and which now appears to rest on firm foundations is their ability to enter into other cells and probably to leave them without injury to themselves or to the "host" cell. Trowell (1958) stated his acceptance of this capacity of the lymphocyte and cited descriptions by several authors, including our now early one (Andrew and Andrew, 1945). A striking instance is given by Ioachim and Furth (1964) who describe a tissue culture of a thymoma in which lymphocytes entered epithelial cells, underwent multiplication by mitosis within epithelial cytoplasm, and then were able to leave their "host" cells.

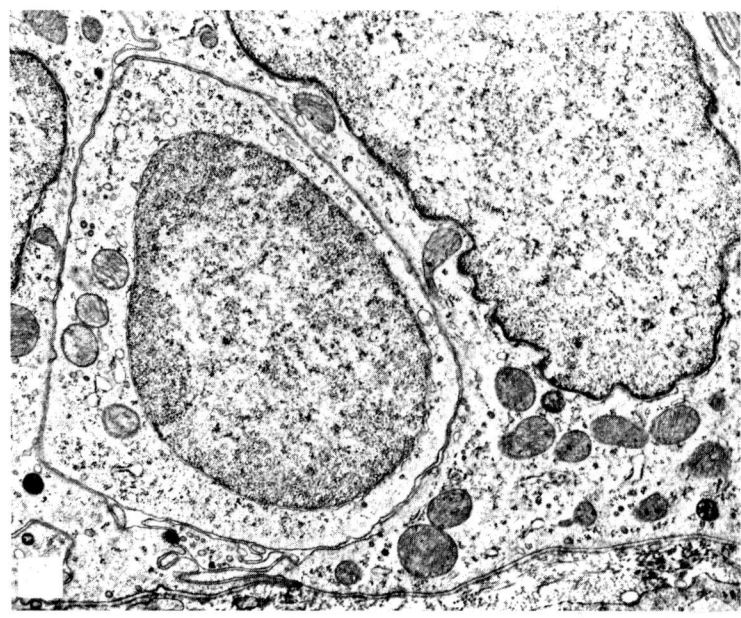

Fig. 13–1. A lymphocyte that has wedged itself between two cells of the intestinal epithelium. Duodenum of rat. Buffered osmic acid, Epon, Karnovsky's stain. ×16,200. (After Andrew, 1965.)

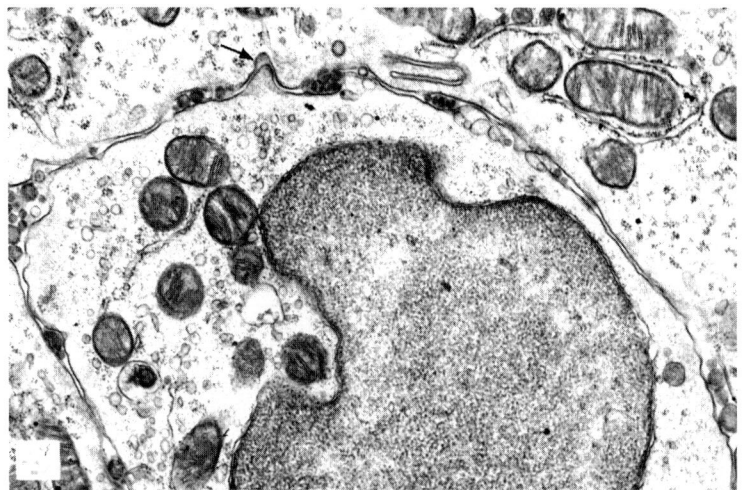

Fig. 13–2. A lymphocyte between two epithelial cells in the duodenum of a C57BL mouse. Lymphocytes in the epithelium show an enlargement of the cytoplasmic body and eventually seem to lose many of their mitochondria and ribosomes, undergoing a transformation to a type of "clear cell." A protrusion of lymphocyte cytoplasm is fitted to an invagination of the epithelial cell (arrow). Buffered osmic acid, Epon, Karnovsky's stain. ×34,200. (After Andrew, 1965.)

Of great importance is the now massive evidence of the role of lymphocytes and lymphoid tissue in the immune reactions of the body. Neonatal thymectomy is followed by an extremely severe impairment of the development of immunological capacity. The thymus itself does not produce antibodies nor develop germinal centers or increased numbers of plasma cells following immunization; not is it in the path of recirculation of the lymphocytes. The thymus, however, has the highest rate of production of lymphocytes of any of the organs (Kindred, 1940). It is thought that its effect on development of the immune reactions of the body may be partly by production of a hormone which stimulates maturation of the other lymphoid tissues (Miller, 1964, 1965) and partly by "seeding" of lymphocytes (the "thymocyte" and small lymphocyte apparently are practically the same type of cell) to other organs (Kellum et al., 1965; Friedenstein and Goncharenko, 1965).

It is true that thymocytes may have important functions in the normal body or under conditions of stress other than their influence on the lymphoid tissues. Thus, Craddock (1964) concludes that they may transfer to cells of regenerating liver either nucleotides or desoxyribonucleic acid needed for growth and division of these cells.

Newer knowledge of the role of the thymus has helped to make it possible to distinguish between two kinds of lymphoid tissue: (1) thymus-dependent, which includes the lymph nodes and spleen; and thymus-independent, which includes the lymphoid tissue of appendix and tonsil (Good and Papermaster, 1964). In fact, the appendix and (in birds) the bursa of Fabricius, appear to some extent to act like the thymus and if they are removed at the time of thymectomy, the resulting immunological depression is more intense than if thymectomy alone is performed.

These few facts and concepts concerning the lymphocytes and lymphoid tissue add to the significance of the age changes which occur in them and which, as we shall see, are often striking ones. These changes may conveniently be described for the lymphoid tissue in its various forms and as it occurs in the individual organs which are composed wholly or in part by such tissue.

Nonorganized or Diffuse Lymphoid Tissue

Small collections of lymphocytes are found in many places in the body. These have been designated as peritoneal, pericardial, peribronchial, perivascular, subpleural, according to their site. While in general they are considered as transitory structures which may come and go, certain regions show them more conspicuously and more consistently in older individuals. This is true of the lymphoid accumulations about the hepatic trinity in the periportal canal of the liver (Andrew, Brown, and Johnson, 1943; Andrew, 1962) (Fig. 13–3). Wallbach (1929) investigated the etiology and development of such cell masses in the liver. According to him, they represent compensatory growth of splenic tissue brought about when such tissue is needed in larger amounts. He believes them to be colonies arising from lymphocytes and reticular cells which come to the liver through the portal circulation from the spleen. They can be produced experimentally by an increase in the amount of foreign albumen, by infections, and by arsenic poisoning.

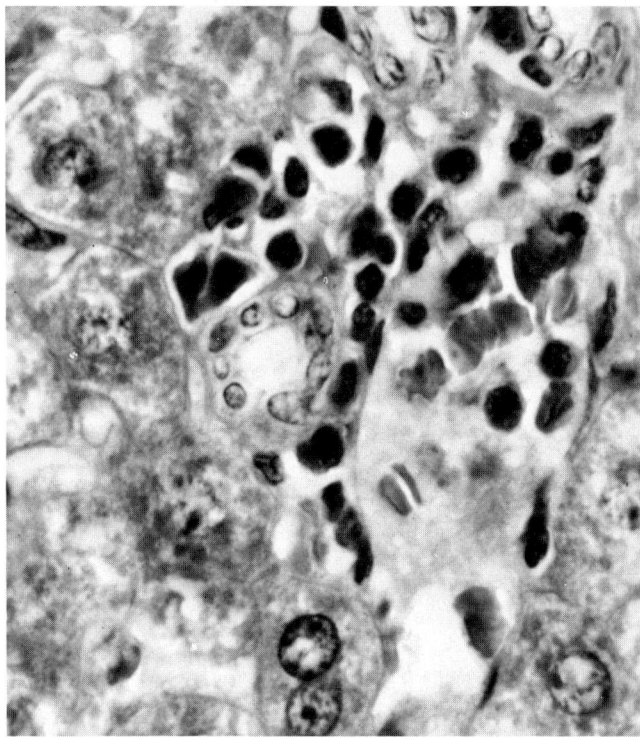

Fig. 13-3. Ectopic lymphoid tissue in the region of the portal trinity of an old rat, a 33-month-old McCollum strain animal. Harris' hematoxylin and eosin. ×1352.

The colonies grow in size both by the accumulation of new cells and by cell division. Their growth apparently leads to destruction of liver cells as the colonies replace portions of hepatic tissue.

In the parotid glands of the rat, accumulations of diffuse lymphoid tissue and occurrence even of small solid nodules amid the glandular tissue were seen as common occurrences in senile animals (Andrew, 1949). Again, this seems to indicate probable colonization and local proliferation. The same phenomenon has been found both in parotid and submandibular glands in many older human subjects (Andrew, 1950). Kingsbury (1945) pointed out the occurrence of lymphatic tissue and its association with degenerative processes in the small glands of the mouth, larynx, and laryngopharynx in the cat. In his material, as in ours, lymphoid tissue usually was found in close relationship to the ducts and vessels in the connective tissue.

Solitary and Aggregated Lymphatic Nodules

While the accumulations of lymphatic tissue which we have described above occasionally show nodular formation, usually this is not the case. In many parts of the body, however, particularly in relation to the mucosal surfaces of the gastro-intestinal and respiratory tracts, fully developed nodules are found in abundance, often showing a darker mass of small lymphocytes peripherally and a light center which has been called a germinal center or a reaction center.

Few data are available on age differences in these solitary nodules. In fact, it is not known whether these structures are in general constant elements or whether they may disappear and be replaced at intervals through life. Ivy (1952, p. 510) says (referring to the appendix): "Its content of lymphoid tissue is said to diminish with age as is generally true of lymphoid tissue throughout the body and in *the alimentary tract*."

The same lack of knowledge is found in regard to the lymphoid tissue of the lower part of the respiratory tree. Macklin and Macklin (1942) said in regard to the lymphoid tissue in the tunica propria of the bronchi and bronchioles: "Doubtless senile atrophic changes involve this layer, along with the others." On the other hand, the evidence for the lung itself, as cited by these authors from Miller (1924, 1937) is that of a very substantial increase in amount of lymphoid tissue in the peribronchial, perivascular, septal, and subpleural location. There is evidence that new lymph nodes are organized from some of this lymphoid tissue, for such nodes occur further down the bronchial tree as age advances.

Williams (1939) described solitary lymph nodules as a rather common occurrence in the bone marrow from various parts of the human skeleton, including the femur, humerus, ribs, and vertebrae of different individuals. These follicles show no clear centers. They are formed more often in persons over 40 years of age. In 202 persons over 40 they were found in 32 per cent of the cases while in 28 persons under 40 they were found in 10 per cent (3 persons). Apparently, they have no pathological significance.

In the next stage of organization of the lymphoid tissue we see aggregations of the lymphoid nodules. These aggregations include the Peyer's patches of the ileum and the symmetrically arranged nodules of the vermiform appendix. Several studies on the lymphatic tissue of the appendix at different ages have been made. According to Bernardo-Comel (1940), the maximum amount of lymphoid tissue in this organ is attained between the ages of 13 to 17 years and maintained until the age of 20. With advancing age the amount decreases, beginning at the tip and progressing toward the cecum. The clear centers diminish in size relatively more rapidly than the nodules as such. In the 60 to 70 year age group the lymphoid tissue is found in greatly reduced quantity and consists of isolated elements near the proximal end, seldom showing clear centers.

Stefanelli (1936) found that generally there are no follicles present after the age of 75 years, although in certain cases the appendix is still high in lymphatic tissue.

Hwang and Krumbhaar (1940) found the weight of the lymphatic tissue highest in the first decade, decreasing to a minimum in the seventh decade but then seeming to continue at the same low level in still older persons.

In considering age changes in the lymphatic tissue in the appendix it should be pointed out that this organ undergoes other very definite degenerative changes in later life. In adults over 50 years of age the lumen is said to be obliterated in one-half of the cases (Ribbert, 1893). Changes in the lymphoid tissue in the wall of the appendix need not, then, be entirely representative of a general trend in the walls of the various other portions of the alimentary tract.

The Tonsils

The organs known as tonsils represent a stage beyond the aggregates of lymphoid nodules. They consist of aggregations of such nodules but they show a higher degree of organization and also bear a more definite relationship to surrounding structures, both through the possession of a frequently well-developed connective tissue capsule on the deep aspect and through a peculiar relationship to the overlying epithelium on the superficial aspect. Waldeyer's ring of lymphatic tissue about the pharynx includes the three types of tonsils, the palatine tonsils, the pharyngeal tonsil or adenoids, and the lingual tonsils, as well as continuous masses of diffuse lymphatic tissue between them.

The tonsils present deep crypts of the epithelial covering with stratified squamous epithelium lining the palatine and lingual tonsils, and pseudostratified ciliated columnar epithelium lining the pharyngeal tonsils. The epithelium is distinguished by a migration of enormous numbers of wandering cells, lymphocytes and polymorphonuclear leucocytes, into it. Frequently this process is so intense that the epithelium takes on a vacuolated reticular appearance and it becomes difficult to distinguish the boundary line between it and the underlying tissue.

While it is customary to speak of the migration through the epithelium as though it were an activity of the leukocytes, Kelemen (1943b) offers evidence that the transfer may be a passive one, that fibrillar cages consisting of argyrophilic networks grow up into the epithelium enclosing masses of lymphocytes and other cells (see Fig. 13-4). Such fibrillar projections often touch even the most superficial layers of the epithelium. In the deeper parts of the epithelium these structures are slender and tube-like but they expand to cyst-like formations as they proceed upward. Finally they reach the lumen and the leukocytes are discharged to pass along threads of fibrin to the mouth of the crypt.

Hieronymus (1933) studied the palatine tonsils from 100 human autopsy subjects. He found a rapid rise to a maximum size at about the twelfth year, a maintenance of a fairly constant size until 35 to 40 years of age, then another leveling off until about 70 years, when a second decline sets in. The clear centers of the nodules do not appear until a few months after birth, according to Pol (1923), third to twelfth month; Barnes (1923) "by the sixth month," and Kniachetsky (1899) sixth month. They are at their maximum development about the time of puberty.

In regard to the histological and cytological details of aging of the tonsils, several workers have made contributions. Wessel (1933) found a great decrease in the numbers of lymphocytes migrating through the epithelium in later life.

Kelemen (1943a) has described the palatine tonsil in the sixth decade of life. Involution is, he says, an atrophic process. There is no sign of a change of the cells or tissues of the tonsils to a functionally less active form but only a decrease in number of the elements. The capsule remains well defined beyond even the last remnant of the organ. No senile fibrosis occurs within the tonsil, nor are formations of bone, cartilage, or cysts any more common in older than in younger adults. However, he does describe the last stages as showing only scattered accumulations of lymphocytes in a scar-like tissue. The crypts tend to be erased.

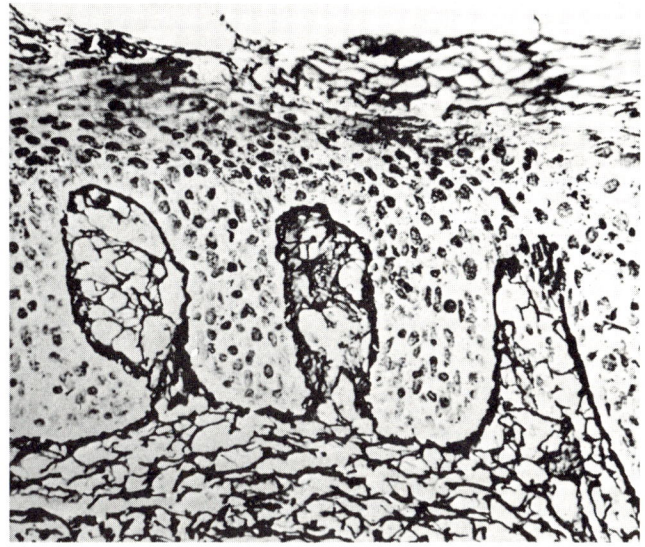

Fig. 13-4. Human palatine tonsil showing projection of reticular meshworks into the epithelium. These appear to carry lymphocytes into the epithelium and liberate them there. (After Kelemen, 1943.)

In a further study Kelemen (1945) made clinical examinations of the tonsils of 309 older patients, including 50 men and 50 women in each of the seventh, eighth, and ninth decades of life and an additional 9 patients from 90 to 103 years of age.

While a steady decline in size of the tonsils in old age was noted, complete absence was rare. The fossa always remained as a more or less spacious niche. Characteristic involutionary forms include (1) the narrow vertical ridge of tonsillar tissue, (2) the large drop-like remnant in the lower part of the niche and suspended by a pedicle-like structure, and (3) the small crater surrounding a last remaining cryptal opening. Local inflammations, tonsillar and peritonsillar, are infrequent but when they occur they take the same course as in other age groups.

Kelemen believes that there is a time of high tonsillar activity with even a "slight possible temporary enlargement" around the sixth decade, corresponding to the similar period before puberty. He concludes that the tonsils frequently are of clinical importance in older persons.

In regard to the pharyngeal tonsils ("adenoids"), the data are more scanty. It has been generally accepted that they are largest in children. Earlier authors give varying figures for the specific age of highest development. Todd (1936) studied them by use of the standard roentgenogram. He found them developing at about 12 months, increasing in size until 3 years, usually remaining stationary until adolescence, and then undergoing a gradual regression. In old age the region may show sparse aggregations of lymphoid tissue scattered over the mucous membrane, according to an early author (Megevand, 1887).

The lingual tonsils, according to Rossing (1940) are both the last to develop and the last to undergo regression. The pharyngeal tonsils are first and the palatine tonsils intermediate in these respects, so that Rossing believes that he has demonstrated a *cranio-craudad progression* in development and regression of the elements of Waldeyer's ring.

The Thymus

The thymus is an organ in which the two major elements are epithelium and lymphocytes (Hammar, 1936). Gregoire (1935) has shown how in grafts of thymic tissue the epithelial portion becomes compact, then is invaded and reticularized by lymphocytes of the host. By irradiating the thymus before transplanting he was able to destroy the autochthonus thymocytes or lymphocytes. He does not believe that the thymocytes of the normal thymus, then, are derived from the epithelial elements, but he thought that he had proved their character as "true" lymphocytes of mesodermal origin. On the other hand, Bailiff (1949) says that in regeneration of the thymus the early reconstruction of the thymocytic portion is carried out almost exclusively by a process of epithelial cell transformation. Murray (1947) believed that he had seen and recorded cinematographically in pure culture of rabbit thymus epithelium a differentiating mitosis in which the two daughter cells from an epithelial cell were, respectively, another epithelial cell and a lymphocyte.

Ackerman and Knouff (1964) have given convincing evidence of the origin of lymphoblasts from the thymic epithelium in the embryo of the hamster.

Hammar (1926, 1936) carried out the most intensive investigation of age changes in the thymus. In a study of the thymus glands from 345 human autopsy cases (1936) he showed that the human thymus grows rapidly through fetal life, continues to increase in absolute weight up to a maximum at the sixth to tenth year, with a slow decline after 16 to 20 years of age.

There has been a somewhat general impression that the thymus undergoes complete or almost complete degeneration, but this does not seem to be the case. Boyd (1932) says that after 11 years of age the weight of the thymus decreases slowly to about the same weight it had at birth (15 g.). Boyd studied 207 cases of death from accidental causes within 24 hours of injury. The cortex began involution at 4 years in her group, while the medulla, with the Hassall's corpuscles, began involution at puberty and the slow replacement of thymic tissue by connective tissue and adipose tissue continued into old age.

In early life the cortex contains many closely packed lymphocytes. As age advances and the cortex atrophies, the distinction between it and the medulla becomes less marked. Finally cortical tissue often appears only as darker areas in the narrowed medullary strands. The great degree of variation in the histological picture and in the degree of involution at different ages must, however, be stressed.

Hassall's corpuscles are seen first in the fetus as a few small rounded bodies 12 to 20μ in diameter. They increase rapidly both in number and in size. By the end of puberty some reach a diameter of 800μ. They seem to decrease gradually in size, so that from the seventh decade none surpasses 300μ in diameter. They decrease also in number as the destruction of both cortical and medullary parenchyma proceeds.

The weight of the thymus as given in deaths due to various diseases probably is low as compared to the normal figure for the given age, since lymphatic tissue tends to be reduced rather rapidly in inanition. In fact it has been said that the large thymus of status thymo-lymphaticus may be simply a normal thymus which, due to the sudden death, has not had an opportunity to decrease in size.

Studies of age changes in the thymus have been made on several of the lower animals. Soderlund and Backman (1909) studied 80 rabbits of known age and found that, as in man, the thymus lost weight with age in spite of infiltration by adipose tissue. Great individual variation was present. After the age of 2 years the cortex was practically indistinguishable from the medulla.

Hassall's corpuscles in the rabbit increase in size and number up until sexual maturity, then decrease, but at the age of 2 years are still large and numerous, according to Syk (1909). Lindberg (1910) found that the first appearance of spermatozoa in the male rabbit and of fertility in the female rabbit coincided with the beginning of involution of the thymus gland and with the time of maximum number of lymphocytes in the blood.

The thymus of the rat appears to reach its maximum weight at the time of rapid increase in the size and development of the gonads. According to Donaldson (1924) the maximum is at 82 days, while at 400 days the thymus weighs only 14 per cent of this maximum. As in man, the thymus seems to persist through life, undergoing a slow atrophic change. Here too the atrophy involves chiefly the cortex, the lymphocytes being replaced by fat and connective tissue.

In the guinea pig the thymus reaches its maximum size when the animal weighs 200 to 300 g. (Paton, 1926). Involution begins at the time of beginning of sexual maturity (1929, Jolly and Lieure). In the horse, Shimpei (1936) found that involution begins in the cortex with atrophy of the lymphoid tissue and replacement by connective tissue. The medulla shows the beginning of involution by an increase in size and number of Hassall's corpuscles and by fatty infiltration. Eosinophils, originally found only in the medulla, are later found in the cortex as well. Frequently they are in a degenerative condition.

A sex difference in the thymus of the mouse has been described by Masui and Tamura (1925), the gland of the nonpregnant female averaging 0.038 g. as against 0.021 for the male. Schirber (1930) found a heavier thymus in the female then in the male goat. In the 462 goats he studied the thymus in females weighed 30.5 g. at birth, 35.5 g. at 3 years, and fell to an average of 5.4 g. at 5 to 11 years. The thymus in males weighed 23.4 g. at birth, 23.0 g. at 3 years, and 11.9 g. at 5 years.

Waschinsky (1925) found, however, a greater weight for the thymus in male than in female pigs. He studied glands of 110 pigs. Here again maximum weight occurs at the age of beginning sexual maturity—6 months. Krupski (1924) found that in cattle the weight of the thymus increases for 7 to 8 weeks after birth and is in full regression by 8 to 12 months. In dogs atrophy begins at 2 to 3 months, according to Hammar (1936).

The thymus is one of the few organs in which some study of age changes has been made in classes of vertebrates other than the mammals. Hammar (1936) states that in fishes the involution of the thymus begins coincidentally with the arrival of sexual maturity. Mucoid and glandular masses appear and the perivascular connective tissue is thickened and may become hyaline. Lymphocytes decrease in number, particularly in the cortex. Picchio (1933) found an appearance of cavities of large size in the thymus of the fish *Lophius budagassa* and a loss of distinction of cortex and medulla with advancing age.

In frogs of 50 g. and more Hammar (1936) also found mucoid areas in the thymus. The cortex was thinned in older (heavier) frogs and the pigment and perivascular connective tissue increased in amount. Cysts were seen but they occurred far down in the weight scale.

In birds involution of the thymus seems to begin relatively late. The organ was of about the same size in all of the hens studied, although in older birds the cortex did become thinner and developed epithelium-lined lumina (Hammar, 1936). Even at 7 to 12 years, however, the medulla remained well developed, although the cortex was much reduced. Riddle and Frey (1924) found regressive changes in the dove thymus, beginning as early as 3 months of age.

The Lymph Nodes

Hellman (1921) was able to demonstrate 200 to 500 nodes in the mesentery alone. The number of lymph nodes in the adult human body probably amounts to several thousand. Lymph nodes appear in the fetus first as primary groups of axillary, inguinal, cervical, and retroperitoneal chains. Later, secondary groups including the epitrochlear, popliteal, intra-mesenteric, and para-aortic appear.

There is considerable variation of opinion in the literature as to when the lymph nodes attain their maximum size. Many authors have stated that they attain maximum size in childhood, but others specify this as maximum relative size referred to body size. Gundobin (1906) stated that maximum size is attained during infancy. It has been generally agreed that the nodes at all ages retain their ability to respond to infection by proliferation of both reticuloendothelial cells and lymphocytes.

It would seem that a study of age changes in lymph nodes of laboratory animals would avoid some of the complications necessarily involved in the study of human autopsy material. Hellman (1914) made a thorough study of the nodes from various regions of the body from 100 rabbits, divided into 12 age groups. The study is, however, primarily quantitative, and histological and cytological features are not described to any great extent. Hellman found an increase in the mass of the nodes up to 5 months of age, then a decrease. Peculiarly, however, a second growth period occurred for the cervical and popliteal nodes, with a final decline beginning after 1 year. The nodes of the oldest rabbits ($3\frac{1}{2}$ years) weighed only about half that found in the stage of their maximum development.

The idea that lymph nodes in young animals are larger than those in old is accepted by Ellenberger and Baum (1915), who dealt with domestic animals in general.

Denz (1947) studied over 300 human lymph nodes collected from 150 autopsies, and in addition a small number of normal glands obtained by biopsy. About a third of the glands were from accident cases, another third from acute medical and surgical cases, and another third from cases of chronic disease. Over 200 of them were from the deep cervical and axillary groups. The rest were mesenteric, axillary, or bronchial. Several fixing and staining methods were used and over 50 of the nodes were studied in serial sections, models being constructed in some cases.

Denz (1947) devoted a good deal of his discussion to consideration of the normal structure of lymph nodes in general. In superficial nodes, such as the inguinal, according to him, the lymphoid tissue consists of a thin layer covering a large fibrous hilum. In deep nodes, such as the deep cervical, there is a greater mass of lymphoid tissue and trabeculae from the relatively small hilum penetrating it, carrying the blood vessels.

Denz finds germinal centers appearing during the first years of life and beginning to retrogress at puberty. He believes that if cyclic changes occur in such centers in the absence of inflammation, they must be gradual and unimportant.

The lymph nodes show differences in age dependent upon their anatomical position and function, although both superficial and deep nodes reach a maximum development during childhood and decrease in size after puberty. In the inguinal nodes the germinal centers and the trabeculae seem to be poorly developed at all ages. After puberty, the cortical tissue decreases in amount. In senility the cortical tissue consists of islands surrounded by medullary tissue. Although there is a shrinking of lymphoid tissue, the size of the node does not change greatly as the hilar connective tissue growth generally compensates for the change in the lymphoid tissue.

The composition of the hilar tissue appears to depend on what is available as replacing tissue in the individual. In adipose subjects, for instance, adipose tissue replaces the lymphoid tissue. In the spare individual fibrous tissue is more common. In starvation or wasting diseases the hilum becomes cystic. Denz says (p. 585), "The connective tissue is greatly reduced and in its place great lymph-containing spaces are found, but the lymphoid tissue is still peripheral in arrangement."

During childhood the deep cervical nodes show well-developed germinal centers and trabeculae. While there is some decrease in size after puberty due to retrogressive changes in the germinal centers, the late changes are less marked than in the inguinal nodes. In extreme old age the deep cervical nodes may resemble those of the fetus in their simplicity of structure, for the cortex and medulla may again appear as continuous units, without nodules, and trabeculae are almost lacking.

A local inflammatory stimulus, at any period of life, may cause the node to take on again the characteristics seen in childhood.

The deep cervical nodes from a series of 100 Wistar Institute rats were studied by Andrew and Andrew (1948). They found that the age differences in the nodes are great enough that one can identify with considerable accuracy, by histological study, nodes from young, middle-aged, and senile rats (Figs. 13–5, 13–6). In the 21-day animals the nodes show a great preponderance of cortex over medulla. The clear centers of the nodules are only beginning to develop at this age. The young mature and middle-aged rats show a gradually increasing size of the medulla with a concomitant decrease in width of the cortex. In the senile animals the medullary sinuses frequently extend almost to the capsule.

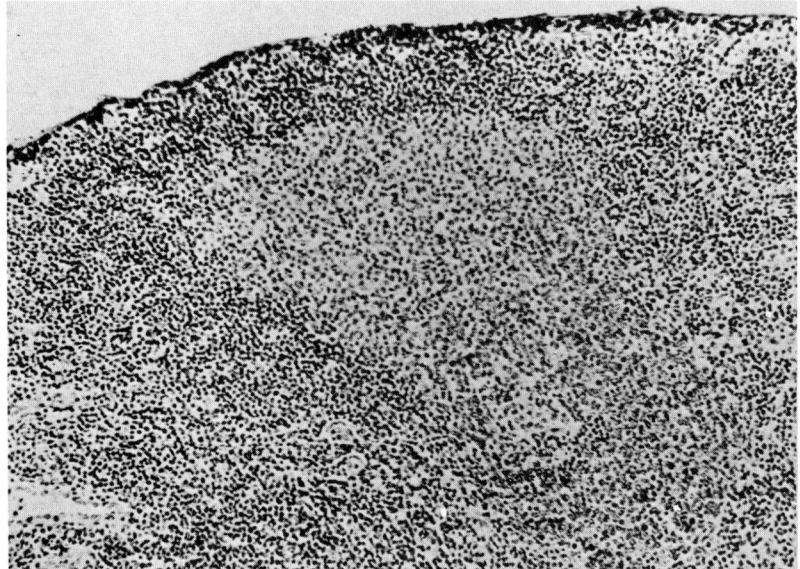

Fig. 13-5. Cortex of a cervical lymph node of a rat of middle age, 300 days old. A generally compact appearance is seen. There is a large reaction (germinal) center present in the center of the field. In Figs. 13-5 through 13-11 the stain employed was Harris' hematoxylin and eosin. ×120. (After Andrew and Andrew, 1948.)

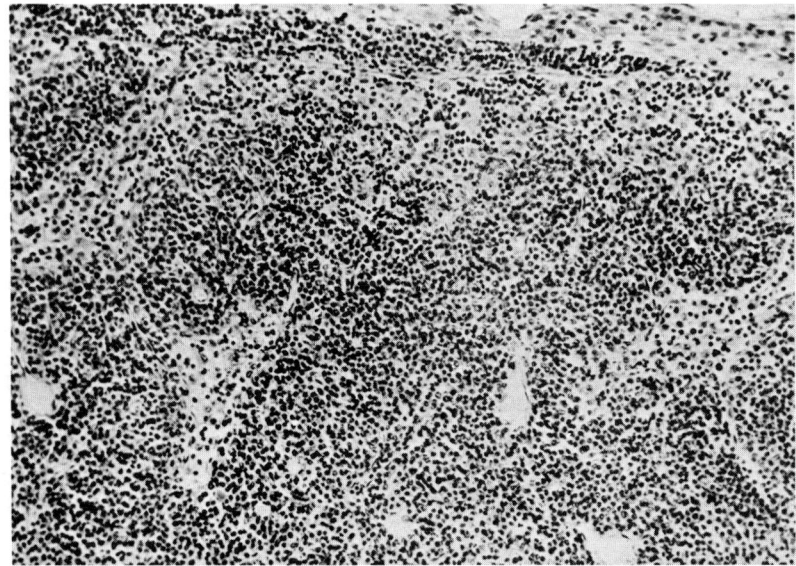

Fig. 13-6. Cortex of a cervical lymph node of a senile rat, 900 days of age. The aspect is relatively loose. Many sections of vessels are present. Reaction centers are lacking. In the cortex in senile rats many of the cells resemble plasma cells rather than typical small lymphocytes. ×120. (After Andrew and Andrew, 1948.)

A conspicuous feature of many of the nodes in senile animals (about 37 per cent) is the presence of large fluid-filled cavities, evidently arising due to atrophy of the lympho-reticular tissue. However, in spite of such atrophy, the mean weight of the nodes in the group of oldest specimens is significantly higher than that in the younger adult rats. This seems to be due to a hypertrophic change in these organs occurring along with the alterations which, by themselves, would lead to atrophy.

The cellular details of the nodes of senile rats differ from those of younger animals both in the cortex and in the medulla. While in the cortex of the younger rats the closely packed cells are definitely small lymphocytes, in senile animals the majority of them frequently are more properly designated as plasma cells, at least in regard to the greater abundance of the cytoplasm and the eccentricity of the nuclei.

In younger rats the medulla shows relatively few free macrophages, most of the cells being stellate in form and connected to the reticular framework by their extended processes (Fig. 13-7). The sinuses generally are well bridged by reticulum. In the senile rats the sinuses are much less frequently bridged by reticular cells and fibers. The number of free macrophages is much increased. The cells are large, rounded, and show many indications of functional activity (Fig. 13-8). Frequently most of the cell-body is filled by large globular masses of deep brown pigment. In many cells large, clear vacuoles occur. In others the bodies of whole lymphocytes or neutrophils or masses of nuclear debris are seen. The numbers, size, and appearance of functional activity of these cells in senile animals are so distinctive as to enable one to identify a node from a senile animal by examination of a single high-power field.

Fibrosis is never of great extent in the nodes of these rats. In a number of the senile nodes, however, some fibrotic change in seen, particularly near the hilum where masses of plasma cells seem actually to coalesce, fibroblasts then appearing in these masses and fibers being laid down. Neither fibrosis nor adipose tissue invasion appears to play any important role in the filling in of spaces left by atrophy of the lymphoid tissue in the rat.

The reaction centers (Fig. 13-9) show phenomena of degeneration of lymphocytes and of phagocytic activity by macrophages in normal rats of all ages, but the number of such centers is decreased in old age.

The findings of Andrew and Andrew (1948) for the rat nodes agree well in many respects with those of Denz (1947) for the human nodes. In both cases a decrease, particularly of the cortical tissue, was noted with advancing age. This decrease is recognized on a histological basis but in the human nodes we do not have weights that might tell us whether a compensatory hypertrophy is occurring, as seems to be true in the rat.

In the human nodes, according to Denz, the type of tissue which replaces lymphoid tissue seems to vary in accordance with what is available. Thus, in obese individuals, it is adipose tissue; in spare individuals it is fibrous tissue; while in inanition there is no replacing tissue and parts of the lymph node show large lymph-containing spaces. The last-named condition is characteristic of the senile rats in the series of Andrew and Andrew. Yet in these animals adipose tissue was abundant in some organs studied, as in the parotid glands (Andrew, 1949).

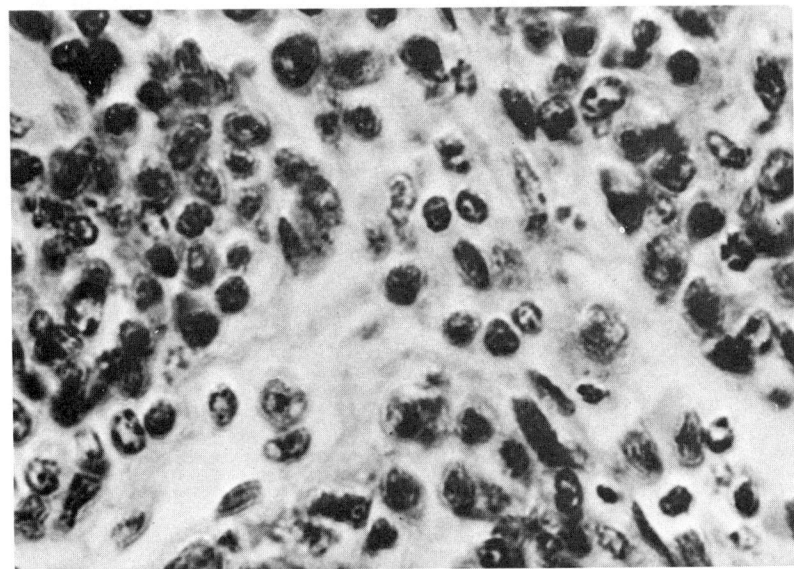

Fig. 13-7. Medulla of a cervical lymph node of a 300-day-old rat. There are few free macrophages present and the sinusoids are relatively narrow. ×360. (After Andrew and Andrew, 1948.)

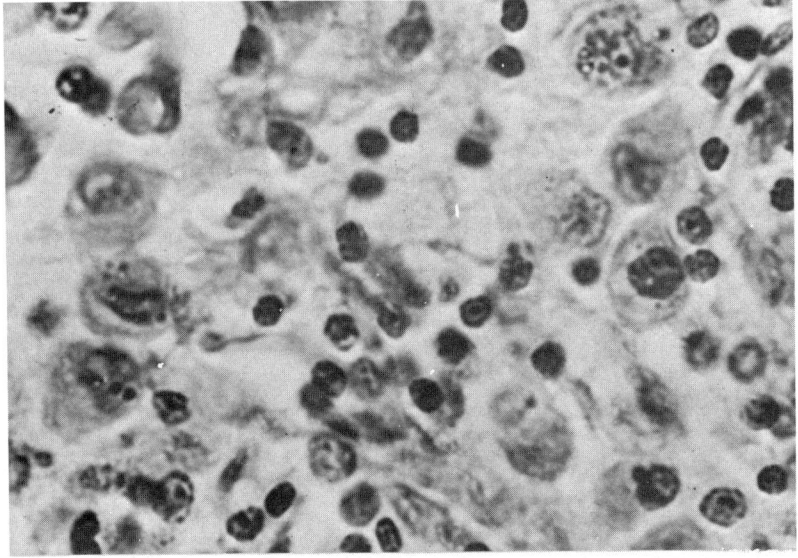

Fig. 13-8. Medulla of a cervical lymph node of a 900-day-old rat. There are many large free macrophages which contain pigment, nuclear debris and vacuoles. The sinusoids are wide. ×360. (After Andrew and Andrew, 1948.)

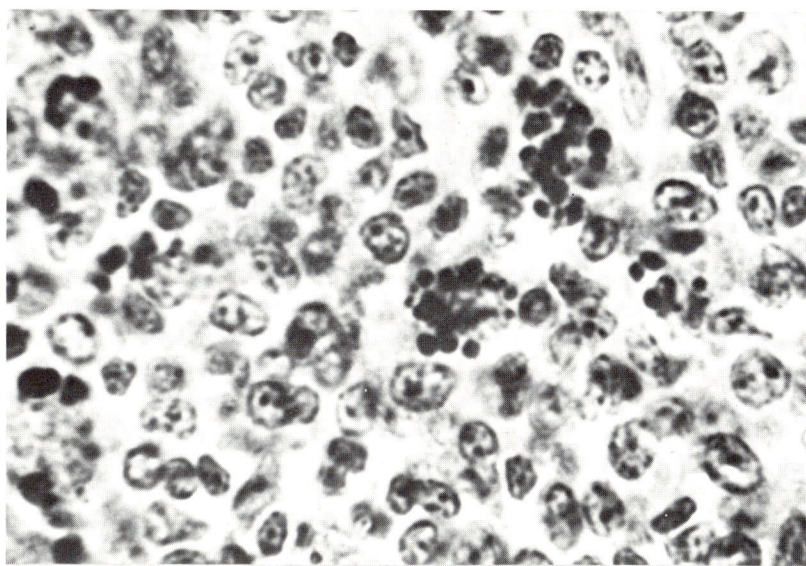

Fig. 13–9. Reaction center in cervical lymph node of a 900-day-old rat. Two macrophages laden with nuclear fragments ("tingible Körper") are seen. Large numbers of degenerating lymphocytes and a great deal of phagocytosis are seen in the reaction centers at all ages, but the centers themselves decrease in number with age. ×1000. (After Andrew and Andrew, 1948.)

Cellular changes both in cortex and medulla are conspicuous in the rat. In man not much attention seems to have been paid to these details, and the problem would be more complicated in autopsy material than in the laboratory animal.

The Spleen

The Malpighian bodies of the spleen are generally well-defined nodules of lymphoid tissue. Nevertheless, the peculiar arrangement of much of the lymphoid tissue in the spleen, scattered diffusely as a component of the red pulp and existing as adventitial sheaths for the extratrabecular arteries, makes a quantitative study of this tissue at different ages a difficult matter.

Hellman (1926) carried out a quantitative study on the spleens of 100 human subjects, ranging from newborn to 84 years of age. All of his cases represent accidental sudden death, suicide, or murder, and it is of interest to note that it required almost 20 years for him to collect the specimens. The subjects died within 12 hours of injury, many of them instantaneously.

Hellman was interested primarily in the weight relationships of the red and white pulp, the secondary nodules, capsule, trabeculae, and blood vessels. He found the average weight of the spleen increasing up to an age between 20 and 30 years and showing a definite decrease only after 50 years of age. The lymphoid tissue (follicular pulp in this instance) increased until directly after puberty (16 to 20 years) at which time the age involution begins. The decrease seemed to continue until the most advanced age. At the age of 5 years the lymphoid tissue amounts to 22 per cent of the weight of the spleen, at 10 years to 17 per cent, at 20 years to 12 percent, at 60 years to 7 per cent, and at 84 years to only 5 per cent. It must be noted, however, that there were in Hellman's series only 5 cases over 50 years of age. It should be noted also that the figures given represent percentages only and must be taken in connection with the total spleen weights in order to arrive at the true amount of follicular lymphatic tissue in the spleen at any given age.

Hellman found the secondary nodules best developed at 1 to 10 years and their average number highest at this time. By the age of 20 years they are already much less conspicuous. Their fullest development, therefore, is somewhat in advance of the fullest development of the lymphoid tissue of the follicles.

The connective tissue (in this instance including the capsule, trabeculae, and blood-vessels) shows some increase in percentage weight with age.

In the 21 to 30 year group it constitutes 6.91 per cent of the total weight. In the group over 50 years old, it is 11.07 per cent of the total weight.

A similar study was made by Hwang, Lippincott, and Krumbhaar (1938), using spleens from a larger number of persons but again only from persons dying violent deaths shortly after the violence occurred and excluding all spleens showing any signs of disease. Microscopic fields of the organ were projected and the outlines of the Malpighian follicles and of the intervening tissue were drawn on paper, the areas then being measured with a planimeter. Knowing the spleen and body weights, it was then possible to get an approximate weight for the follicular lymphatic tissue in the spleen and also its ratio to the body weight at the different ages. In this way 300 cases of violent death distributed in 9 age groups from birth to 70 years and over were analyzed. As in Hellman's work, the percentage of lymphatic tissue was found to be low in the first year (4.5) although the relative number of follicles was then at a maximum. The maximum percentage of lymphatic tissue is reached in the first decade (10.8 per cent), again agreeing with Hellman. However, whereas in Hellman's study the percentage then fell steadily throughout life, their curve dropped sharply to 7.7 per cent in the 11 to 30 age groups, then fell more gradually in the next two age groups. In the 50 to 70 year group a slight rise occurred, while in the oldest group (all cases over 70) the figure fell to 5.8 per cent. These authors believe that a possible increase of lymphatic tissue in late middle age, associated perhaps with increased resistance to various infections, is indicated by their figures.

Krumbhaar and Lippincott (1939) also made a study of the total weight of the human spleen at different ages. They pointed out the great variation in figures found by earlier authors. Schridde (1923), for example, gave as the normal weight of the spleen 115 g. while Hyrtl (1846) had set it at 250 to 300 g. Rossle and Roulet (1932) studied spleen weights in 802 subjects, 569 males and 233 females. They found a maximum weight for males in the third decade (169.1 g.) and for females in the fourth decade (153.7 g.). In the seventh decade average weight of the spleen was about 112 g. and for all persons over 70 years of age about 103 g. Spleens from 2 persons over 90 years of age averaged only 65 g.

Krumbhaar and Lippincott (1939) analyzed 2,000 routine autopsy cases, omitting those which at autopsy or during histological examination showed any considerable splenic changes, particularly such as might tend to splenomegaly. The subjects were divided into 18 semi-decades, ranging from birth to 95 years.

Significantly heavier spleens were found in the 16 to 20 and 21 to 25 age groups than in any other groups. However, the lower weight level of the 26 to 30 years age group was maintained with little change up to the sixty-fifth year, after which a sharp loss occurred. The authors say that while body weight seems to have nothing to do with the peak weight from 16 to 25, it is "presumably a factor" in the final decline.

Krumbhaar and Lippincott also assembled data on spleen weights of 2,000 persons dying violent deaths. They state that they realize that in such cases there are many unknown factors and even when full allowance is made for the possible inclusion of undetected abnormalities, the normal spleen must be recognized as one of the most variable organs in size and weight. The graphs of the mean spleen weights of the 2,000 disease deaths and of the 2,000 violent deaths show a considerable degree of parallelism. The former probably average too high, they say, and the latter too low, so that the middle curves in the graph, representing the data for all 4,000 cases, probably come nearer to giving the true picture.

The ratio of the spleen weight to the body weight in all 4,000 cases is highest in the earliest age group. From there on the slope of the ratio is slightly but "not significantly" down, except for a plateau in the groups between 51 and 65 years where the ratio is significantly higher than those of the age groups on either side. This again would seem to be an indication of a probable increase in lymphatic tissue at this period of life. The very oldest spleens were light in weight, but while fibrosis, arteriosclerosis, and atrophy (low weight) are the rule in the spleen of the aged, the lymphatic tissue may persist in considerable amounts even in extreme old age.

These authors summed up the results of their studies on the weight of the human spleen and on the percentage of lymphatic tissue in it at various ages as follows: "the spleen appears to grow parallel with the growth of the body during childhood, but reaches its maximum weight (170 g.) earlier (i.e., in the 16 to 20 age group as opposed to 26 to 40 years for the body weight). From the ages of 26 to 65, its absolute weight remains approximately unchanged, though, compared with the average body weight, the spleen weight falls slightly, except for the 51 to 65 age group where it rises significantly. After 65, both spleen and body weight fall steadily, the spleen weight more than the body weight, so that in the very old, weights less than 100 g. are frequent.

The lymphatic tissue in the Malpighian follicles, low at birth, quickly rises to a maximum percentage early in the first decade. It then drops, most sharply in the second decade, until the age of 50, when there is a distinct increase lasting until about 65. Thereafter, it falls again eventually to reach a percentage similar to that of early infancy.

Both the spleen weight and the lymphatic tissue percentages show a marked individual scatter.

Andrew (1946) studied the spleens of 100 Wistar Institute rats, from 21 days up to 1170 days of age. In the senile group (animals 800 to 1170 days in this study) there were 35 animals.

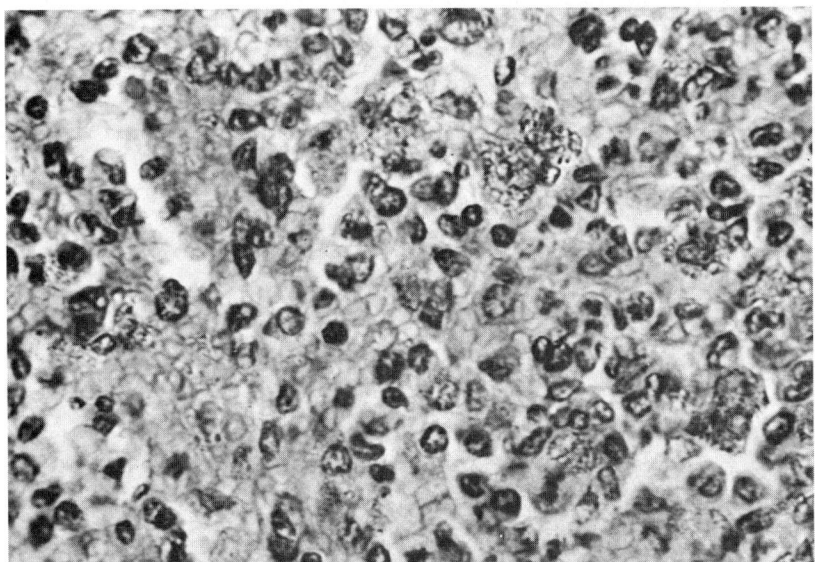

Fig. 13-10. Red pulp of the spleen of a rat 600 days of age. The sinusoids are inconspicuous and the appearance is compact. Macrophages, of which many are seen in the field, contain a granular, golden-yellow pigment. ×580. (After Andrew, 1946.)

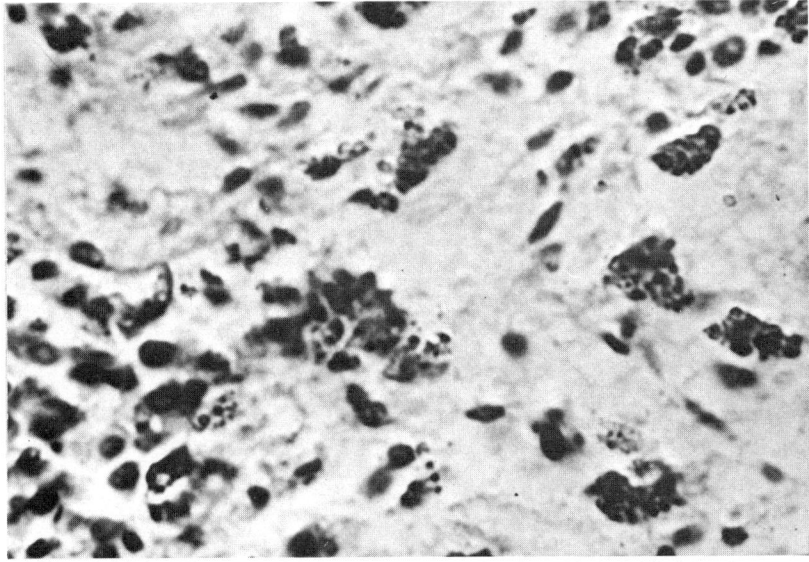

Fig. 13-11. Red pulp of the spleen of a senile rat, 1,060 days of age. The tissue has a loose appearance. The pigment in the macrophages is of a deep brown color, much darker than in younger animals. ×580. (After Andrew, 1946.)

Qualitative as well as quantitative differences in the spleens from rats of different ages were found. In the immature rats (21 days) Malpighian follicles are in process of formation. No reaction centers are seen. In young rats (50 to 200 days) reaction centers are numerous and well-marked. They persist in the rats of middle age (300 to 726 days). In senile animals reaction centers are lacking in almost all cases (32 out of 35 animals for the several large sections studied).

There is a decrease in amount of follicular tissue in old rats and a loss of distinctness in the separation of the two zones of the follicle as well as of the demarcation between red and white pulp. The follicles of senile animals often show several sections of thick-walled artery rather than the single section, which may indicate either a branching or a tortuosity of the central artery.

Senile rats present a conspicuous change in the red pulp consisting in a transformation from a predominantly compact, reticular type of tissue to a predominantly sinusoidal type (Figs. 13-10 and 13-11). This change would seem to be brought about by an actual metaplasia of "reticular" cells to "endothelial" cells with formation of new sinusoidal spaces from the open ends of the old ones.

The spleen of 21 day rats shows no pigment-containing macrophages. They make their appearance first in 50 day rats in this series and increase in numbers per volume of spleen up into middle age. The variation in number in senile spleens, however, is very great. In spleens in which the architecture of the white pulp is largely destroyed they are scarce.

In senile animals the red pulp shows more degenerating pycnotic and fragmenting lymphocytes than those seen in young animals. These are often phagocytized by macrophages. Plasma cells are more abundant in old rats.

Megakaryocytes are plentiful in the spleens of young rats, undergo a sharp decrease in number at an early age but persist in fair numbers in senility. A megakaryocytic splenomegaly was seen in a number of old rats (17 per cent of the 35 animals).

Range in weight of the spleen in old rats is greater than in any of the younger groups. While a number of the senile spleens seem to be atrophic, the average weight actually is *greater* than that for the young and middle-aged animals.

REFERENCES

Ackerman, G. A. and R. A. Knouff, 1964. The epithelial origin of lymphocytes in the thymus of the embryonic hamster. Proceedings of the American Society of Hematology. Blood, vol. 24, p. 838 (Abstract).

Andrew, W., 1946. Age changes in the vascular architecture and cell content in the spleens of 100 Wistar Institute rats, including comparisons with human material. Amer. J. Anat., vol. 79, pp. 1-74.

――― 1949. Age changes in the parotid glands of Wistar Institute rats with special reference to the occurrence of oncocytes in senility. Amer. J. Anat., vol. 85, pp. 157-198.

――― 1950. A comparative study of senile changes in the salivary glands of man and of the rat. J. Geront., vol. 5, pp. 385-386.

――― 1965. Comparative Hematology. New York: Grune & Stratton, 200 pp.

――― and N. V. Andrew, 1948. Age changes in the deep cervical lymph nodes of 100 Wistar Institute rats. Amer. J. Anat., vol. 82, pp. 105-166.

――― H. M. Brown, and J. B. Johnson, 1943. Senile changes in the liver of mouse and man, with special reference to the similarity of the nuclear alterations. Amer. J. Anat., vol. 72, pp. 199–221.

Bailiff, R. N., 1949. Thymic involution and regeneration in the albino rat, following injection of acid colloidal substances. Amer. J. Anat., vol. 84, pp. 457–510.

Barnes, H. A., 1923. The Tonsils, Faucial, Lingual, and Pharyngeal. St. Louis: C. V. Mosby.

Bernardo-Comel, M. C., 1940. Il tessuto linfoide dell appendice umana nelle varie età della vita. Arch. Ist. Biochem. ital., vol. 9, pp. 199–222.

Boyd, E., 1932. The weight of thymus gland in health and disease. Amer. J. Dis. Child., vol. 43, pp. 1162–1214.

Cooper, E. H. and D. Inman, 1963. The ultrastructure of transformed lymphocytes. In Proceedings IX Congress European Society of Hematology, p. 11 (Abstract).

Denz, F. A., 1947. Age changes in lymph nodes. J. Path. Bact., vol. 59, pp. 575–593.

Donaldson, H. H., 1924. The Rat: Data and Reference Tables, 2d ed. Wistar Institute of Anatomy.

Ellenberger, W. and Baum, H., 1915. In Handb. d. vergleich. Anat. d. Haustiere, Hirschvald, Berlin.

Gowans, J. L., McGregor, D. C., and D. M. Cowen, 1962. Initiation of immune responses by small lymphocytes. Nature, London, vol. 196, pp. 651–655.

Hammar, J. A., 1936. Die normalmorphogische Thymusforschung im letzten Vierteljahrhundert. Barth, Leipzig.

Hellman, T., 1926. Die Altersanatomie der menschlichen Milz. Ztschr. f. Konstitutionslehre, vol. 12, pp. 270–415.

——— 1921. Die Bedeutung der Sekundärfollikel. Beitr. z. path. Anat. u. z. allg. Path., vol. 68, pp. 333–363.

——— 1914. Die normale Menge des lymphoïden Gewebes beim Kaninchen in verschiedenen postfetalen Altern. Upsala läkaref. förh. Suppl., pp. 1–408.

Hwang, J. M. S and E. B. Krumbhaar, 1940. Amount of lymphoid tissue of the human appendix and its weight at different ages. Amer. J. Med. Sci., vol. 199, pp. 75–83.

——— Lippincott, S. W., and E. B. Krumbhaar, 1938. The amount of splenic lymphatic tissue at different ages. Amer. J. Path., vol. 14, pp. 809–819.

Hyrtl, J., 1846. Lehrbuch der Anatomie des Menschen. F. Ehrlich, Prague.

Ioachim, H. and J. Furth, 1964. Intrareticular cell multiplication of leukemic lymphoblasts in thymic tissue cultures. J. Nat. Cancer Inst., vol. 32, pp. 339–359.

Ivy, A. C., 1942. Digestive System. Chapter 10, pp. 254–301. In Cowdry's Problems of Aging, 2d Ed., Baltimore: Williams & Wilkins.

Jolly, J. and C. Lieure, 1929. Sur l'involution du thymus. Compt. rend. Soc. de biol., vol. 102, pp. 762–764.

Kelemen, G., 1945. Clinical observations on the palatine tonsil. Ann. Otol., Rhin., and Laryng., vol. 54, pp. 421–436.

——— 1943a. Clinical observations on the palatine tonsil in the aged. Ann. Otol., Rhin., and Laryng., vol. 52, pp. 419–444.

——— 1943b. Pathway of the tonsillar lymphocyte. Arch. Otolaryng., vol. 38, pp. 433–444.

Kingsbury, B. F., 1945. Lymphatic tissue and regressive structure, with particular reference to degeneration of glands. Amer. J. Anat., vol. 77, pp. 159–188.

Kniachetsky, 1899. Über die Tonsillen der Kinder. Dissertation, St. Petersburg.

Krumbhaar, E. B. and S. W. Lippincott, 1939. Postmortem weight of "normal" human spleens at different ages. Amer. J. Med. Sci., vol. 197, pp. 344–358.

Krupski, A., 1924. Über die akzidentelle Involution der Thymusdrüse beim Kalb. Schweiz. Arch. f. Tierheilk., vol. 66, pp. 14–21.

Lindberg, G., 1910. Zur Kenntnis der Alterskurve der weissen Blutkörperchen des Kaninchens. Folia. hematol., vol. 9, pp. 64–80.

Masui, K. and Y. Tamura, 1925. The effect of gonadectomy on the weight of the kidney, thymus and spleen of mice. Brit. J. Exper. Biol., vol. 3, pp. 207–223.

Maximow, A. A., 1928. Cultures of blood leucocytes; from lymphocyte and monocyte to connective tissue. Arch. Exp. Zellforsch., vol. 5, pp. 169–268.

Megevand, J. A., 1887. Contribution à l'Étude Anatomorphologique des Maladies de la Voute de Pharynx. Thesis, Geneva.

Moe, M. and O. Behnke, 1962. Cytoplasmic bodies containing mitochondria, ribosomes, and rough surfaced endoplasmic membranes in the epithelium of the small intestine of newborn rats. J. Cell Biol., vol. 13, pp. 168–171.

Murray, R. G., 1947. Pure cultures of rabbit thymus epithelium. Amer. J. Anat., vol. 81, pp. 369–411.

Nowell, P. C., 1960. Phytohemagglutinin: An initiator of mitosis in cultures of normal human leukocytes. Cancer Res., vol. 20, pp. 462–466.

Picchio, T. S., 1933. Richerche sul timo di Lophius budagassa Spin. e. sulle sue modificazione nell'adulta. Arch. ital. Anat., vol. 31, pp. 549–568.

Pol, R., 1923. Zur Funktionsfrage der lymphadenoiden Organe, inbesondere der Tonsillen. Verhandl. d. deutsch. path. Gesellsch., vol. 19, pp. 286–289.

Porter, K. A. and E. H. Cooper, 1962. Recognition of transformed small lymphocytes by combined chromosomal and isotopic labels. Lancet, vol. 2, pp. 317–319.

Riddle, O. and P. Frey, 1924–1925. The growth and age involution of the thymus in male and female pigeons. Amer. J. Physiol., vol. 71, pp. 413–429.

Rossing, F., 1940. Histologische Unterlagen zum kraniokaudalen Entwicklungsgang des Waldeyerschen Rachenringes. Ziegler's Beitr., vol. 105, p. 17.

Rossle, R. and F. Roulet, 1932. Mass und Zahl in der Pathologie. Julius Springer, Berlin.

Schirber, A., 1930. Ein Beitrag Zur Anatomie und Rückbildung des Thymus bei der Ziege. Dissertation, Veter, Berlin.

Schridde, H., 1923. Die blutbereitenden Organe. In L. Aschoff's Pathologische Anatomie, 6th ed., Fischer, Jena.

Shimpei, E., 1921. Über die Involution der Thymusdrüse des Pferdes. Japan. J. Med. Sci., vol. 1 (quoted by Hammar, J. A., 1936).

Soderlund, G. and A. Backman, 1909. Studien über die Thymusinvolution. Die Altersveränderungen der Thymusdrüse beim Kaninchen. Arch. mikr. Anat., vol. 73, pp. 699–725.

Stefanelli, C., 1936. L'Appendicite acuta nei soggetti de età nel vecchio. Policlinico (sez. chir.), vol. 43, pp. 644–651.

Sukiennikow, W., 1903. Topographische Anatomie der bronchalen und trachealen Lymphdrüsen. Dissertation, Berlin.

Syk, I., 1909. Über Altersveränderungen in der Anzahl der Hassalschen Körper nebst einem Beitrag zum Studium der Mengenverhältnisse der Mitosen in der Kaninchenthymus. Anat. Anz., vol. 34, pp. 560–567.

Tanaka, Y., L. B. Epstein, G. Brecher, and F. Stohlman, Jr., 1963. Transformation of lymphocytes in cultures of human peripheral blood. Blood, vol. 22, pp. 614–629.

Todd, T. W., 1936. Integral growth of the face. I. The nasal area. Internat. J. Ortho. and Oral Surg., vol. 22, pp. 321–332.

Trowell, O. A., 1958. The lymphocyte. Int. Rev. Cytol., vol. 7, pp. 236–293.

Wallbach, G., 1929. Untersuchungen über die Ätiologie und die Genese der mesenchymalen Zelleherde in der Leber als Beitrag zu einer Lehre der mikroskopische sich darstellenden Zellfunktionen. Ztschr. f. d. ges. exper. Med., vol. 68, pp. 569–620.

Waschinsky, G., 1925. Über den Thymus des Schweines. Dissertation, Veter., Berlin.

Wessel, O., 1933. Die Lymphozytendurchwanderung durch das Epithelium der Tonsillen. Dissertation, Rostock.

Williams, R. J., 1939. The lymphoid nodules of human bone marrow. Amer. J. Path., vol. 15, pp. 377–384.

Wolfe-Heidegger, G., 1939. Zur Frage der Lymphocytenwanderung durch das Darmepithel. Ztschr. f. mikr. anat. Forsch., vol. 45, pp. 90–103.

Yoffey, J. M. and F. C. Courtice, 1956. Lymphatics, Lymph and Lymphoid Tissue. Cambridge: Harvard University Press. 510 pp.

14
The Respiratory System

The lungs, as would seem natural in relation to their functional properties, are rich in elastic connective tissue fibers.

Rather unexpectedly, however, there seems to be a definite increase in amount of elastic tissue with age in these organs (Bickerman, 1952; Blumenthal *et al.*, 1964).

The elastic fibers of the lungs, furthermore, do not show some of the qualitative changes seen in such fibers in other locations. In the arteries, for example, there is an age-related increase of a yellow pigment, a fluorescent compound, associated with the elastic fibers; this is not the case with the lungs. While some calcification may occur with age in pulmonary elastic tissue, there certainly are not the spectacular calcific changes so often seen in the large arteries in old age.

The reason for the increase in amount of elastin and elastic tissue may be found in the stimulus of continuous rhythmic stress (Krahl, 1964; Bunting, 1939; Dyson and Decker, 1958).

There is an increase in the amount of diffuse lymphoid tissue along the branches of the respiratory tree and around the small blood vessels (Figs. 14–1 and 14–2).

An electron microscope and histological study of the musculature of the vessels and bronchial tree within the lung was made by Heppleston (1961). He used normal CBA mice with an age range of 2 to 27 months. Quantitative variations in the smooth muscle of the bronchial tree were found but appeared to be on the basis of individual variation. No consistent alteration with age was seen in the size, number, or fine structure of muscle in intrapulmonary arteries or veins or in the bronchial tree.

The author extrapolates these findings to the conclusion that the reduced amounts of muscle in the bronchial tree which often are seen in chronic pulmonary emphysema cannot be accounted for "by age alone." Thus, it would seem desirable to have such studies on the human lung.

One interesting, but not age-related finding in this study is the presence of large amounts of cardiac muscle in the intrapulmonary veins, beginning abruptly in the venous system about midway between the hilus of the lung and the pleura.

Sirtori (1964) has demonstrated differences between alveolar cells of the lungs of young and old mice (15 days and "over 1 year" old). In old mice the lamellar inclusions contain a smaller number of lamellae. Cells with lamellar inclusions

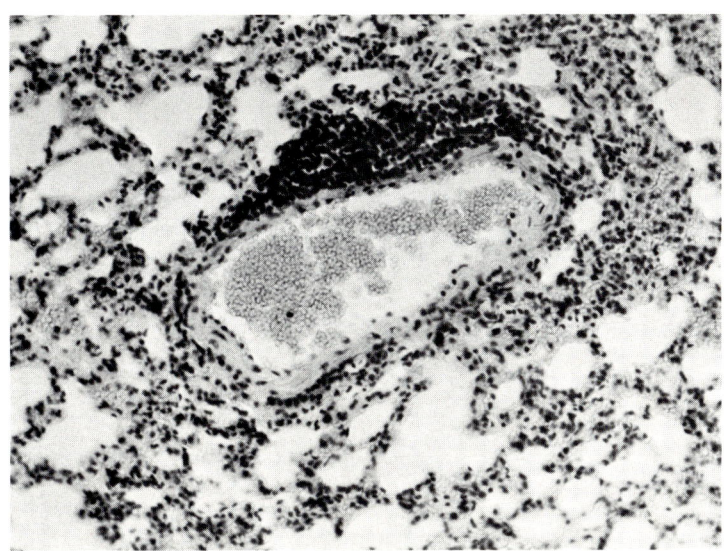

Fig. 14–1. Lung of a senile C57BL mouse, 808 days of age. The large vessel shows a dense aggregation of lymphocytes on its upper border. Such masses of ectopic lymphoid tissue generally are, as in this instance, eccentrically located. Harris' hematoxylin and eosin, from serial section of an entire lung. ×125.

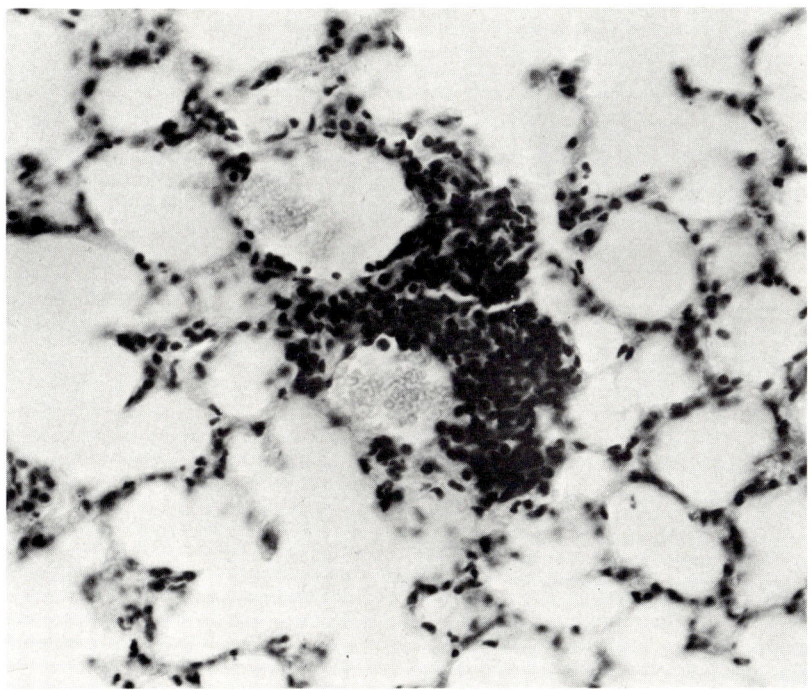

Fig. 14–2. Lung of the same animal presented in Fig. 14–1. Lymphocyte masses here are present out among the alveoli themselves. Harris' hematoxylin and eosin, from serial section of an entire lung. ×250.

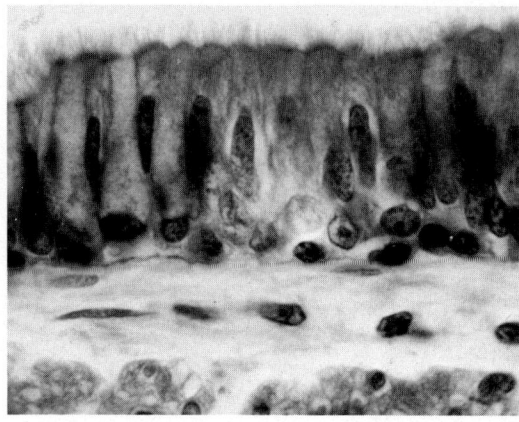

Fig. 14-3. Portion of lining of a bronchus. The cilia of the pseudo-stratified epithelium help to keep the respiratory passages clear throughout life ×800.

seem to be selectively damaged by influenza virus and to a greater degree in old than in young animals. Sirtori believes that this great susceptibility to influenza virus can lead to greater susceptibility to cancer when the lungs are exposed to carcinogens. This author has also found what he considers to be "precancerous" changes in normal epithelium of nipple, of endometrial glands, and of the cervix of the uterus.

Boucek et al. (1961), in an attempt to study the fibrous proteins of the lungs with advancing age, found it difficult to separate elastin and collagen. However, the data of their chemical analyses indicate that elastin and collagen may tend to bind together in a process of copolymerization. Such alteration would lead to an increase in crystallinity and a decrease in elastic properties of the pulmonary fibrous proteins.

REFERENCES

Ashoff, L., 1938. Zur normalen und pathologischen anatomie des Greisenalters. Urban und Schwarzenberg. Berlin.

Basilevich, I., 1939. Clinical observations on a group of twelve centenarians. The Brit. Encycl. of Med. Practice, vol. 9, p. 528.

Bickerman, H. A., 1952. The respiratory system in the aged. Chapter 22, pp. 562-613 in Cowdry's Problem of Aging, 3rd ed., edited by A. I. Lansing, Baltimore: The Williams & Wilkins Co.

Boucek, R. J., N. L. Noble, and A. Marks, 1961. Age and the fibrous proteins of the human lung. Gerontologia, vol. 5, no. 3, pp. 150-157.

Brock, B. C., 1947. Anatomy of the Bronchial Tree. London: Oxford University Press.

Christopherson, J. B. and M. Broadbent, 1934. A new method of approach in certain respiratory disorders in elderly persons. Brit. J. Physical Med., vol. 9, p. 5.

Heppleston, A. G., 1961. The musculature of the aging mouse lung. J. Geront., vol. 16, pp. 106-109.

Hollender, A. R., 1944. Influence of aging on the nasal mucosa. A histopathologic study. Ann. Otol. Rhin. and Laryng., vol. 53, p. 595-596.

——— 1944. Histopathology of the nasal mucosa of older persons. Arch. Otolaryng., vol. 40, pp. 92-100.

Korenchevsky, V., 1942. Natural relative hypoplasia of organs and the process of aging. J. Path. Bact., vol. 54, pp. 13-24.

Krahl, V. E., 1965. A four-dimensional approach to the study of pulmonary anatomy and physiology. The William Snow Memorial Lecture. Am. Rev. Resp. Dis., vol. 42, pp. 228-237.

Miller, W. S., 1924. The pulmonary lymphoid tissue in old age. Amer. Rev. Tuberc., vol. 9, pp. 519-524.

―――― 1947. The Lung, 2d ed., Springfield, Ill.: Charles C Thomas.

Monroe, R. T., 1951. Chap. 4, Diseases of Old Age. Cambridge: Harvard University Press.

Rolleston, H., 1932. Medical Aspects of Old Age. London: Macmillan.

Saxton, J., 1949. Pathology of senescent animals. Conference on Problems of Aging. Transactions of the Eleventh Conference. Josiah Macy, Jr. Foundation, New York.

Sirtori, C., 1964. Relationship between cancer and senile changes in the lung. Electron microscopy study. Gerontologia, vol. 9, no. 4, pp. 239-248.

Smith, C. G., 1942. Age incidence of atrophy of olfactory nerves in man. J. Comp. Neurol., vol. 77, pp. 589-595.

15
The Digestive System

Mouth, Salivary Glands, and Pharynx

The mouth and lips of the older individual tend to be drier than those of the young. The amount of salivary secretion and also the concentration of its carbohydrate-splitting enzyme are decreased (Meyer, Spier, and Neuwelt, 1940).

Schram (1933) described the human submandibular gland at various ages. He studied the submandibular glands of 26 subjects and the sublingual glands of 37 subjects. Even before the seventh decade he found some fat present in many of the glands. After the age of 60 all of the glands contained abundant fat. Larger amounts were seen in the sublingual than in the submaxillary glands.

Yamaguchi (1924) observed an increase in fat content of the salivary glands in various pathologic conditions and in senility. He found the fatty infiltration in general to be greatest in the submandibular gland, next in the parotid, and least in the sublingual gland. His studies were made on frozen sections prepared by various fat stains. The fat in the gland-cells gave a positive Smith reaction for neutral lipoids. He believed that the fatty infiltration of the gland cells, on the basis of histologic observations at least, would seem to have no inimical effect on the secretory activity of the cells.

In the parotid gland of the rat, the process of fatty change seems to be one of frank fatty degeneration with destruction of cells, alveoli, and even whole lobules of secreting tissue (Andrew, 1949). The histologic picture would indicate a severe loss of function in this gland in old age in the rat (Fig. 15-1). The submandibular gland, however, exhibits practically no fatty change with age.

A second important feature described as occurring in the salivary glands of senile individuals is the presence of peculiar large cells, the oncocytes (Fig. 15-2). These cells were first described in the human glands by Hamperl (1933), and later in the rat (Andrew, 1949). In our material they possessed a distinctive cytoplasm of a foamy, reticulated type.

Hamperl expressed the belief that the honeycombing of the cytoplasm is a transitional stage in the development of the fully differentiated granular cytoplasm.

The nuclei are often pycnotic, but sometimes vesicular with a very large nucleolus. The nuclei of either type may be found in process of amitotic division.

In the rat, both in the parotid and submandibular glands, the oncocytes occur in such large numbers in senility that study of only one or a few low-power fields makes it possible to tell whether the section is from a young or old animal.

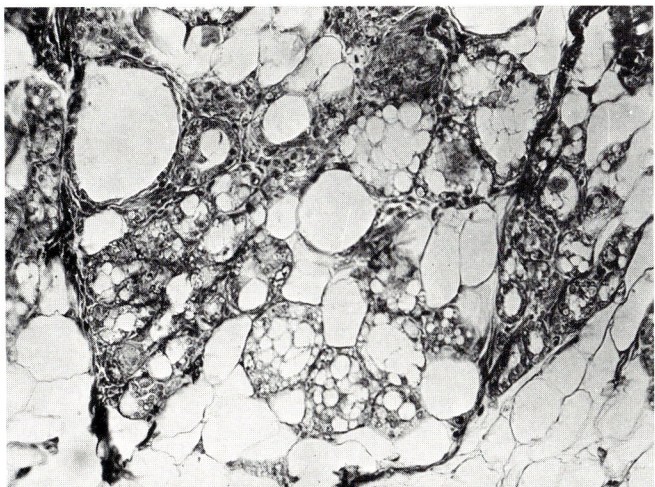

Fig. 15–1. Parotid gland of a 700-day-old Albino Wistar Institute rat. Much of the glandular tissue in this field is involved in a process of fatty degeneration. Harris' hematoxylin and eosin. ×100.

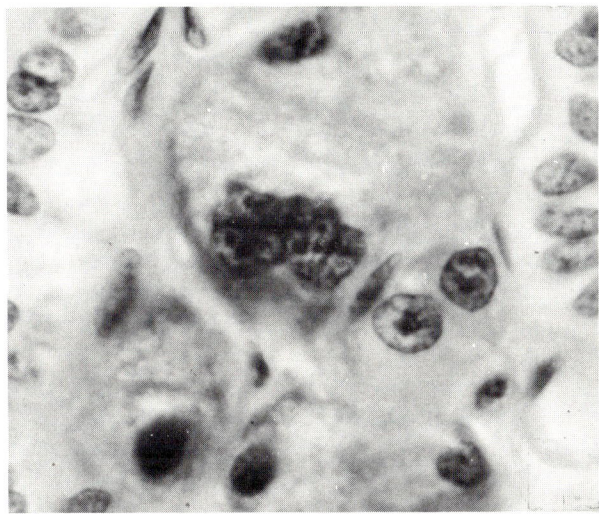

Fig. 15–2. Oncocyte with a large, multilobular nucleus of a monstrous type in a parotid gland of a 1000-day-old female Albino rat. ×1000.

A third feature of the senile salivary glands in the rat is the infiltration of the interlobular and interalveolar connective tissue by lymphocytes and mast cells and the penetration of such cells into the epithelium of the ducts. This infiltration was almost constantly present in sections of old glands and did not seem to be a part of an inflammatory process in any usual sense of the term.

We have studied human salivary glands in surgical specimens obtained from the Ellis Fischel State Cancer Hospital at Columbia, Missouri (Andrew, 1952).

The surgery involved was the radical neck operation for removal of carcinoma in or near the oral cavity.

While some objections might be raised in regard to drawing conclusions from the study of the salivary glands of such patients, even greater ones could be raised in relation to most autopsy material where a widespread disseminated group of changes may be expected. In regard to these specimens, according to the opinion of L. V. Ackerman (1952) the only really serious objection would be in cases in which previous irradiation had been given and the salivary glands were included in the field. In choosing the paraffin blocks we therefore avoided such specimens.

The tissue was sectioned at 6μ and several sections mounted on each of 10 slides. Four slides were stained with Harris' hematoxylin and eosin, 2 with Heidenhain's iron hematoxylin and eosin, 2 with Mallory's triple connective tissue stain, and 2 with hematoxylin-eosin-azure, particularly for lymphocytic and other leucocytic elements.

Thirty-one glands were represented in this study, 16 parotid and 15 submandibular. Unfortunately, it was not possible in all cases to have both glands from the same individual. There were 10 subjects, however, for which both glands were studied, leaving 11 for which only one or the other gland is represented (Table 15–1).

Since 10 or more large sections of each gland were available for study, it was believed that a fair sampling for the characteristics of the individual organs was obtained, although ideally it would have been better to have the entire gland or a number of pieces for study. It is true, however, that the features seen in small areas of a section generally seemed to obtain throughout the specimen or the greater part of it.

Study of the sections showed that a great amount of fat is present in the majority of the parotid glands, more so than in the submandibular glands, but that all of the latter also have varying amounts of fat. It seemed advisable to set up some partly objective criterion for the amount of fat present and this was done by indicating the degrees as + (less than 10 per cent of the section occupied by fat), + + (10 to 25 per cent), + + + (25 to 50 per cent), and + + + + (50 per cent or more). With the amount of difference noted between the two types of glands (Table 15–1) this grading system proved very useful.

It will be noted that 4 of the parotid glands had 50 per cent or more of parenchyma replaced by fat and 7 have 25 to 50 per cent replacement, the remaining 4 having lesser degrees. None of the submandibular glands had over 25 per cent replacement, although 11 were in the 10 to 25 per cent group, and the remaining 5 showed less than 10 per cent replacement.

The question of whether the replacement means an actual loss of secreting parenchyma through fatty degeneration of the cells is important. There were a few clear-cut instances in which the histologic picture would allow it to be said definitely that this is the case. In some areas, however, the alveolar cells seem to be in process of degeneration (Fig. 15–3) with the appearance of fat spaces or "cells" replacing individual alveoli.

To investigate the mechanism of fatty change further, a number of sections of parotid glands of senile rats were re-examined. A very significant conclusion can be drawn from such sections, namely, that the fat locules result from a fusion of in-

Table 15-1. *Some features of the parotid and submandibular glands in older human subjects. Relative degrees of fatty degeneration, lymphoid infiltration, and parenchymal cell change are indicated by 0 (none) and + to ++++ (slight to marked).*

Pathology Number	Age in Years and Sex	Fatty Degeneration		Lymphoid Tissue		Fibrosis		Aberrant Epithelial Cells	
		Parotid	Submandibular	Parotid	Submandibular	Parotid	Submandibular	Parotid	Submandibular
472110	54 M	++++	++	0	++	0	0	0	0
47159	60 M	++++	++	0	+++	0	+	0	0
471902	65 M	+	++	0	++	0	+	0	0
471123	65 M	++++	++	+	++	0	0	+	+
472205	66 M	++++	++	+		0	+	0	
46827	68 M	++++		+		0		0	
4821	69 M	++++		0		0		0	
462103	69 M	++++	++	0	0	0	0	0	0
461509	69 F	+++		0		0		0	
48420	71 M	+++		+		0		0	
45643	71 M	++++	+++++	0	+++++	0	0	0	++
48604	72 M	++++	++++	0	++++	0	0	0	++
481648	72 M	+++	++++		+++		0		+
47549	72 M		++	0	0		0	0	
481640	74 M	+++	+++	0	0	0	++	0	+
471973	74 M		++		0		++		0
472215	75 M	++++	+++	+	+++	0	+	0	0
481271	76 M								
451019	76 M	++	++	+	+++	0	0	0	+
471863	77 M				+		0		0
45468	83 M		+		++		0		0

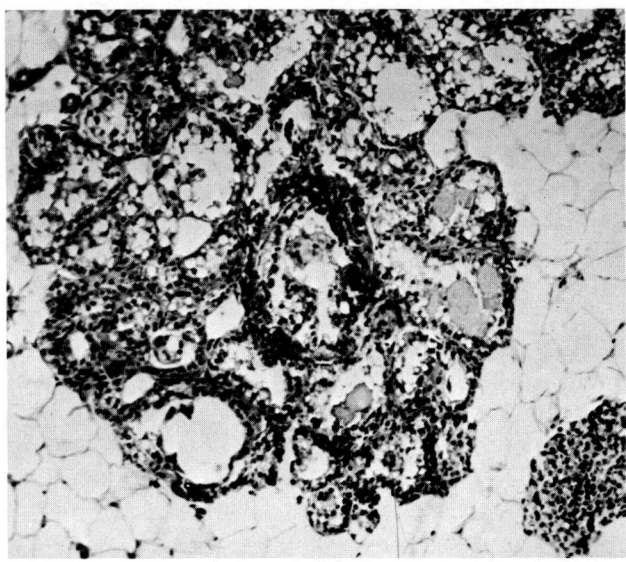

Fig. 15-3. Parotid gland of a senile Albino rat, 900 days of age. Large areas of parenchyma have been destroyed by fatty degeneration of alveoli. Mallory's aniline blue connective tissue stain. ×100.

creasingly larger droplets. The fat droplets within individual gland cells fuse to form larger ones and these eventually fuse with those of neighboring cells. The fat locule or "cell" seems to be truly of multicellular origin in this process (Fig. 15-4).

For the submandibular gland, since fatty change is almost nonexistent in the senile rat, it was not possible to obtain a clear picture. The histologic evidence in the human gland could be taken as favoring a simple fatty infiltration more than a truly degenerative change.

It seems a matter of fundamental importance to pathology to recognize the type of fatty change seen so clearly in the parotid gland of the senile rat and less so in man. Other instances of this hitherto undescribed type of replacement of parenchymatous glandular elements by fatty tissue should be sought for.

The second feature to which particular attention was paid is the presence or absence of aberrant cells in the salivary glands, either in the alveoli or in the ducts. In Table 1 it is seen that only 1 parotid gland and 6 submandibular glands show such cells. Study of the parenchyma in this series of human glands, then, does not present any such striking criteria of senility as seen in the rat.

The aberrant cells present a varied picture in this human material. Nuclear peculiarities, the division of the nucleus into lobes or individual adherent vesicles, are the more outstanding features both in the parotid and submandibular glands.

It would be difficult, in fact, to characterize the cytoplasm of such cells other than to say that it varies more in appearance than it does in normal cells. It is light and wide-meshed in some aberrant cells but deeply staining in others.

In a 66-year-old male subject, examination of the parotid gland showed well-developed sebaceous glands. Such glands were found in considerable numbers in

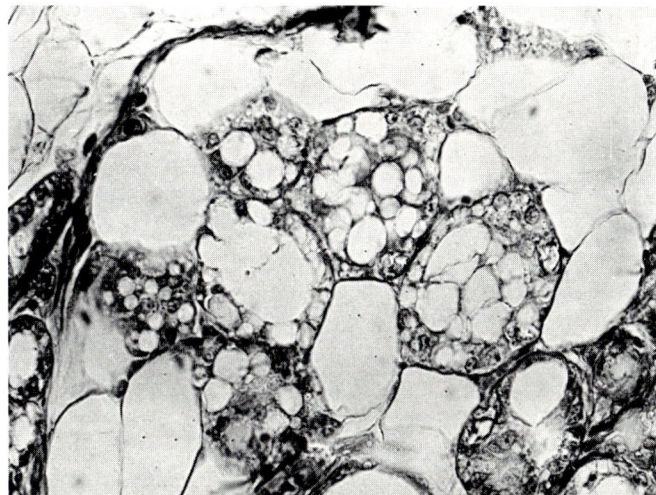

Fig. 15-4. Detail of gland seen in Fig. 15-3. Fat droplets appear in individual epithelial cells but appear to coalesce from one cell to another, eventually filling the volume of the entire alveolus and expanding its framework. ×400.

this parotid and also in the submandibular gland of the same individual, but none were found in the other subjects. They apparently arise by a metaplasia of portions of the striated ducts. Their resemblance to oncocytes, when seen only in part, is striking (Fig. 15-5).

The third feature to which attention was paid was the infiltration of the glandular tissue by lymphocytes or the presence of accumulations of such cells in the interlobular connective tissue. Infiltration of the ducts is present almost constantly but varies in degree in different areas of the same section. Accumulations of lymphocytes are seen in 6 of the parotid glands and in 14 of the submandibular glands (Table 1). Such accumulations consist primarily of small lymphocytes and often occupy almost all of the interlobular space outside of the ducts and vessels. In some of them reticular cells also seem to be present.

The lymph nodes within the glands in many of these specimens show the presence of inclusions of epithelial tissue which, as far as can be seen, are isolated amid the lymphoid tissue. They seem to be in tubular and alveolar form and show a very distinct basement membrane formed of collagenous fibrils (Fig. 15-6). Similar epithelial inclusions were described by Carere-Comes (1938). In lymph nodes of 3 out of 30 human subjects in which different groups of nodes were studied, none were cancerous. Andrew and Andrew (1948) have described epithelial inclusions in the deep cervical nodes of Wistar Institute rats, in which they are more common in younger animals.

Fibrosis is not by any means a marked feature in the senile salivary glands. None was found in the parotid glands but it did occur to a varying degree in 6 of the 16 submandibular glands. It occurs in these glands as a phenomenon of lobular distribution, i.e., certain individual lobules are markedly affected while nearby ones may be entirely spared. The connective tissue which separates the alveoli

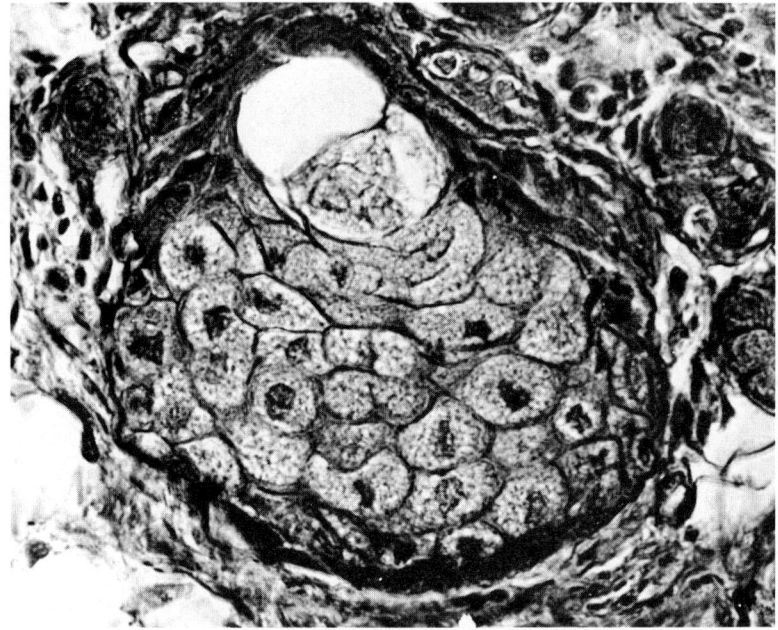

Fig. 15–5. A sebaceous gland which has developed in the submandibular gland of a 66-year-old male subject. Tracing of serial sections reveals these glands to be formed by metaplasia of salivary ducts. ×1000.

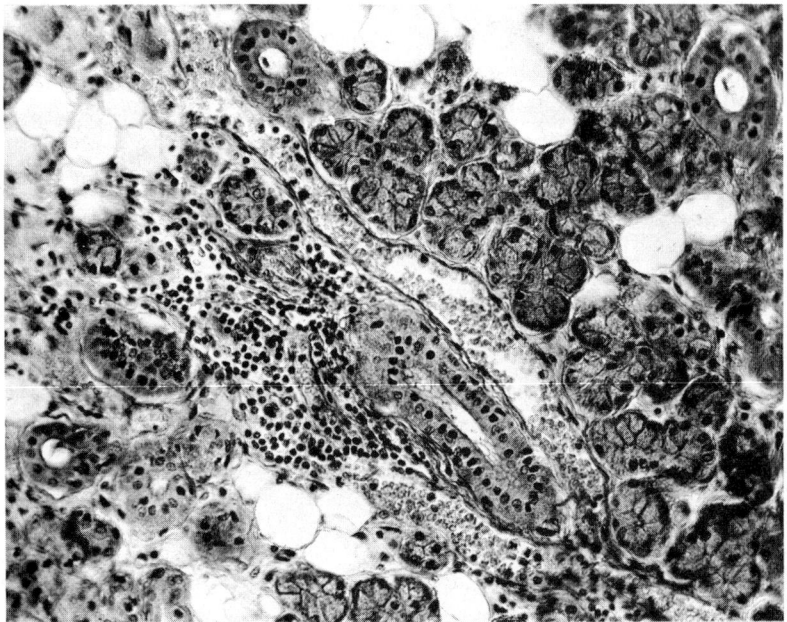

Fig. 15–6. Accumulation of lymphocytes in the interlobular connective tissue of the submandibular gland of a 72-year-old man. ×300.

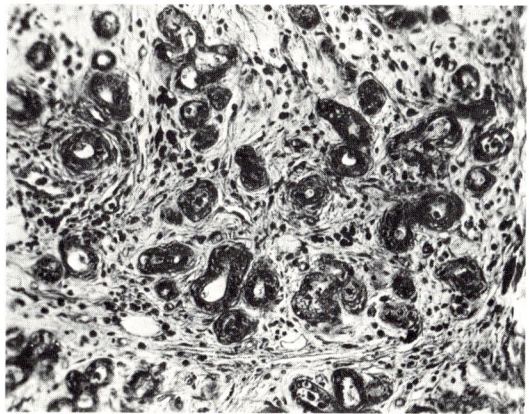

Fig. 15–7. An unusual case of marked fibrosis in the submandibular gland of a 60-year-old man. Fibrotic change is not common even at more advanced ages. Malloy's aniline blue connective tissue stain. ×300.

widely from one another (Fig. 15–7) consists of fine collagenous fibrils and of many fibroblasts of a primitive "embryonal" aspect.

Kurtz (1961) attempted to study the finer cytological changes in the salivary glands, both by means of classical staining methods and electron microscopy. He found the mitochondria in secretory cells of the parotid gland of young rats to occur as rods, filaments, and spheres with an even distribution and with only slight variations in appearance from cell to cell. In old rats most of the mitochondria were spheroidal and varied greatly in number and distribution from cell to cell. Occasional cells of strange shape and unusual distribution of mitochondria were seen.

In young rats the Golgi apparatus in secretory cells of the parotid gland was a delicate network of uniform strands, while in the senile animal it consisted of variously formed, frequently coarse clumps. Osmophilic droplets were seen in the Golgi preparations in the old animals.

In serous secretory cells of the submandibular glands of young rats the mitochondria were seen as short, plump rods evenly distributed in the cytoplasm. The Golgi apparatus, as in cells of the parotid, consisted of evenly distributed delicate strands. In senile rats the mitochondria were generally reduced in diameter, some being very narrow threads, and varied greatly in appearance from cell to cell. The Golgi apparatus showed variations in thickness of strands with loss of their integrity and an irregular distribution. Osmiophilic droplets also were seen in the Golgi preparations.

In his electron microscope studies Kurtz (1961) added new details concerning the alterations with age. In the submandibular gland the matrix of these organelles appeared more electron dense and the cristae "were not easily seen." The mitochondria, again, seemed to be of smaller diameter than in young rats. The interior of the nuclei seemed more dense and the nuclear membrane frequently was wrinkled. For the endoplasmic reticulum (ergastoplasm) of the serous cells, the smooth flowing appearance of the young was changed to a rather coarse, uneven one, chiefly by fragmentation of the membranes at short intervals. No statement was made on abundance of ribosomes.

Serous secretory cells and cells of the intralobular ducts showed scattered clumps of pigment, a substance completely lacking from the young gland. The pigment occurred in rounded or ovoid masses, which consisted in turn of spheroidal bodies of variable size and density. Kurtz says that these structures do not resemble typical intracellular droplets of fat and that it is not clear whether the great electron density is due to "inherent density" of the substance or to impregnation with osmic acid. No apparent adverse effect of the pigment on the cell was seen here, in contrast to the condition in the pancreas, where presence of pigment is often accompanied by cytoplasmic aberrations.

In the parotid gland globules of pigment were seen, but here they could be identified in the young rats also, although they were present to a far greater extent in the senile rats. Kurtz believed that these bodies represented the material described earlier by Andrew (1949) and identified with the light microscope as fat; hence, that the "fat" may in fact be a lipid-containing pigment. In the parotid secretory cells, as in the submandibular the appearance of a well-ordered architecture in young specimens is altered in old ones to an appearance of a coarser type in which loss of order and integrity is seen.

The subject of the teeth in relationship to the aging process is important but so much a part of a special field that we do not attempt here a description of these changes.*

Functional changes in the oral cavity affect the mobility and strength of the tongue and the facility of swallowing.

That this may be based to some extent on muscle atrophy seems probable, although one of the few studies on age changes in these structures (Bucciante and Luria, 1934) found an actual increase in size of muscle fibers. There was also, however, an increase in amount of interstitial connective tissue in at least some areas, so that even with hypertrophy of individual muscle fibers there may be a decrease in total bulk of muscle tissue.

The lymphoid tissue of the three pairs of tonsils—the pharyngeal tonsils or "adenoids" of the nasopharynx, the conspicuous palatine tonsils between the fauces, and the lingual tonsils on the back of the tongue—contains large lymphoid nodules which, through much of life, show well-developed germinal or reaction centers. The epithelium over these nodules shows a massive infiltration with lymphocytes especially dense in the walls of the deep invaginations or crypts.

Wessel (1933)† studied quantitative relations of the infiltrating lymphocytes and found that they decreased in number with increasing age.

Pol (1923)† gave particular regard to the germinal centers. According to him, they do not appear until the third to the twelfth month of postnatal life and reach maximum development about the time of puberty, waning after that time.

The palatine tonsils of subjects of 100 autopsies were described by Hieronymus (1933). He found a rapid growth to a size which reached its maximum in the

---

* For a good recent description, see A. E. W. Miles, "Aging in the Teeth and Oral Tissues," Chapter 21, pp. 353–397. in Bourne, *Structural Aspects of Aging*, New York: Hafner, 1961.
† See Chapter 13 References.

twelfth year. Size remained fairly constant until middle life (35–40 years), declining then but reaching a second plateau after a few years, to undergo a final decline beginning at about 70 years.

The most comprehensive studies of the palatine tonsil have been made by Kelemen (1943).* He found in old age a marked atrophic process in which the lymphoid tissue may disappear almost entirely. The crypts tend to lose their character by one of two kinds of happenings: (1) they may undergo an eversion, by which the epithelium which had lined the crypt becomes a surface epithelium; (2) they may show a narrowing of the lumen and, as this becomes obliterated, their walls become a part of the surrounding connective or surviving lymphoid tissue.

Esophagus and Stomach

A "chronic atrophic gastritis" is described as a common finding in subjects over 50 years of age (Hebbel, 1949). In this condition there is a thinning of the mucous membrane, a leucocytic infiltration, and frequently a change in the character of some of the gastric glands, so that the more complicated ones of the body and fundus of the stomach "dedifferentiate" to the simpler pyloric type which lack parietal cells. The histological change in the glands may account for the achlorhydria of gastric secretion seen frequently in the elderly. Stout (1945) emphasized the possible relationship between these changes and the incidence of gastric carcinoma.

The occurrence of benign gastric polyps increases in old age in the dog (Goodpasture, 1918) and in the human subject (Thompson and Oyster, 1950).

There appears to be little known about age changes in the esophagus.

The Small Intestine

Ivy and Grossman (1952) cite a "general belief" that with advancing age there is some atrophy of smooth muscle and of mucosa in both small and large intestine.

In our own studies on the small intestine of C57 Black mice (Andrew and Andrew, 1957) we found some conspicuous changes with age. The lamina propria of younger mice was consistently a highly cellular tissue and the individual crypts of Lieberkühn lay close to one another. In senile animals, a large amount of a rather homogeneous material, apparently amyloid, was found in the lamina propria (Fig. 15–8). The crypts of Lieberkühn appeared to be atrophic and often were widely separated from each other. In the intercryptal tissue of some of the old mice there were masses of basophilic material containing cells in lacunae and resembling small spicules of cartilage (Fig. 15–9).

* See Chapter 13 References.

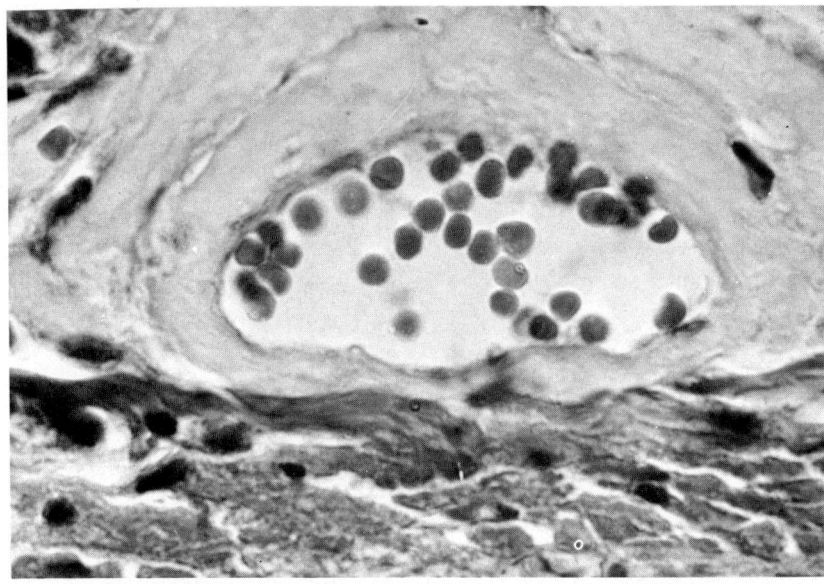

Fig. 15-8. Amyloidosis in the wall of the small intestine of an 818-day-old male mouse of the C57 Black strain. Much of the submucosa has a hyaline aspect and the entire wall of the blood vessel shown is involved in the process. Harris' hematoxylin and eosin. ×1000.

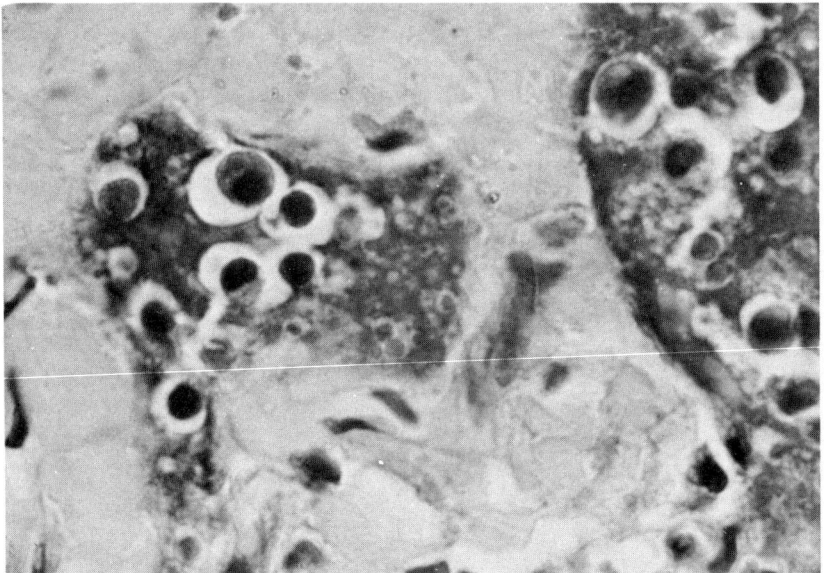

Fig. 15-9. Calcification in the wall of the small intestine of the same animal presented in Fig. 15-8. Patches of basophilic material appear in the midst of the amyloid and the contained cells lie in large clear spaces. Harris' hematoxylin and eosin. ×1000.

Alkaline Phosphatase in Epithelium and Connective Tissue

The alimentary system, with its function of preparation and absorption of the varied foodstuffs needed by the tissues, may well be thought to be an important subject for study of the aging process. Some recent histochemical and histological studies indicate that definite changes do occur in the walls of the tract.

The phosphatases are a group of enzymes that liberate inorganic phosphate from phosphoric esters. Alkaline phosphatase, which acts at pH 8.6, is generally present in bone, kidney, muscle, blood, and intestinal wall.

A study of histochemical features of the intestine of strains CE, 129 and Swiss mice was made by Suntzeff and Angeletti (1961). They used five levels: duodenum, ileum, proximal colon, sigmoid colon, and rectum. The reaction for alkaline phosphatase was found to be readily evidenced in the small intestine of young mice and was highly positive after 20 minutes of incubation in the glycerophosphate medium. The activity occurred in the cytoplasm of the epithelial cells but was localized chiefly in the "brush border." While the distribution of activity was the same at all ages, its intensity "seemed to decrease with aging."

Radiographic studies on human small intestine have given evidence of a hypertonic condition in many older persons (Jungman and Cosin, 1948) and of a "coarser" outline of the mucous membrane in older persons (Protis and Kind, 1952).

The Large Intestine

The increased incidence of diverticulosis and diverticulitis in older persons gives evidence for a weakening of the wall of the large intestine. Kocour (1937), however, found in 700 autopsies that diverticula increase markedly after the age of 40, but that their numbers remain relatively constant during the seventh and eighth decades.

Ribbert (1893), studying the human appendix, found the lumen to have become obliterated in one half the subjects over 50 years of age. The process of obliteration seems to be a fibrotic one and to begin at the tip and proceed toward the base. Prior to fibrosis, the originally abundant lymphoid tissue of the organ has begun to decrease in amount.

Pancreas

Early authors noted some age changes in the pancreas.

Thus Balò and Ballon (1929) described a metaplasia of the ducts in human material, by which their epithelium changed from a columnar to a squamous type. Goodpasture (1918) stressed the great variation of nuclear and cytoplasmic structure in the acini of the pancreas of senile dogs.

Andrew (1944) described a change in the pancreas of senile rats which was of great prominence in animals of the Gray Norway strain and much less so in others (Albino). The undifferentiated intralobular duct cells underwent proliferation, with later formation of lumina. Thus large numbers of duct-like structures with wide lumina came to occupy much of the volume of individual lobules (Fig. 15-10), apparently causing a degeneration of the secretory acini and of the islets of Langerhans. The dilation of the newly formed cyst-like structures proceeded to so great an extent that the lining epithelium became a simple squamous type. In the lumen of some of them a fibrous-appearing substance resembling keratin was seen. The total effect of this process seemed to be a highly deleterious one and it might be classed as a tumor formation. It did occur, however, in all of the senile rats of Gray Norway type which were studied.

Andrew (1944) also confirmed the metaplastic change of the epithelium in ducts of the pancreas of senile human subjects and emphasized the large amount of "adipose tissue" change in the lobules in old age.

Kurtz (1961) found little change in the ultrastructural aspects of secretory cells of the pancreas with age, although he noted that variations in general appearance from cell to cell in the senile rat are certainly marked. In the mitochondria of the senile pancreas the matrix seemed less electron-lucent and the cristae extended further inward from the peripheral membrane. The endoplasmic reticulum shows a change by which its membrane portion becomes difficult to discern in many places and the ribosomes are the only prominent portion, maintaining some degree of orderly arrangement even when the membrane is no longer present or visible. The Golgi apparatus did not show definite changes.

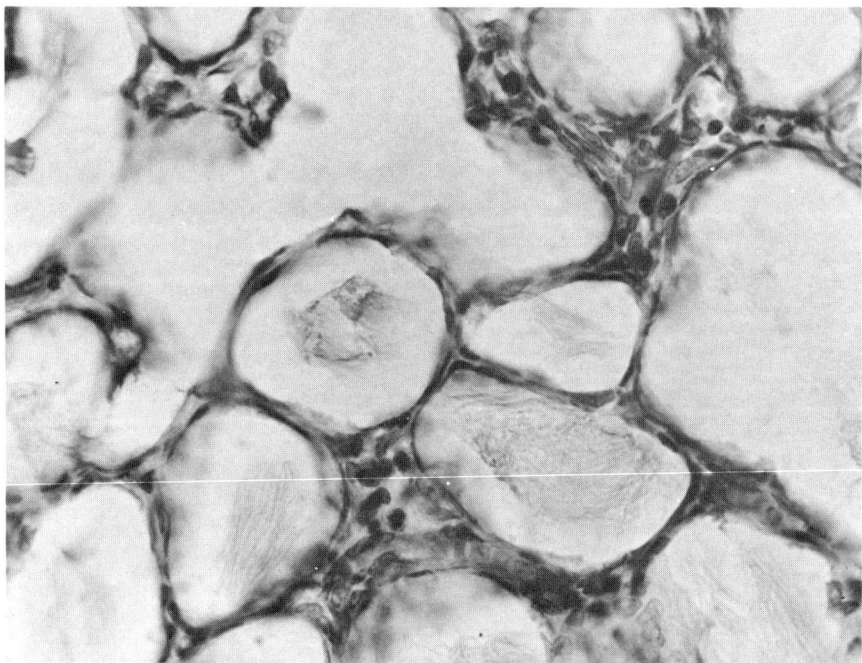

Fig. 15-10. Multiple cyst-like locules in the pancreas of a 981-day-old male Gray Norway rat. The locules arise from duct-like structures in which the cuboidal epithelium undergoes a metaplasia to simple squamous type. Some locules contain a substance resembling keratin. The alveoli no longer are seen. ×380.

Pigment, in the form of irregularly shaped particles, was not uncommon in cells of the pancreatic acini of senile rats. The particles ranged from 0.1μ to over 1.0μ in diameter. Kurtz emphasizes the high degree of electron density of the pigment "greater than that of fat droplets"—which also were seen occasionally. Large amounts of pigment seemed to lead to disorganization of the endoplasmic reticulum.

Liver

The liver, the largest gland in the body, has a wide variety of functions and its parenchymal cells perform duties in relation to the general metabolism, food handling, and excretion, in addition to their clearly glandular function in secretion of the bile.

One might expect to see fibrotic changes in such an organ if such changes in glands are a general phenomenon in old age. Actually, however, the livers of senile animals and man seldom show a conspicuous fibrosis (Fig. 15-11). Study by means of silver impregnation methods, which demonstrate the fibrous skeleton of the liver all the way from the coarse collagenous fibers to the fine reticular network along the sinusoids, does reveal, however, moderate degrees of fibrotic change. Hinton and Williams (1968) studied four stocks of laboratory mice by these methods. These are the C57BL, DBA, RF, and TS strains. The mean life span of these strains, so far as known is close to 700 days (Storer, 1966). Animals of all strains were fed on Purina laboratory chow. They were sacrificed by digital compression of cervical spinal cord.

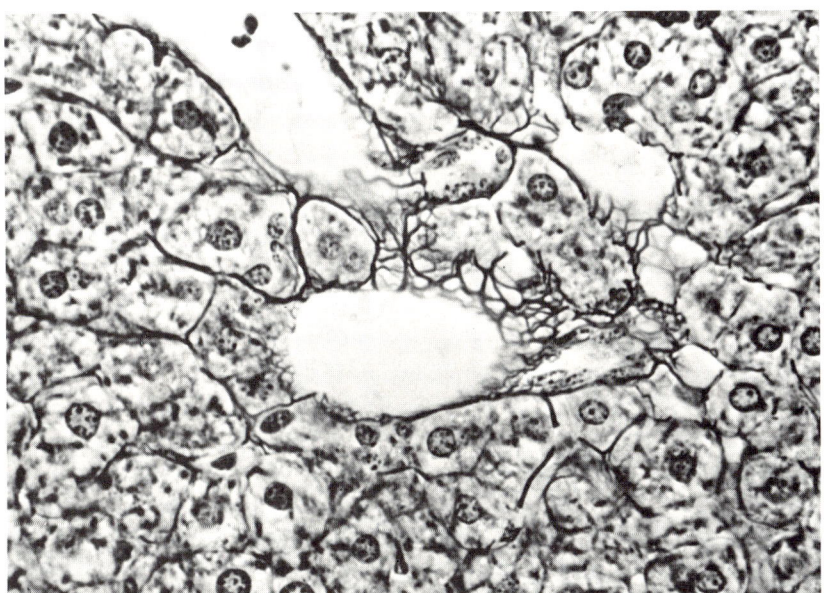

Fig. 15-11. Liver of a 70-day-old C57 Black male mouse. The nuclei are spheroidal and generally uniform in size and appearance. Gomori's stain. ×900.

These investigators found that they were able to distinguish precisely the central vein and portal areas of lobules by counterstaining adjacent sections with hematoxylin and eosin.

There are three main categories of fibrosis which occurred in "older" animals: (1) subcapsular; (2) perivenous, i.e., about the initial tributaries of the hepatic veins; and (3) intralobular. Of these, the subcapsular is the most common type. It does not consist of an increase in thickness of the capsule itself but results from increments of the perivenous and perisinusoidal reticular tissue in this region. Collagen develops from the reticular fibers in this type more frequently than in the other two.

In the perivenous fibrosis thin, irregular tracts of reticulum extend from the walls or the perimural regions, generally from the central veins. As such tracts become more extensive, central veins are linked to central veins by them, thus partitioning off rather large masses of parenchyma, with the portal areas as centers. Formation of collagen is rare in this kind of fibrosis.

The third and least common type of fibrosis is the intralobular. This consists of an increase in the perisinusoidal reticular fibers. The changes are irregular in different lobules as far as the amount, distribution and extent of the reticular tissue is concerned. Collagen appears not to be formed in this type.

"Significant" collagenesis, indeed, seems to be found only in subcapsular fibrosis in mice of 52–68 weeks of age. Fibrosis of all three types is most common in the C57BL stock, next in the DBA, and slightly less in the RF.

These findings of some actual "fibrotic change," as shown by special staining, are of considerable interest. Since the animals of 52–68 weeks of age are so much younger than many of the C57BL stock with which we have worked in relation to other features of the liver (364–476 days, as compared with 700–800 days of age), we may look forward to future studies by these workers or others on fibrotic changes in mice of more advanced age.

While the changes in amount of connective tissue relative to parenchyma do not appear to be great ones, there are important alterations in the parenchyma itself.

By microspectrophotometric determinations Swartz (1956) found that ploidy classes vary with age in human liver. Only diploid nuclei were found until the age of 6 years, by 11–14 years a tetraploid class was established and at the age of 20 an octoploid class. After this age three DNA classes persisted through life.

The numbers of binucleate cells at different ages was studied by St. Aubin and Bucher (1952) by the method of cell suspension. They found only 7 per cent in the embryo, the highest count at 5 weeks—58 per cent—and 35 per cent in the old animal.

The occurrence of aberrant cells with giant nuclei in the liver of man and of rodents is a common finding in old age. Invaginations of the nuclear membrane of liver cells (Fig. 15–12) seem to lead on to the formation of intranuclear inclusions (Fig. 15–13) which actually represent entrapped masses of cytoplasm.

A comparison of various quantitative aspects of the mitochondria of Sprague-Dawley rats of different ages has been made by Kment et al. (1966). They used 22 5-month-old animals and 20 20 to 21-month-old animals of both sexes. A total of 21,361 mitochondria were studied with the electron microscope. For the

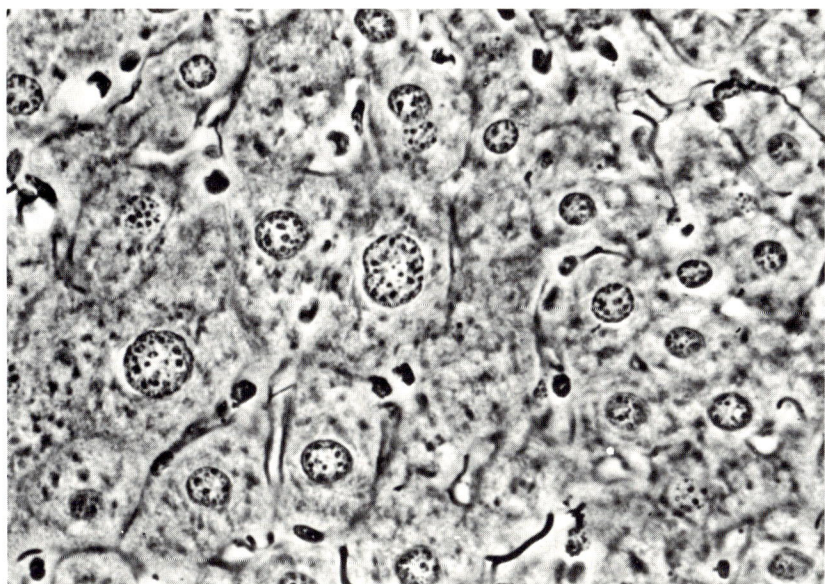

Fig. 15-12. Liver of a senile C57 Black male mouse, 767 days of age. Some nuclei are very large and of aberrant shape. Gomori's stain. ×900.

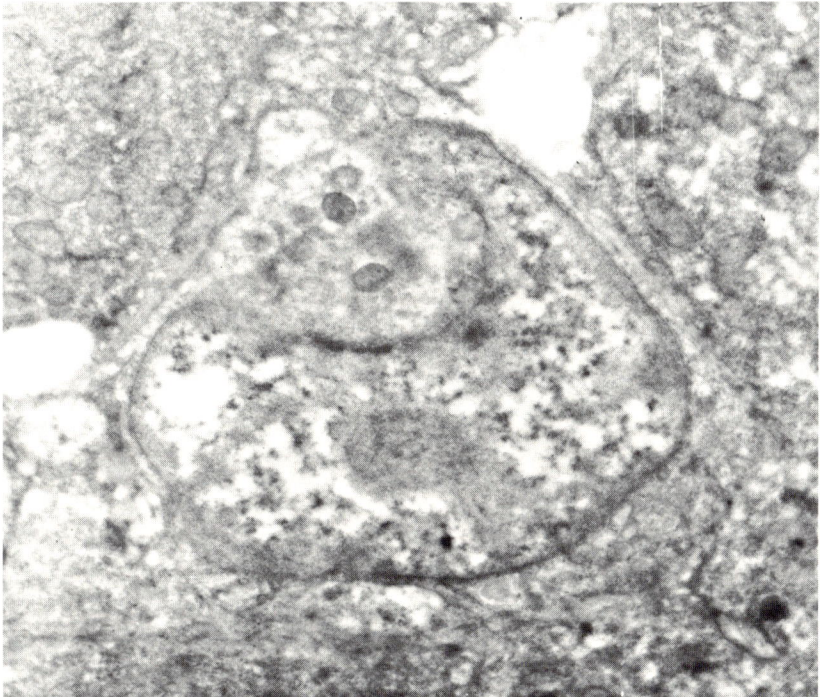

Fig. 15-13. An invaginating nucleus in the liver of a senile C57 Black mouse, a 728-day male. In the deep pocket formed by infolding of the nuclear membrane some cytoplasmic elements are being engulfed. (After Andrew, 1962.)

older animals the number of mitochondria per unit cell cut was significantly higher while cut areas of individual mitochondria were significantly smaller than for young rats. The total cut area of mitochondria in per cent of the cell cut area showed no significant difference between young and old animals (the "old" animals here were hardly "senile").

In a correlated chemical and anatomical study (Andrew, Shock, Barrows, and Yiengst, 1959)* the liver, which gives relatively little histological evidence of loss of units (cells), showed also no consistent change in proportion of extracellular to total water nor in the amount of the intracellular components.

In a survey over any area of a section as extensive as one lobule, individual hepatic parenchymal cells or small groups of such cells in senile animals will show variations in some of the cell organelles. One type of alteration is an invagination of the nuclear membrane. Such invagination has been recognized as a step in the formation of intranuclear inclusions (Kleinfeld, Greider, and Frajola, 1956) which are of relatively common occurrence in senile liver (Fig. 15-14B). They often contain cytoplasmic elements, such as mitochondria and fragments of endoplasmic reticulum.

Some hepatic cells in old mice show a marked change in the endoplasmic reticulum itself. The channels contain an opaque material and give the appearance of a network of rather coarse trabeculae.

Scattered cells and groups of cells show changes in structure of the mitochondria and reduction in their size and number (Andrew, 1962).

Nucleoli in some cells in senile mice showed a clear center (Fig. 15-15) as though a cavitation had occurred.

Actual degeneration of nuclei is seen in a few cells. Such necrobiotic change is rare in the senile livers described and is thought to be due to vascular changes, since some small vessels do show marked changes in their walls.

Quantitative studies on liver parenchyma of autopsy subjects have been carried out (Tauchi and Sato, 1962). Their 74 subjects ranged from 19 to 96 years of age.

No difference in *size* of the cells was found after maturity but apparently there is a decrease in *number*. However, size of nuclei, nucleo-cytoplasmic ratio, degree of *irregularity* in size of cells and of nuclei, and number of binucleate cells are greater in old age.

The decrease in number of hepatic cells is followed by increasing binuclearity of the cells, later replaced by an increase in volume of individual nuclei. These authors believe that some extracellular inhibitory factor slows the replacement of hepatic cells.

The relation of the more *generalized* changes with age to the increased tendency of tissues to hyperplasia and neoplasia is a subject about which some important data are being obtained. Essner and Masin (1966) studied the liver in C3H male mice 19 months of age, giving attention to the fine structure of the parenchyma of the organ in general and of 10 hyperplastic nodules of spontaneous origin. In general, the structure of parenchymatous cells in the nodules did not differ greatly from that in normal liver. In most areas glycogen was present and

---

*See Chapter 11 References.

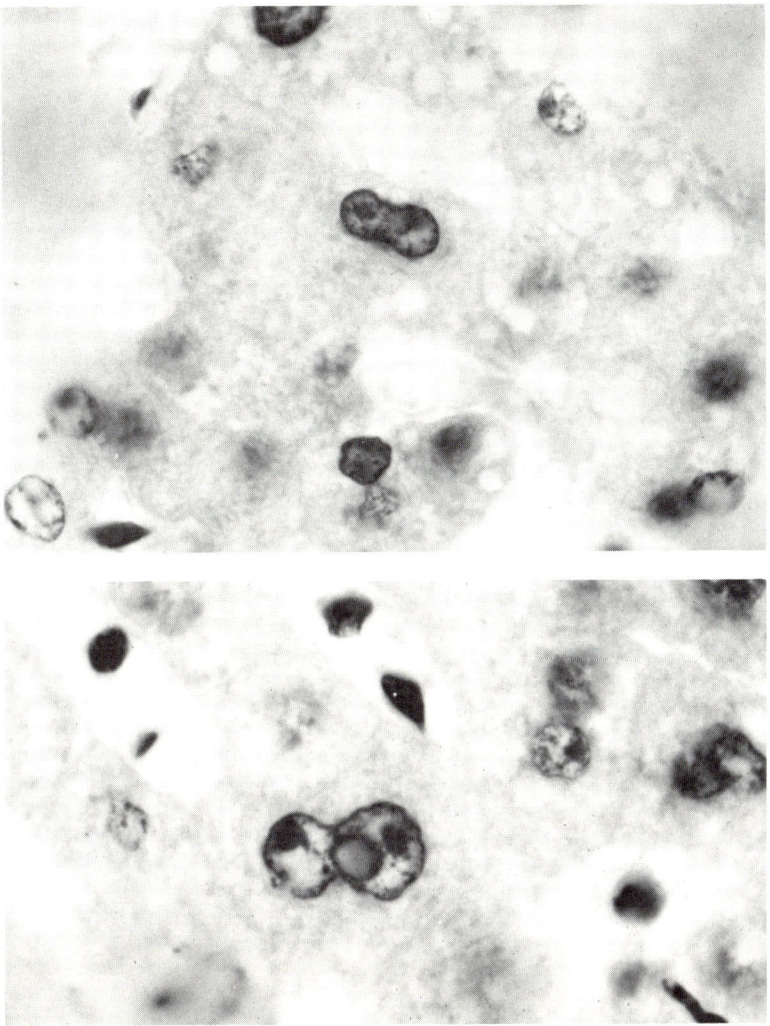

Fig. 15-14. Amitotic division of nuclei in the liver of senile mice. (A) A nucleus of apparently normal size and aspect undergoing division. ×1000. (B) A large nucleus, with an intranuclear inclusion, undergoing division. ×1000.

microbodies were found. In various areas of each nodule, however, there was a definite alteration of the rough endoplasmic reticulum, the cisternae of which were dilated. The usual regular pattern of the endoplasmic reticulum was disturbed and individual cisternae were randomly oriented. A homogeneous material of low opacity had accumulated in many of these cisternae. It contained fine tubular elements (Fig. 15-16).

In livers of senile mice of this same strain which showed *no gross evidence* of nodules, the hepatic cells often presented a similar, although not identical, picture to that seen in the nodules. In these cells of the general parenchyma the individual cisternae had transformed into cylindrical structures containing tubular elements

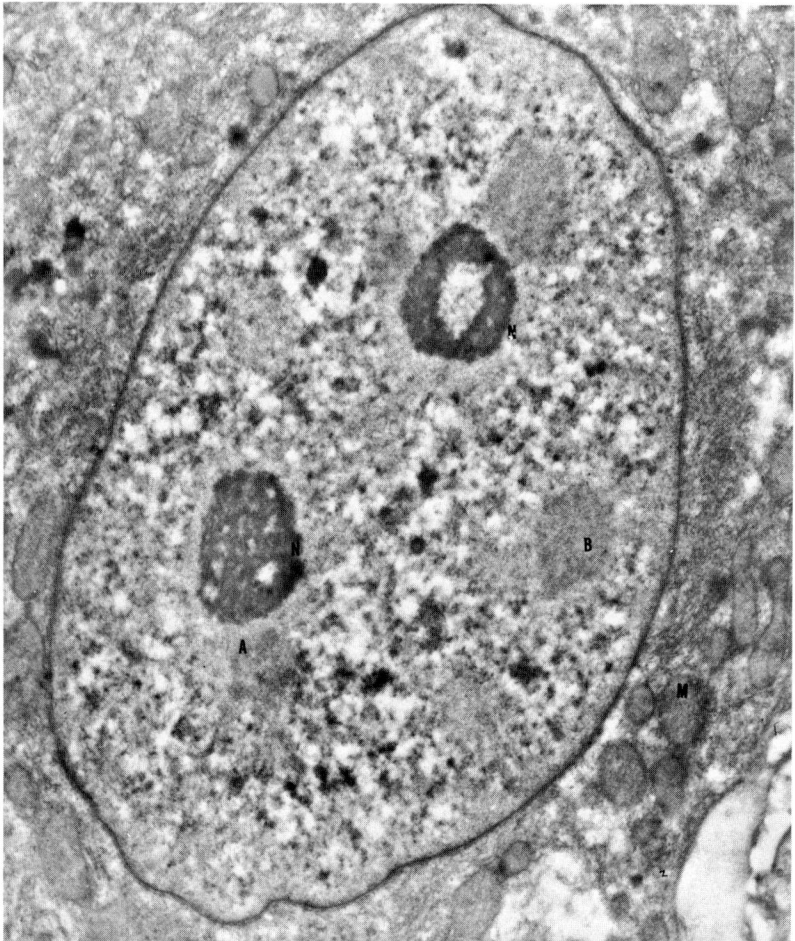

Fig. 15-15. Nucleolar change in liver of a 728-day male C57 Black mouse. The nucleolus in the upper part of the field shows a clear center. Several "nucleolus-like" bodies, of lesser density, are present. ×16,500. (After Andrew, 1962.)

similar to those seen in the cells of the nodules, as well as numerous opaque particles similar to the liposomes of the Golgi saccules.

Essner and Nasin (1966) point out that the whole of the liver of senile mice appears to be highly prone both to hyperplasia and neoplasia. They suggest that the generalized changes seen in the rough endoplasmic reticulum may be a manifestation of this increased susceptibility. Foulds (1963) has suggested that in the mammary gland a similar *generalized* susceptibility to preneoplastic and neoplastic change occurs.

The mitochondria of spontaneous nodules in old mice also showed conspicuous changes. In many of them there appeared a core of fibrillar material (Fig. 15-17), while abnormal enlargement of the whole mitochondrion and changes in the cristae mitochondriales were seen. Some individual mitochondria were encircled by elements of endoplasmic reticulum, either normal or dilated.

The Digestive System 167

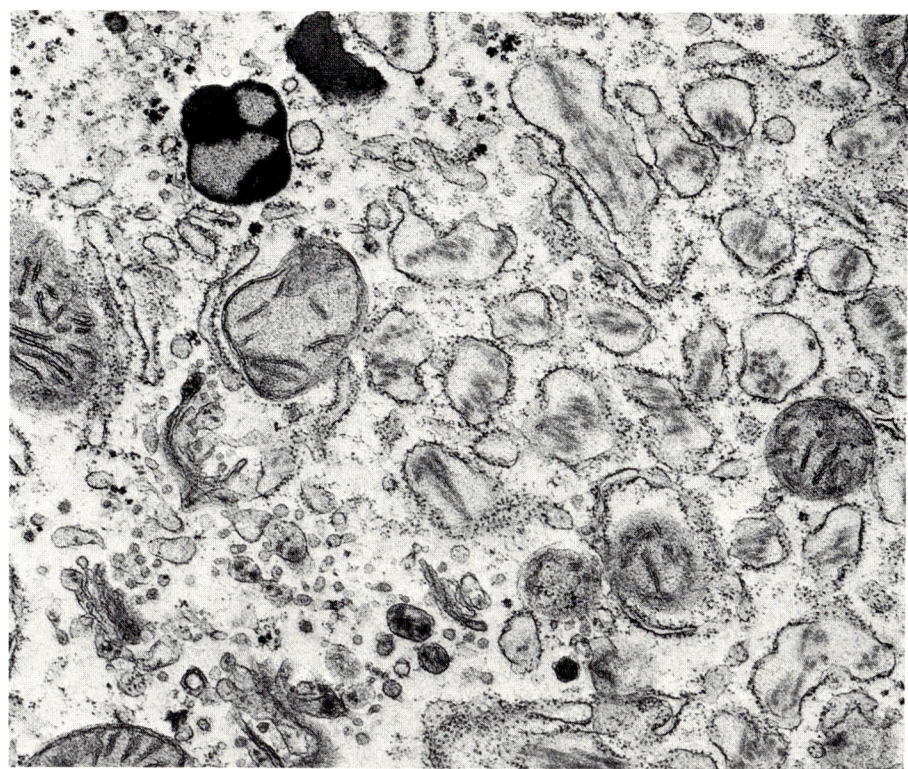

Fig. 15–16. Alterations in the endoplasmic reticulum in liver cell from a hyperplastic nodule in an old C3H male mouse. There is dilatation of the cisternae and they contain a fibrillar material. ×20,200. (Courtesy of Dr. E. Essner, with permission of E. Essner and E. J. Nasin, 1966.)

Tauchi *et al.* (1966) have shown striking age changes in mitochondria of the liver in senile rats which had been treated with thyroxine (refer to Fig. 15–18). Often they grew to be "giant" mitochondria, several microns in diameter. A similar, but less marked, change occurred in liver mitochondria of treated young animals. Bizarre shapes of mitochondria were most prominent in the liver of thyroxine-treated senile rats. The majority of the enlarged mitochondria seemed essentially normal in relation to density of matrix and numbers and structure of the cristae. Some of them, however, did show true "swelling" with diluted matrix and partial or complete loss of the cristae.

Recently Tauchi and Sato (1968) have made a careful study of human liver tissue obtained at the beginning of upper abdominal surgery. Fifty-two subjects, with an age range of 21–79 years and who showed no abnormality in routine liver tests, and whose liver samples were microscopically free of pathological change, were selected. Specimens were embedded in epon. Sections were cut and stained with uranyl acetate and lead.

Measurements of area and circumference of the mitochondria of hepatic parenchymal cells were made on electron micrographs.

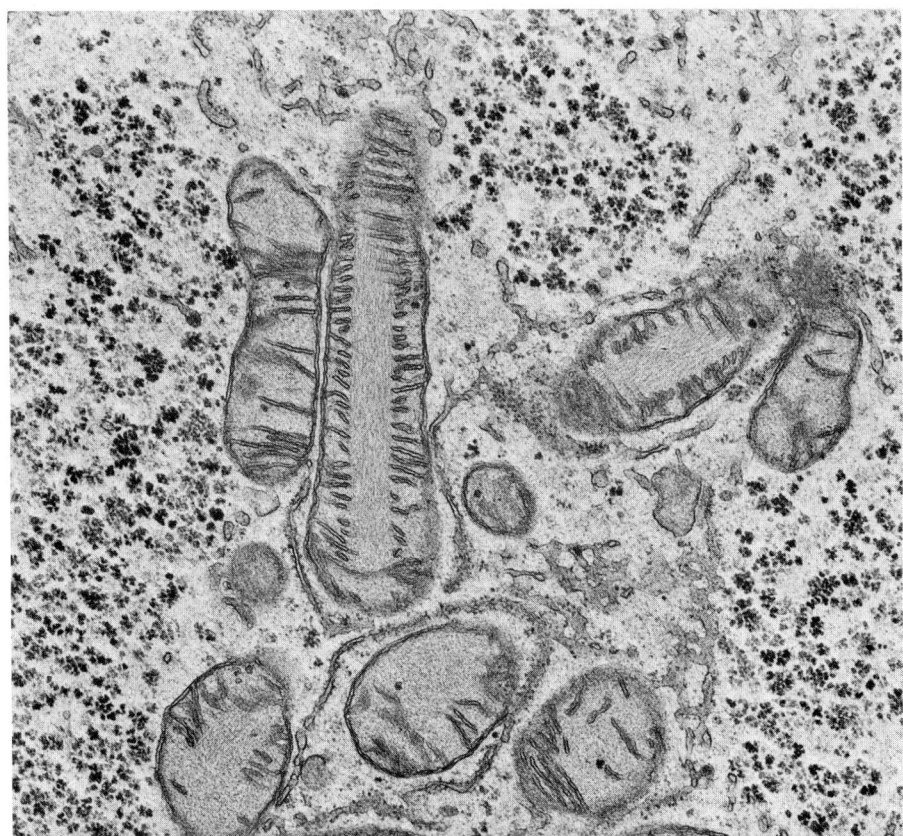

Fig. 15–17. Alterations of mitochondria in a hyperplastic nodule of the liver of an old C3H mouse. A core of fibrillar material is seen within the matrix. ×20,200. (Courtesy of Dr. E. Essner.)

The mitochondria decrease in number but increase in size, especially after 60 years of age (Figs. 15–19, 15–20).

Histochemical studies revealed no definite difference in succinic dehydrogenase activity in relation to age.

The hepatic parenchymal cells of the fetus and young child apparently contain no lipofucsin whatever. This "aging pigment" is found, at least in some cases, in small quantities in adult human subjects and is more common in old persons. According to Weinbren (1952) cells newly formed during regeneration of the liver are devoid of pigment. Among the Amphibia, as we have seen, there is a great increase in the amount of pigment in the liver of *Ambystoma mexicanum*. The relation between lipofucsin in the hepatic parenchymal cells and the function of the liver in salvaging some components of pigment from the breakdown of red blood corpuscles requires further study.

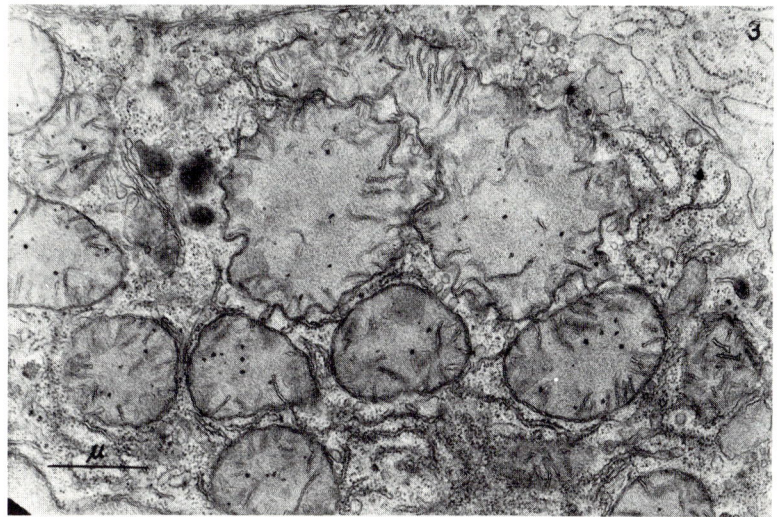

Fig. 15–18. Giant mitochondria of bizarre form in a 26-month rat treated with thyroxine. ×20,200. (Courtesy of Professor H. Tauchi.)

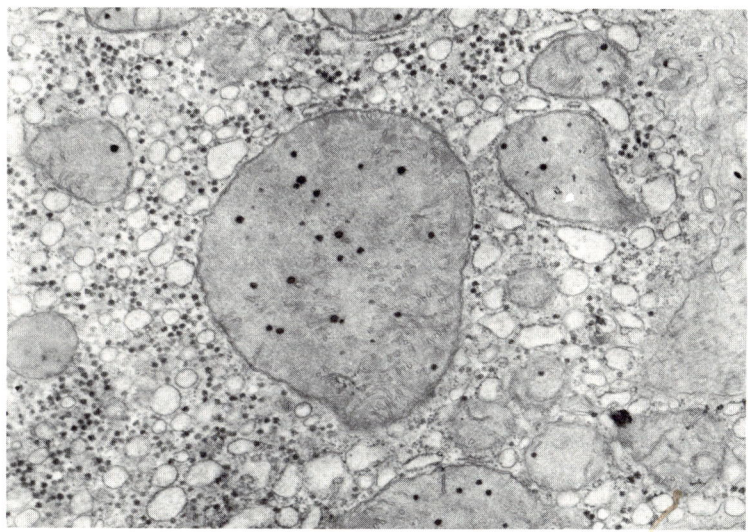

Fig. 15–19. Giant mitochondrion in an hepatic parenchymal cell of a senile human subject. The surrounding mitochondria are of approximately "normal" size. ×20,000. (Courtesy of Professor H. Tauchi.)

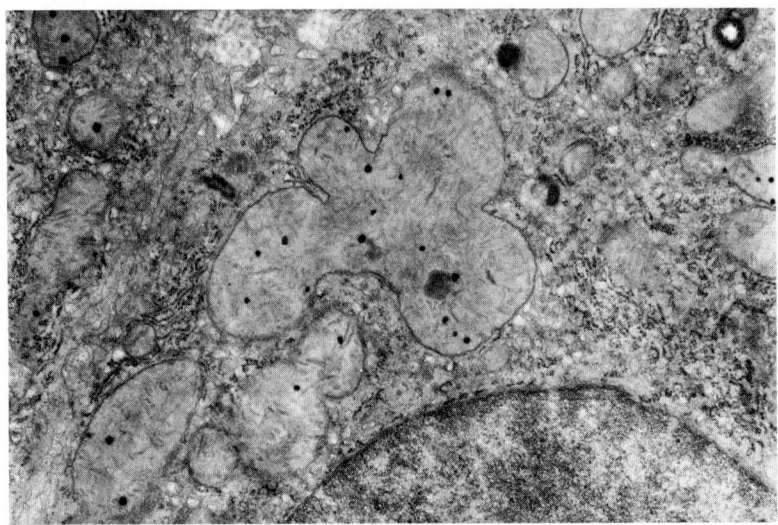

Fig. 15–20. Giant mitochondrion of an hepatic parenchymal cell of a senile human subject. This mitochondrion, besides its abnormal size, is of a bizarre form. ×20,000. (Courtesy of Professor H. Tauchi.)

REFERENCES

Andrew, W., 1944. Senile changes in the pancreas of Wistar Institute rats and of man. Amer. J. Anat., vol. 74, pp. 97–128.

———, 1949. Age changes in the parotid glands of Wistar Institute rats with special reference to the occurrence of oncocytes in senility. Amer. J. Anat., vol. 85, pp. 157–199.

———, 1962. An electron microscope study of age changes in liver of the mouse. Amer. J. Anat., vol. 110, no. 1, pp. 1–18.

———, R. H. Behnke, and Y. Shimizu, 1966. Variations in cell population of intestinal lamina propria to age. Gerontologia, vol. 12, pp. 129–143.

Andrew, W., and N. V. Andrew, 1957. An age involution in the small intestine of the mouse. J. Geront, vol. 12, pp. 136–149.

Baló, J. and H. C. Ballon, 1929. Metaplasia of basal cells in ducts of pancreas; its consequences. Arch. Path., vol. 17, pp. 27–43.

Bucciante, L., and Sluria, 1934. Trasformazioni nella struttura del mused; volontari dell'vomo nella senescenza. Arch. Ital. Anat. Embriol., vol. 33, pp. 110–187.

Coons, A. H., E. H. Leduc, and F. M. Connoly, 1965. Studies on Antibody Production. I. A method for the histochemical demonstration of specific antibody and its application to a study of the hyperimmune rabbit. J. Exp. Med., vol. 102, pp. 49–60.

Dokov, V. D., 1958. Uber Die Vom Alter Und Ernährungsregime Bedingte Zusammensetzung Der Zellelemente in Der Propria Des Verdauungs-Systems Beim Schaf. Academie Bulgare Des Sciences. Comptes Rendus. V. 11, no. 2, Mar.–Apr. 1958. pp. 145–148.

Essner, E. and E. J. Nasin, 1966. Ultrastructure of spontaneous, hyperplastic nodules in livers of aging mice. Electron Microscopy, vol. II (Proceedings of Sixth Int. Congress for Electron Microscopy, Kyoto, Japan), pp. 755–756.

——— and A. B. Novikoff, 1960. Human hepatocellular pigments and lysosomes. J. Ultrastruct. Res., vol. 3, pp. 374–391.

Fifield, L. R., 1927. Diverticulitis. Lancet, vol. 1, p. 277.

Foulds, L., 1963. Biological organization at the cellular and supercellular level. Edited by R. J. C. Harris, pp. 242–250, New York: Academic Press.

Goodpasture, E. W., 1918. An anatomical study of senescence in dogs, with especial reference to the relation of cellular change of age to tumors. J. Med. Research, vol. 38, p. 127.
Guiss, L. W. and F. Stewart, 1943. Chronic atrophic gastritis and cancer of the stomach. Arch. Surg., vol. 46, pp. 823–843.
Hardt, L. L., F. Steigmann, and G. Milles, 1948. Gastric polyps. Gastroenterology, vol. 11, p. 629.
Hausemann, D., 1896. Über die Entstehung falscher Darmdivertikel. Virchow's Arch., vol. 144, p. 400.
Hebbel, R. 1949. The topography of chronic gastritis in otherwise normal stomachs. Amer. J. Path., vol. 25, pp. 125–141.
Hinton, D. E. and W. L. Williams, 1968. Hepatic fibrosis associated with aging in four stocks of mice. J. Geront., vol. 23, pp. 205–211.
Jordan, H. E., 1931. The pigment content of the liver cells of *urodeles*. Anat. Rec., vol. 48, pp. 351–365.
Kleinfeld, R. G., M. H. Greider, and W. J. Frajola, 1956. Electron microscopy of intranuclear inclusions found in human and rat liver parenchymal cells. J. Biophys. and Biochem. Cytology, vol. 2, no. 4, Part 2, Suppl., pp. 435–439.
Kocour, E. J., 1937. Diverticulosis of the colon. Amer. J. Surg., vol. 37, pp. 433–436.
Kurtz, S. M., 1961. Ageing changes in the salivary glands and pancreas. Chapter 7, pp. 71–84 *in* Structural Aspects of Ageing, G. H. Bourne, ed., London: Pitman Medical Publishers.
Lubarsch, O., 1922. Über das sog. Lipofuscin. Virchow's Arch., vol. 239, pp. 491–503.
Maljatzkaja, M. J., 1933. Über die Speicherungsvörgänge von Vitalfarbstoff in der Darmivand. Z. Zellforsch., vol. 18, pp. 110–131.
Moeschlin, S. and B. Demiral, 1952. Antikörperbildung der Plasmazellen in vitro. Klin. Wschr., vol. 30, pp. 827–829.
Nossal, G. J. V. and O. Makela, 1962. Autoradiographic studies on the immune response. I. The kinetics of plasma cell proliferation. J. Exp. Med., vol. 115, pp. 209–230.
Okamoto, Hiroo, 1925. Über die Leber und Milzpigmente der Kröte. Frankf. z. Path., vol. 31, pp. 16–53.
Ortega, L. G. and R. C. Mellors, 1957. Cellular sites of formation of gamma globulin. J. Exp. Med., vol. 106, pp. 627–640.
Polland, W. S., 1933. Histamine test meals, an analysis of 988 consecutive tests. Arch. Int. Med., vol. 51, p. 903.
Schmidt, J. E., 1905. Beiträge zur normalen und pathologischen Histologie einiger Zellarten der Schleimhaut des menschlichen Darmkanales. Arch. mikrosk. Anat., vol. 66, pp. 12–40.
Storer, J. B., 1966. Longevity and gross pathology at death in 22 inbred mouse strains. J. Geront., vol. 21, pp. 404–409.
Stout, A. P., 1945. Gastric mucosal atrophy and carcinoma of the stomach. New York J. Med., vol. 45, pp. 973–977.
Suntzeff, V. and P. Angeletti, 1961. Histological and histochemical changes in intestines of mice with aging. J. Geront., vol. 16, pp. 225–229.
Tauchi, H., F. Adachi, and M. Shamoto, 1966. Effect of age on the cellular changes experimentally produced in the rat liver. Proc. VIIth Int. Congr. Geront., Vienna, Austria, pp. 127–135.
——— and T. Sato, 1962. Some micromeasuring studies of hepatic cells in senility. J. Geront., vol. 17, pp. 254–259.
——— and ———, 1968. Age changes in size and number of mitochondria of human hepatic cells. J. Geront., vol. 23, no. 4, pp. 454–461.
Thompson, H. L. and J. M. Oyster, 1950. Neoplasms of the stomach other than carcinoma. Gastroenterology, vol. 15, pp. 185–243.
Vierordt, H. 1906. Daten und Tabellen für Mediziner, 3d ed., Jena: G. Fischer.
Yiengst, M. J. and N. W. Shock, 1949. Effect of oral administration of vitamin A on plasma levels of vitamin A and carotene in aged males. J. Geront., vol. 4, pp. 205–211.

16
The Urinary System

The Human Kidney

An evaluation of the vasculature of the human kidney at various ages, using corrosion specimens of the renal arterial vessels, was made recently by Hackel (1966). He used preparations from 92 subjects. The arterial vessels show a change from a straighter course to a curved, arcuate, or even helical one. Cross-section area of renal arterial vessels decreases from the sixth decade. The thickness of kidney cortex decreases continuously from the fourth decade and "glomerular density" (numbers of glomeruli), which is in general constant between the first and fifth decades of life, decreases from the sixtieth year.

Concepts of the aging human kidney have been reviewed by Laroche and Mathe (1955). Three main types of gross change are listed as occurring in the senile human kidney: (1) atrophy, (2) capsular thickening and adhesion, and (3) production of surface irregularities. Four main types of histological change are listed: (1) glomerular changes, ranging from congestion to fibrous or hyaline transformation, (2) tubular changes, particularly distension and atrophy, (3) vascular changes, particularly of the small arteries, and (4) interstitial changes, chiefly sclerosis of connective tissue in the cortex.

While the changes listed by various authors as representing true age differences may give some idea of the general "impressions" of pathologists, they by no means represent features that are universally accepted as characteristic of the aged kidney. Howell and Piggott (1948) reported the results of 300 postmortem studies on subjects over 65 years of age. They found no gross renal change in 47 per cent of the cases. The classical "senile" or coarsely scarred kidney was seen in only 14 per cent of these cases. They say: "An organ of this kind was often found below the age of 70, while an apparently normal kidney could be found even at 88. This great variation in the appearance of the kidney in old age is a marked contrast to the descriptions in the text-books."

These authors studied also "20 microscopic slides" from among their senile subjects. They found no correlation within the group between the age of the individual and the proportion of abnormal glomeruli. The lowest frequency of normal glomeruli on any slide was 42 per cent in a subject of 74 years, while the highest was 83 per cent in a subject of 86 years.

Moore (1931) made counts on glomeruli in abnormal human kidneys from individuals of various ages. He found that the kidney of the senile person may have only half to two-thirds as many of these units as does that of the young person. Howell and Piggott (1948) believe that Moore's conclusions may be "somewhat misleading."

Thus, from the anatomical standpoint, studies on the human kidney leave us somewhat uncertain (1) as to whether any changes can be designated as definitely due to aging and use, and (2) as to whether such changes in the parenchyma as have been described as senile changes are always associated with the general changes of arteriosclerosis in the senile organism in a "cause and effect" relation.

It would seem important for a classification of our knowledge of the kidney in old age to turn to a study of some laboratory animals. However, little work has been done in this area.

Up to now, studies on lower animals have often concentrated on the question of the numbers of glomeruli present at different ages. Arataki (1926), working with Wistar Institute rats, found the (left) kidney at maturity (300 days) to have some 31,000 glomeruli, while in animals of 500 days he found only 20,000 glomeruli. Measurements on the glomeruli showed that there is a continuation of growth even in the adult animal. The "average diameter" at birth was 62μ, at 250 days, 111μ, at 350 days, 118μ, while at 500 days it was 124μ.

Moore and Hellman (1930) also describe a decrease in number of glomeruli "with age" in the rat. In their animals of 230 days of age they found an average of 28,000 glomeruli, while in their older animals, which actually were only 320 days old, the count was 18,000–22,000 per kidney.

Rogers (1950) studied the kidneys of aged guinea-pigs. He described gross features such as the presence of multiple cysts and hemorrhages seen frequently in such organs. As microscopic features, his chief findings were fatty infiltration of proximal convoluted tubules.

Kidney of the Rat

Andrew and Pruett (1957) studied kidneys from 50 Wistar Institute rats. These animals ranged in age from 75 days to 1170 days. All of the young and "middle-aged" rats were Albinos, while the senile ones (868–1170 days) were divided between Albino (12) and Gray Norway animals (18).

Tabulated data on the individual animals indicated that the features of the senile kidney were similar in the two strains. Young (75–100 days) and middle-aged animals (300 days) were equally divided between the sexes (10 male, 10 female), while the senile animals were almost so (16 male, 14 female).

Studies included a careful qualitative study of sections of the kidney from each of the 50 animals and a quantitative investigation of the size of the glomeruli. For the qualitative study the animals were divided into 4 groups: Group I consisting of rats of 75–100 days; Group II, 300 days; Group IIIa, 868–983 days; and Group IIIb, 1000–1170 days. It was felt that with an early and a late senile group, it might be possible to detect differences within the senile period.

For the quantitative study they were divided into three groups: Group I, 75–100 days; Group II, 300 days; and Group IIIab, 868–1170 days. For the measurements a calibrated ocular micrometer was used in direct study of the glomerular tufts with the oil immersion lens. Two diameters, the largest and the one at right angles to this in the same plane were measured.

The kidneys of the young rats (75–100 days) in Group I and the majority of those (11 specimens) of the rats in Group II (300 days) showed what we may describe as the typical histological picture of the "normal" kidney (Fig. 16–1). The glomeruli showed no evidence of fibrosis. They had a compact appearance. The lumina of Bowman's capsules and other parts of the nephrons either were entirely clear or, occasionally, showed a very faintly staining, inconspicuous coagulum. The cells of the tubular epithelium, except some that were binucleate, were uniform in appearance when corresponding portions of the nephron in different fields were compared. The walls of the blood vessels showed neither deposits nor tissue change in any of their layers and the connective tissue about them was free of aggregations of lymphocytic cells. In the kidneys of two of the 300-day rats in Group II the lumina of a few tubules contained small amounts of colloid-like material which varied from markedly eosinophilic to lightly basophilic in staining reaction and gave a weakly positive Millon's reaction for proteins. This material apparently represents plasma proteins which have passed the glomerular barrier and precipitated to some extent in the tubules. We referred to it as "colloid," with the understanding that this appellation signifies only a general histological resemblance. In three of the middle-aged rats there was a moderate accumulation of lymphocytes and plasma cells in the adventitia and surrounding connective tissue of the arteries.

The kidneys of the first group of senile rats (group IIIa: 868–983 days) differed from those of young and middle-aged animals not so much in the glomeruli as in the tubules and arteries. In the majority of the animals (Table 16–1) a high proportion of tubules, the various parts of the nephron, and the large and small

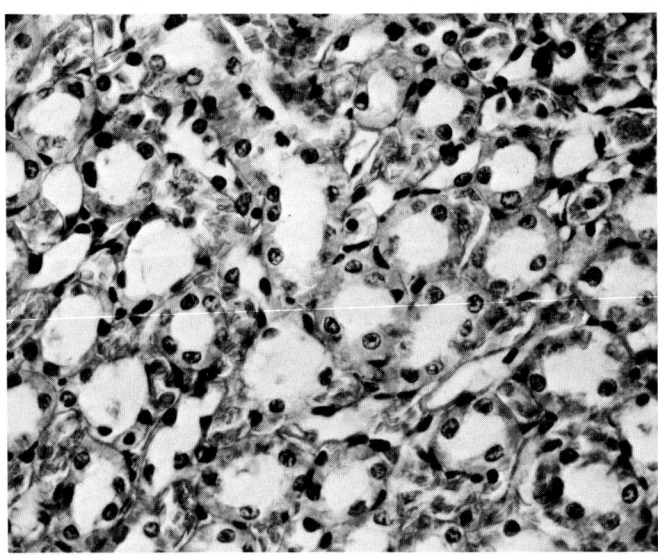

Fig. 16–1. Tubules in the renal medulla of a young female rat. The lumina are generally clear in sections. ×470.

Table 16–1. *Age changes in kidneys of Wistar Institute rats*

Age Group	Malpighian Corpuscles			Tubules		Blood Vessels	
	Slight Fibrosis of Glomeruli (Glomerular Tufts)	Moderate to Marked Fibrosis of Glomeruli	Colloid	Aberrant Cells	Colloid	Basophilic Deposits in Wall	Lymphocytic Infiltration
75–100 days (9 animals)	0	0	0	0	0	0	0
300 days (11 animals)	0	0	0	0	0	0	27.3
868–983 days (19 animals)	44.4	10.5	21.1	89.5	84.2	47.4	100
1000–1170 days (11 animals)	36.4	0	27.3	100	100	54.5	100

The per cent of animals in each age group which showed the described changes is listed in this table. From Andrew and Pruett, 1957.

collecting tubules, showed large masses of colloid (Fig. 16-2). Frequently these masses completely filled the segment of a tubule and even distended it, with resultant flattening and atrophy of the epithelial lining, suggesting that precipitation of proteins had occurred during life and that these appearances were not artifactual, nor due to postmortem change.

In 4 of these 19 rats colloid was found within some of the Bowman's capsules (Fig. 16-3). Here also this material often seemed to have distended the capsule.

Both in the tubules and in Bowman's capsule the colloid frequently contained vacuoles, usually at its periphery. It is interesting to speculate whether such vacuoles indicate an attempt at reabsorption. They may be artifacts, however, as the majority of workers seems to consider the vacuoles in the thyroid colloid to be.

A second feature of the tubules in this senile group was the occurrence of definitely aberrant epithelial cells. These appeared in any portion of the nephron but were not seen in the collecting tubules. The aberrant cell usually is much larger than the neighboring cells and, whatever its location, tends to have a granular, eosinophilic cytoplasm. Its most prominent distinction is the greatly enlarged and often hyperchromatic nucleus (Fig. 16-4). Frequently, a very prominent nucleolus, far larger than that in the ordinary cell, was present (Fig. 16-5). Occasionally a small group of such aberrant cells were seen, but generally they were found singly. Many of these aberrant cells resembled the oncocytes of the salivary glands and other organs.

The conspicuous feature of the blood vessels in this group of senile animals was the presence of a marked infiltration of the adventitia of many of the arteries and of the adjacent connective tissue by masses of lymphocytes and plasma cells. We use the term *infiltration* here as generally descriptive since it is not clear whether

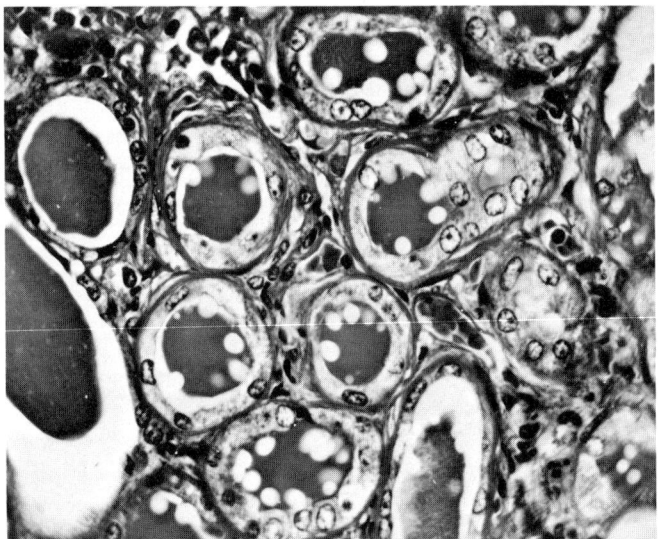

Fig. 16-2. Tubules in the renal medulla of a senile female rat, 1170 days of age. A dense-appearing colloid is seen in the lumina of the majority of tubules. ×470. (After Andrew and Pruett, 1957.)

The Urinary System

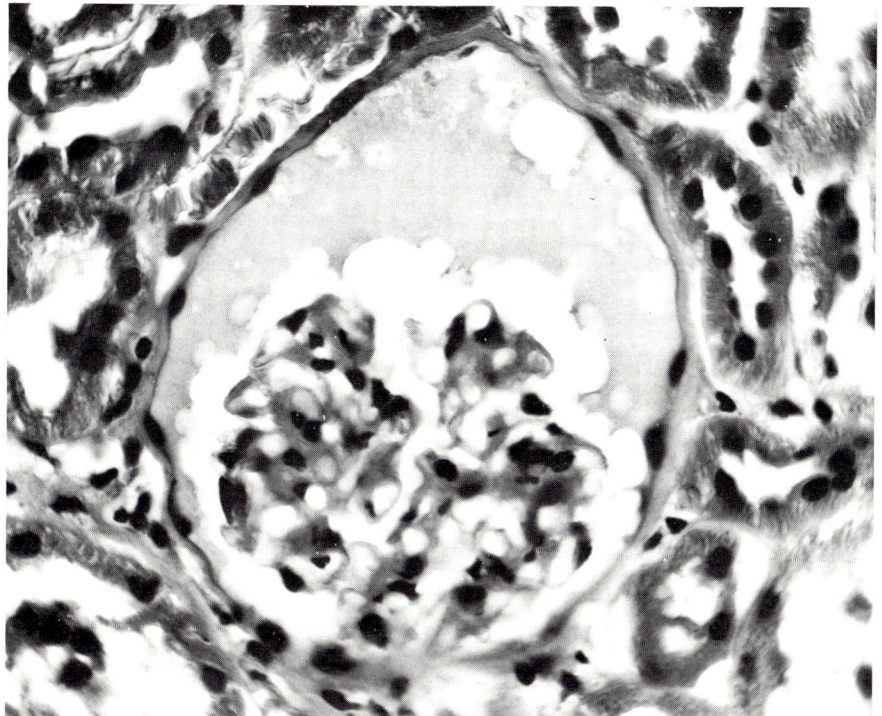

Fig. 16-3. Renal corpuscle of a senile rat, an 869-day female Gray Norway. Old rats of both Albino and Gray Norway types show a thickening of the basement membrane of Bowman's capsule. Frequently, as here, a vacuolated colloid is seen in the lumen and seems to have distended Bowman's capsule. ×570. (After Andrew and Pruett, 1957.)

the cells that migrated through the wall of the vessel near which they lay had come from capillaries in the connective tissue, or had arisen in some other way. The appearance of the cell masses, both from the combination of lymphocytes and plasma cells within them and from the presence of cells which probably are primitive fibroblasts, is one of fairly stable aggregations; little indication of true nodule formation and no germinal (reaction) centers have been seen in them.

In about half of the animals in this group the adventitia of the blood vessels showed deposits of a deeply basophilic amorphous material. The shape of these deposits was irregular and their distribution seemed to follow no regular pattern. They occurred in arteries, arterioles, and in some veins. They are evidently deposits of calcium salts since they give a positive reaction with van Kossa's test for calcium in sections. We have seen them neither in the media nor in the intima.

Many arteries in the senile animals presented "vacuoles" or clear spaces among the muscle cells of the media (Fig. 16-6). They were of varying size, some very minute, others much larger than a smooth muscle cell. Some were rounded, some ovoid, while others were irregular in shape. The elongated ones usually follow the circumferential course of the muscle fibers. With Scharlach R the media showed some droplets of fat-positive nature but they did not seem to cover as large an area as the clear spaces.

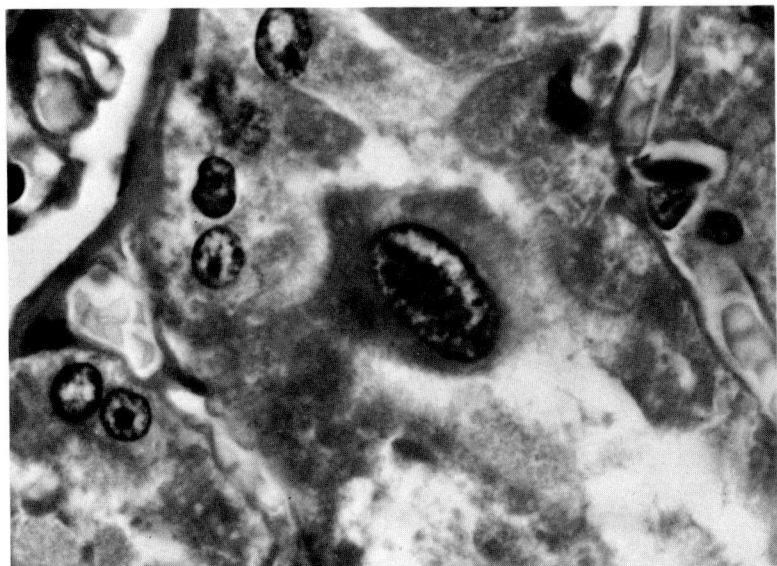

Fig. 16-4. Aberrant cell in the kidney of a 981-day-old Gray Norway male rat. This cell apparently developed in the epithelium of a proximal convoluted tubule. The nuclear change is particularly striking. ×1000. (After Andrew and Pruett, 1957.)

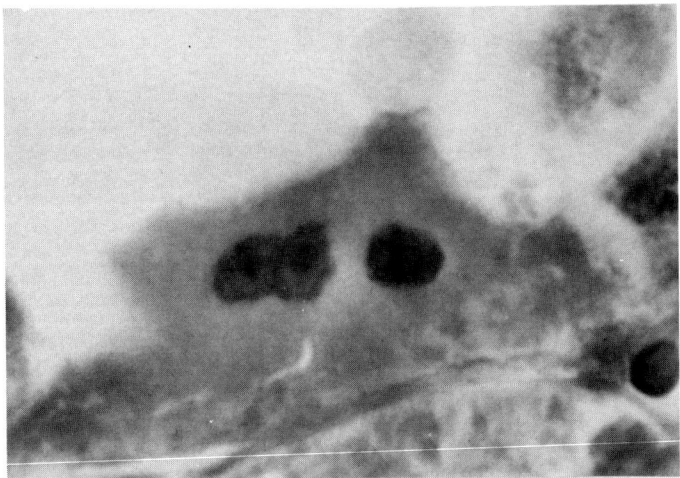

Fig. 16-5. Cells lining a dilated tubule in kidney of a 900-day Albino rat. By their size, large nucleoli, and tendency to amitotic division, these cells may be considered as "oncocytes." ×970. (After Andrew and Pruett, 1957.)

The phenomenon which we see in the media of the arteries, the appearance of clear spaces or "vacuoles," is, interestingly, similar to that described by Kuhlenbeck (1954) for the media of the arteries in the brains of Wistar Institute rats. Such spaces probably represent areas of deposit of lipoids.

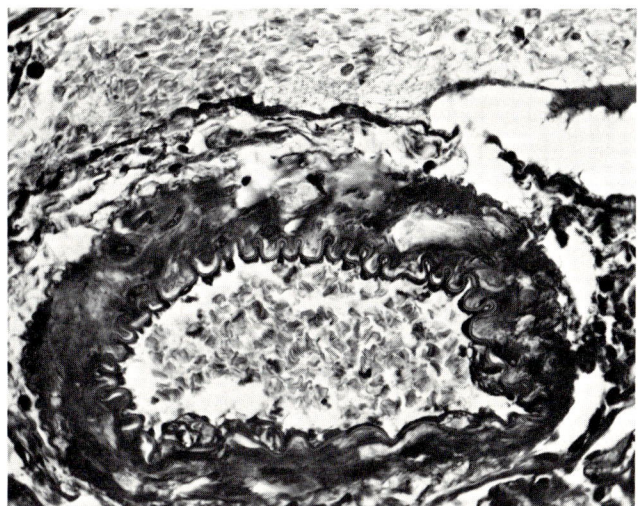

Fig. 16-6. Arciform artery in kidney of a 982-day-old Gray Norway rat. The most noteworthy change in the arterial wall is the appearance of clear spaces or vacuoles in the smooth muscle. (After Andrew and Pruett, 1957.)

On the basis of the description of the senile human kidney, we might have expected to see a large proportion of the glomeruli in various stages of fibrosis and indeed many of them in such an advanced stage as to indicate functional obliteration of the capillary bed of the glomerulus. This was not by any means the picture seen in the senile rats in either Group IIIa (868–983 days) or Group IIIb (1000–1170 days). After the general survey of sections, in which several hundred glomeruli were concerned, 100 glomeruli were counted in each animal and classified in relation to the presence or absence of fibrosis. Figure 16-7 shows a clearly fibrotic glomerulus. Our category of "slight fibrosis" (Table 16-1) relates to the number of fibrotic glomeruli and indicates only 1 or 2 out of the 100 glomeruli as fibrotic. In the category "moderate to marked fibrosis" there were only two of the animals, a 911-day Gray Norway male (20 glomeruli as fibrotic) and a 911-day Gray Norway female (60 glomeruli fibrotic).

Frankly fibrotic glomeruli, then, were rare in the great majority of the animals. In only 2 out of the 19 (10.5 percent, Table 16-1) was fibrosis of glomeruli conspicuous.

There is, however, a difference between the glomeruli of the senile and nonsenile rats which is obvious after a few comparisons of members of the different age groups have been made. This difference, which involves practically every glomerulus of the senile kidney, consists in a greater prominence of the basement membrane, a sharper outlining of the individual capillary loops, and their greater degree of dilation.

The kidneys from the animals in Group IIIb, the older senile group, differed from those in Group IIIa more in degree than in any qualitative way. The features of the senile kidney were accentuated in most instances. Thus colloid occurred in

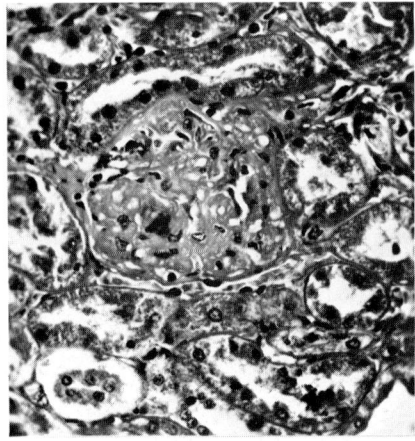

Fig. 16–7. Complete fibrosis of a glomerulus. The appearance of individual fibrotic glomeruli in the rat is similar to that in man but in the series studied by the author, fibrotic glomeruli are not at all common even at advanced age. In this 1170 day Albino female, the oldest of our animals, only 2 per cent of the glomeruli were fibrotic. (After Andrew and Pruett, 1957.)

the tubules in 100 per cent of the rats in Group IIIb and in general seemed more abundant per kidney section. Aberrant cells were present in all animals in this group. The aspect of the arterial walls was similar to that in Group IIIa, with the consistent occurrence of infiltration by lymphocytes and plasma cells and the presence of basophilic deposits in many cases.

The percentage occurrence of fibrotic glomeruli in this group, however, was no greater than in Group IIIa. About a third of the animals (36.4 percent, Table 16–1) showed "slight fibrosis," again indicating no more than one or two fibrotic glomeruli per hundred, while none of these 11 animals showed moderate or marked fibrosis of glomeruli. In Group IIIb, the prominent basement membrane and the dilated appearance of the capillaries were consistently present.

Dimensions of Glomeruli

Measurements on 50 glomeruli from each animal (Table 16–2) showed that the glomerular dimensions in Group I and Group II were closely similar. A test for possible significance of the difference between these groups was made, the difference of the means was divided by the square root of the sum of the two probable errors and the results were referred to a table of significance (Pearl, 1940, p. 472). There was a 50 per cent chance that the difference between the glomerular size in Group I and Group II rats might occur at random—hence the difference between Groups I (75–100 days) and II (300 days) was not significant.

The comparison of the figures for Group IIIab (senile, 868–1170 days) with Group I and Group II, showed even by simple inspection that the mean of the short diameters in the senile specimens is greater than the mean of the long distance in either young or middle-aged specimens. The test for significance indicated that for both short and long diameters there is less than one possibility in 65,000,000,000 that the differences between Group IIIab and Groups I and II respectively would be due to chance alone! The differences in glomerular size between the senile and the young or middle-aged animals, then, were highly significant. These, it must be remembered, are linear measurements and the differences of area or volume would be greater.

Table 16–2. *Age changes in kidneys of Wistar Institute rats*

Size Characteristics of Glomeruli and Combined Weight of Two Kidneys		Group I 75–100 days (9 animals)	Group II 300 days (11 animals)	Group IIIab 868–1170 days (30 animals)
Average size of glomeruli	Short diameter, in μ	79.43 ± 1.093	74.40 ± 1.449	96.45 ± 0.689
	Long diameter, in μ	89.28 ± 0.999	91.30 ± 1.884	117.76 ± 0.809
Range in size within group	Short diameter, μ	66.02 – 79.50	54.82 – 82.80	83.92 – 110.24
	Long diameter, μ	83.90 – 97.16	67.76 – 106.84	103.58 – 133.06
Combined weight of two kidneys, gm		1.989 ± 0.095	2.149 ± 0.127	2.687 ± 0.035

* Average size of glomeruli (glomerular tufts) in different age groups of Wistar Institute rats, as obtained by measurement of short and long diameters of 50 glomeruli in each animal. The probable error is given for each diameter. Kidney weights for each group are included in this table. From Andrew and Pruett, 1957.

REFERENCES

Altschule, M. D., 1930. The changes in the mesonephric tubules of human embryos ten to twelve weeks old. Anat. Rec., vol. 46, pp. 81–91.

Andrew, W. and Pruett, 1957. Senile changes in the kidneys of Wistar Institute rats. Amer. J. Anat., vol. 100, pp. 51–80.

Arataki, M., 1926. On the postnatal growth of the kidney, with special reference to the number and size of the glomeruli (albino rat). Amer. J. Anat., vol. 36, pp. 399–436.

Furno, A., 1909. Richerche Anatomo-Patologiche intorno al Rene Atrofico Senile. Lo Sperimentale, vol. 63, pp. 99–129.

Grafflin, A. L., 1937. Glomerular degeneration in the kidney of the Daddy Sculpin (Myoxocephalus Scorpius). Anat. Rec., vol. 57, pp. 59–79.

——— 1939. Cyst formation in the glomerular tufts of certain fish kidneys. Biol. Bull., vol. 72, pp. 247–257.

Howell, T. H. and A. P. Piggott, 1948. The kidney in old age; preliminary communication. J. Geront., vol. 3, pp. 124–128.

Hackel, von F., 1966. Altersveränderungen der Niere und ihre klinische Bedeutung—Neuere Ergebnisse der Korrosiontechnik. Zeitschr. f. Altersforschung, vol. 19. pp. 221–234.

Kuhlenbeck, H., 1954. Some histologic age changes in the rat's brain and their relationship to comparable changes in the human brain. Confinia Neurologica, vol. 14, pp. 329–342.

Laroche, Cl. and G. Mathé, 1955. Le Rein Senile. *In* Binet and Bourlière, Précis de Gerontologie, Chapter 11, Paris: Masson et Cie, pp. 329–347.

Lowry, O. H. and A. B. Hastings, 1952. Quantitative Histochemical Changes in Aging. *In* Cowdry's Problems of Aging, 3d ed. Baltimore: Williams & Wilkins, pp. 105–138.

Moore, R. A., 1931. The total number of glomeruli in the normal human kidney. Anat. Rec., vol. 48, pp. 153–168.

——— and L. M. Hellman, 1930. The effect of unilateral nephrectomy on the senile atrophy of the kidney of the white rat. J. Exp. Med., vol. 51, pp. 51–57.

Moritz, A. R. and M. R. Oldt, 1937. Arteriolarsclerosis in hypertensive and non-hypertensive individuals. Amer. J. Path., vol. 13, pp. 679–728.

Oliver. J. R., 1952. Urinary System. *In* Cowdry's Problems of Aging, 3d ed. Baltimore: Williams & Wilkins, pp. 631–650.

Rogers, J. B., 1950. Renal pathology in senile guinea pigs. J. Geront., vol. 5, p. 387 (Abstract).

17
The Reproductive System

Male Reproductive System

Production of viable sperm frequently continues into what we would call "old age" in the human species. In half or more of the male population in the sixties and seventies sperm formation is seen in the testes. When degenerative change in the seminiferous tubules occurs, it is accompanied by fibrotic changes. Whether the initial degenerative change is due to alterations in circulating hormones or to an increasing refractoriness of the testis as "target organ" to respond to them has not been determined.

The microscopic changes most often observed in the testes are thickening of the basement membrane of the tubules and increasing amounts of fibrous tissue in the intertubular areas.

Although many cases of fertility are cited even in octogenarians, and also in nonagenarians the changes that occur in some of the secondary sexual organs may be marked. Alterations in the prostate gland are widespread in a man 40 years of age. Nodular hyperplasia is the most common form of change. Appearances here suggest that the primary change is in the fibrous tissue of the stroma and that the epithelium is affected only secondarily. The change in the periurethral tissue unfortunately acts to bring about partial blockage of the passage of urine from the bladder. Moore (1952) has described some of the details of the prostatic changes. Hypertrophy of the prostate gland is still a process the etiology of which is unknown, other than to say that it is due to aging. An upset of the androgen-estrogen balance may be the true cause.

The seminal vesicles of adult man show a yellow pigment in the epithelium. According to early writers the pigment is generally not present before puberty. It probably, then, is related to the function of these glands rather than having a significance such as that of lipofuscin, the "age pigment." However, the granules, ranging from very small dimensions up to a diameter of 8μ, appear to accumulate in the columnar cells as aging occurs. The basal cells remain with little or no pigment.

That these organs remain under hormonal control into advanced age was shown by the experimental studies of Korenchevsky and his colleagues (1953, 1956). Estrogens and androgens affect weight and histological structure even in senescent rats.

184 Man and Mammals

In relation to the epididymis, there is little information as to changes in old age. Faller (1941) made a study of the amount of vitamin C in this organ in the rat. He found a greater concentration in its upper portion in infant and young animals, while in adult and senile animals the concentration was greater in its lower portion.

Prostate Gland

In the 40- to 60-year period the outstanding feature of the prostate is the variation in appearance from one part of the gland to another. The stroma presents atrophic change of the smooth muscle such that this tissue, originally about equal in bulk to the connective tissue, comes to occupy a relatively much smaller volume. The collagenous tissue becomes more homogeneous and denser. The lumina remain large but protrusions into them are less numerous and the height of the epithelium is decreased so that it changes from a columnar to a cuboidal type. Concretions (corpora amylacea, Fig. 17-1) appear. These changes are irregular in their distribution. They may be marked in some parts of the gland while minor or even lacking in other parts.

In subjects over 60 years of age these changes tend to be more widespread or even constant throughout the gland. Complete atrophy of some acini, with

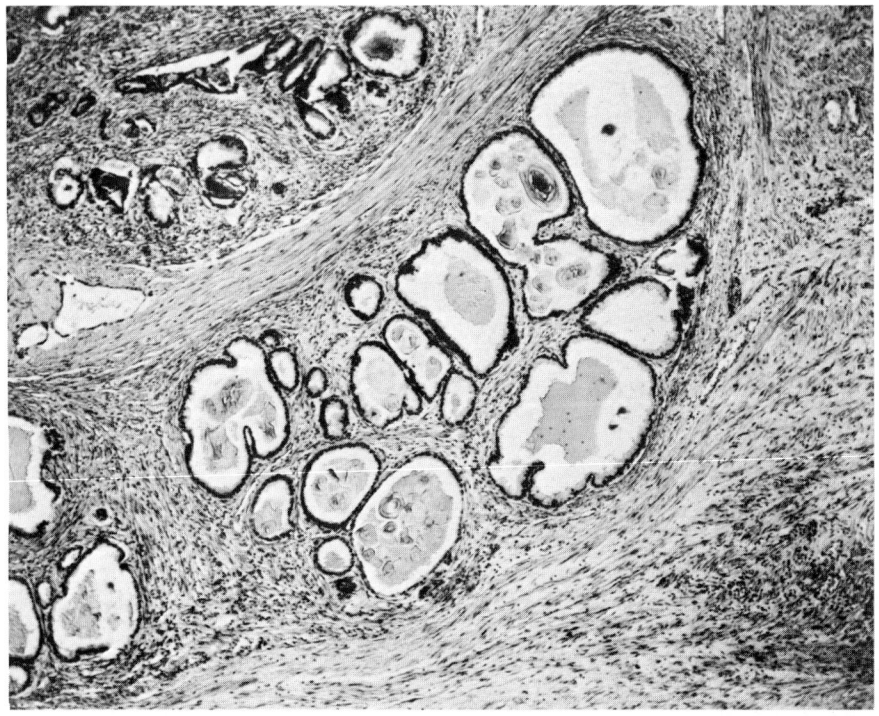

Fig. 17-1. Prostate gland of an 87-year-old man. Hypertrophy of the gland, due primarily to increase of the stroma, is common at advanced ages. ×35.

replacement by connective tissue, occurs. Corpora amylacea becomes numerous, due probably to relative stagnation of the secretion, and are of larger size than formerly.

In actuality, there is much individual variation in the rate at which these changes occur in the prostate gland.

Testes and Ducts

Engle (1952) stated that in his own series of testes and their ducts, derived chiefly from traumatic deaths and from organs with adequate histological fixation, more than half of the testes and ducts of men past 70 had abundant spermatozoa. The seminiferous tubules with developing spermatozoa present the appearance seen in Fig. 17-2. According to Blum (1936), in 165 cases of men over 60, spermatozoa were present in ejaculates of 68.5 per cent of men aged 60 to 70, 59.5 per cent of men 70 to 80, and 48.0 per cent of men 80 to 90 years of age.

According to Stieve (1930) a progressive increase in the elastic tissue of the wall of the seminiferous tubule is an age change and a similar increase occurs in the tunica albuginea of the testis and in the epididymis. Collagenous tissue decreases in amount as elastic tissue increases. According to Stieve, the width of the tubules does not change as long as sperm production is going on, but the tubules tend to change from a round form in cross section to a polygonal one.

The only change in interstitial cells of the testes clearly associated with aging is the appearance and increase of pigment. Oiye (1928) found no changes in numbers of interstitial cells with age.

The Sertoli cells are of great importance for the maturation of the spermatozoa, and their fate during the aging process is of considerable interest. Lynch and Scott (1950) have studied these cells in a series of 168 males ranging from 2 months to 84 years of age. The lipid content of Sertoli cells and of the interstitial cells of Leydig was the special object of their study. The lipid of the Sertoli cells

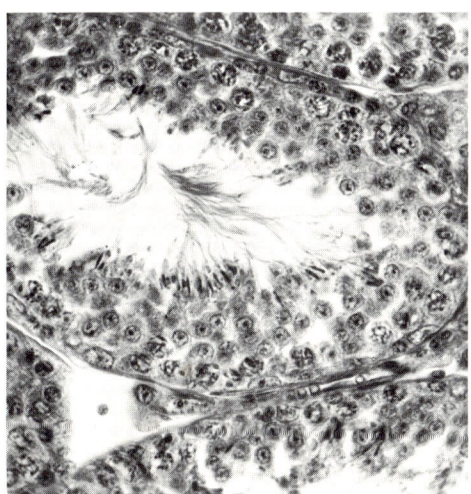

Fig. 17-2. Seminiferous tubules of a mouse. The process of formation of spermatozoa in man and other mammals, while generally slowed in the latter part of the life span, often continues to some degree even in advanced age. ×400.

appeared first at age 15 and increased gradually to a maximum in old age. The lipid of the Leydig cell appeared first at 12 years of age and reached a maximum at 35 years. Thereafter it decreased progressively. The interstitial cell lipid may be an indication of manufacture or storage of the androgen complex, since its curve compares fairly well with the ketosteroid excretion curve given by Hamilton and Hamilton (1948). Increased lipid content of Sertoli cells may be evidence of estrogen production. Whether this is a fact and how it relates to functional changes in the male are important questions, but the fact seems clear that Sertoli cells do show a very definite age-related change.

A significant characteristic of the male reproductive system in advanced age is the presence of pigment. This usually occurs as golden brown spheroidal granules and is found in smooth muscle of the seminal vesicles and prostate, in efferent tubules of the testes, and in the interstitial cells of Leydig. Pigment may actually be excreted to some extent via the efferent tubules as it can be found in their lumina and in the semen.

While the interstitial cells tend to accumulate much pigment with age, no definitive investigations have shown any direct relationship between their numbers and pigment content with the condition of the accessory glands, with the activity of the seminiferous tubules, or with the level of the sex drive of older men.

Female Reproductive System

The period of active cyclic growth and development in the human female reproductive system comes to a close in the menopause, an event to which there is no actual correspondent in the male. Follicles (Fig. 17-3) cease to develop and ovulation ceases. The stroma of the ovary shows fibrotic change and masses of hyalinized collagenous fibers are seen. Some of the epithelial structures, however, including the germinal epithelium and that of the rete tubules of the ovary frequently show proliferative change and may form papilliform projections. Cyst formation, sometimes with secretion of undue amounts of estrogen, is common.

The uterus and vagina, freed from the cyclic stimuli of the ovary, cease their own periodic activity and their mucous linings become relatively stable. The organs decrease in size, and varying degrees of fibrosis occur in their walls. In the vagina the epithelium becomes thin. The reaction in the vagina changes from acid to alkaline and there is a lowered resistance to infection. There is a drying process with some keratinization of the vaginal outlet and the vulva. These changes often lead to irritation and vulvitis.

Information on ovarian aging in older mammals is scanty. The general impression is that involution is a slower processs than in the human (Eckstein, 1955) and that hormonal influences and growth of follicles still are found in senile animals, as in the monkey (Krohn, 1955). This is true for the mouse (Bloch and Flury, 1957), but the number of ova liberated is small and many of these are abnormal in old animals (Boot and Muhlbock, 1954). It appears, however, that even when there is no longer any evidence of estrogenic stimulation, the ovaries

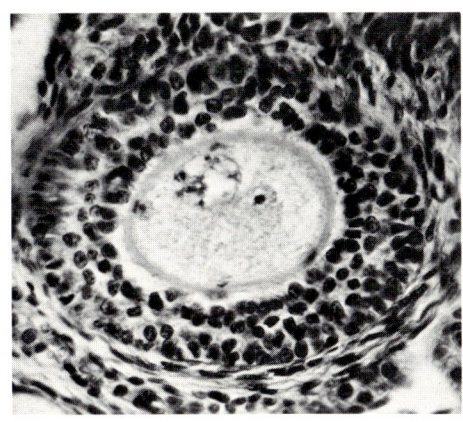

Fig. 17-3. Ovum undergoing atresia in rat ovary. The atretic changes of follicles and ova represent one type of degenerative change which occurs without regard to age and is found in very young as well as older individuals. ×400.

of senile mice may be reactivated by administration of pituitary hormones (Zondek and Aschheim, 1927). Thus, relative to the life span, the process of senescence of the ovary in the mouse appears to be slower and more protracted.

It is interesting to note that in the ovaries of some senile mice there is a great proliferation of interstitial cells which, like hypertrophic stromal cells in the human ovary, produce estrogenic hormones.

In such mice the vagina shows a phenomenon of continuous stimulation, a finding often seen in the C57 Black strain (Thung et al., 1956).

The ovaries of mammals in general do not show as marked involutional changes as do those of the human female. In the mouse, at least, there is less accumulation of residual structures such as corpora albicantia, corpora fibrosa, and sclerosed blood vessel walls. There are differences, primarily quantitative in nature, in the proliferating germinal epithelium in the human or the mouse. These differences may be conditioned by the dense stroma in the human ovary. In any event, the papillary projections occur in women but not in mice, while in mice the ingrowing ducts and cysts are associated with the formation of anovular follicles. Proliferations of downgrowing epithelium in the mouse are much more extensive than in the human ovary and resemble tumor formations. Transitions to a granulosa type of cell are common.

Some features of tissue change with aging of the human uterus were investigated by Woessner (1963). He found the wet weight, collagen, and elastin increasing to a maximum by age 30, remaining constant for the next 20 years, then declining in the 50-65 year subjects to values about ½ of the maximum.

Bacterial collagenase effected a complete digestion of uterine collagen at all ages. The rate, however, is much slower in old age.

Postmenopausal atrophy of the ovaries and uterus (Fig. 17-4) may be of such a degree that in a woman of 70 or over the uterus is a much reduced fibrous body, while the ovaries may be difficult or impossible to identify at surgery or autopsy. Chronic infection of atrophied vagina and uterus occur frequently. There may be intense irritability of vulva or vagina. Administration of estrogen may bring about a change of the thin vaginal membrane to a thick one and accumulation of glycogen in the epithelium, with a change from an alkaline to an acid reaction of the vaginal secretion. The questions of when and how such administration should be made, and of the complications in relation to it, are beyond the scope of a work on anatomical changes with age.

In the human female the highest incidence of cancer of the mammary glands and of the cervix and body of the uterus is during the fifth and sixth decades, in the time immediately before and immediately after menopause. Why this is true is difficult to say, but from animal experimentation, a link with the stimulation by the sex steroids seems reasonable.

Cyclosenility

In an important paper Sirtori (1969) has pointed out the resemblance of the involutional changes in cells of the endometrium during the premenstrual phase, to those of senility and has used the term *cyclosenility* to describe them. He emphasizes the existence of "segmentary" damage involving only some of the organelles of the cells and emphasizes that this accounts, among other things, for the segmentary response which may be obtained with hormone therapy, as in the administration of estrogens (see Figs. 17–5, 17–6, and 17–7).

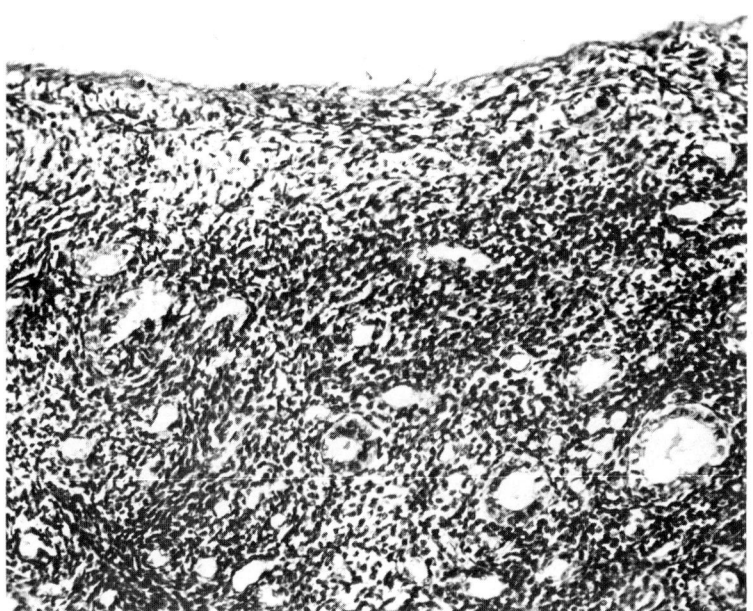

Fig. 17–4. Atrophic endometrium of the uterus of a 64-year-old woman. This subject also had suffered from prolapse. The surface epithelium is missing in places and the glands are dilated and tend toward cyst formation. ×125.

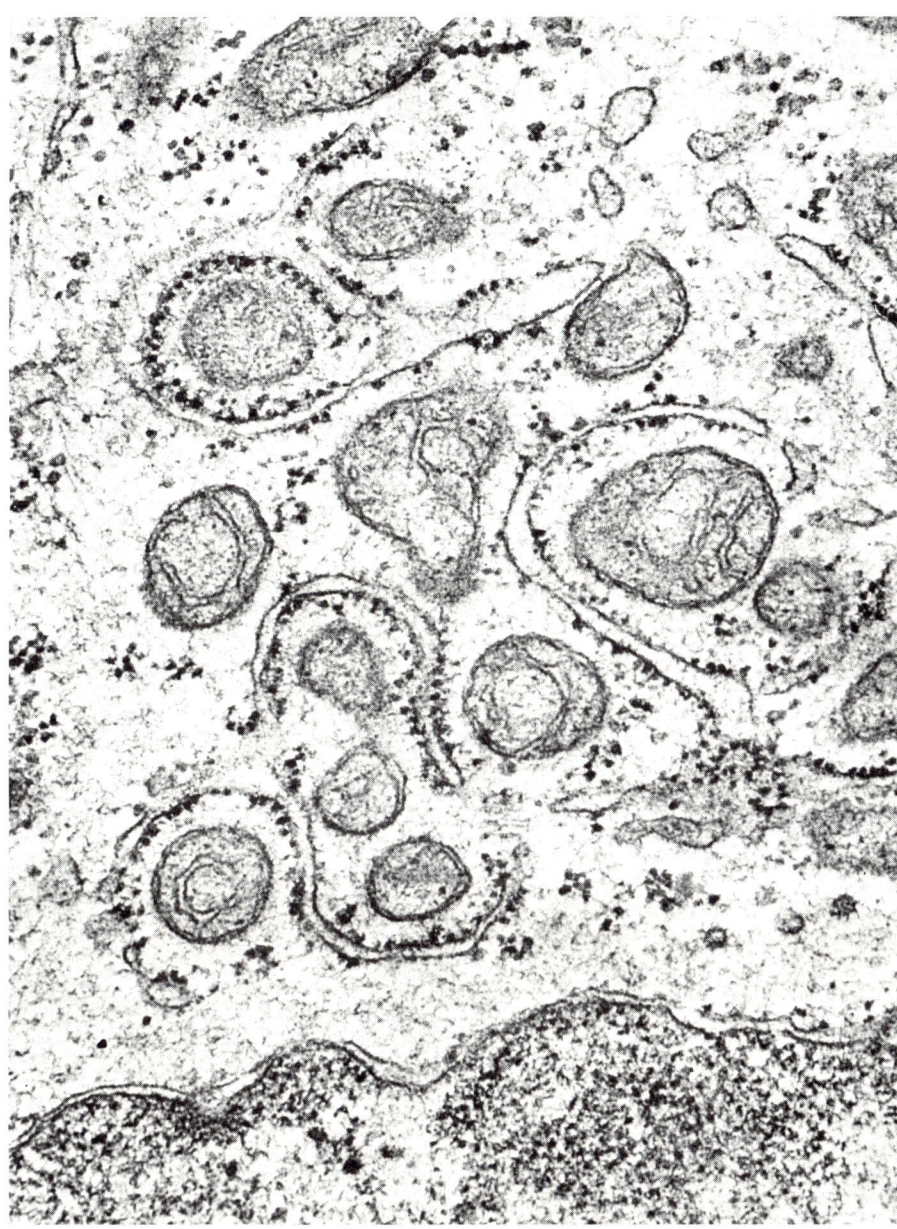

Fig. 17-5. Mitochondria of endometrial epithelium of a human subject in the latter part of the follicular phase. The micochondria are of "normal" size and appearance. Rough endoplasmic reticulum is prominent. ×40,000. (Reprinted with permission from C. Sirtori, 1969. Some ultrastructural aspects of human uterine physiopathology: cyclosenility, senility, antisenile, and antitumoral therapies, and contraceptive drugs. Proc. Int. Symp. Ob. and Gyn. Carlo Erba Foundation, Milan.)

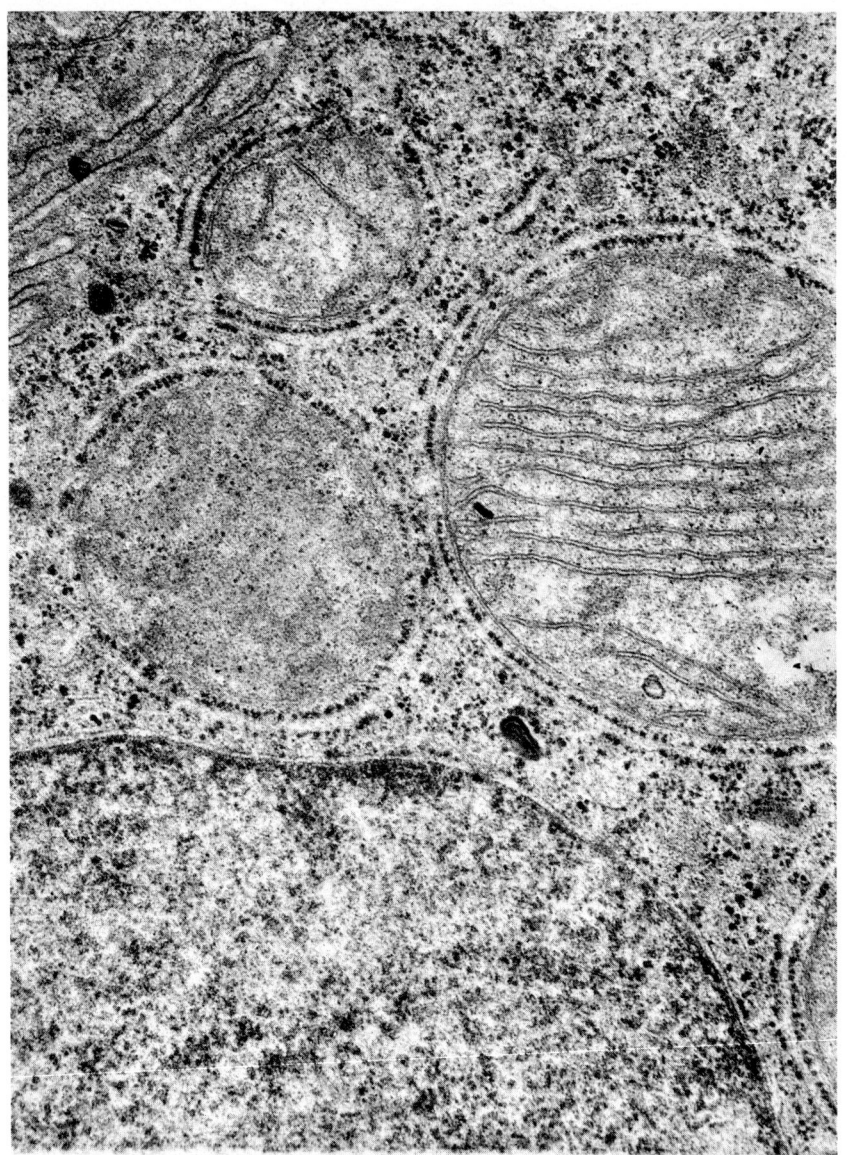

Fig. 17-6. Mitochondria in the luteinic phase. Many of these mitochondria are immensely hypertrophied. The one on the left is undrgoing degenerative change. ×45,000. (Reprinted with permission from C. Sirtori, 1969. Some ultrastructural axpects of human uterine physiopathology: cyclosenility, senility, antisenile, and antitumoral therapies, and contraceptive drugs. Proc. Int. Symp. Ob. and Gyn. Carlo Erba Foundation, Milan.)

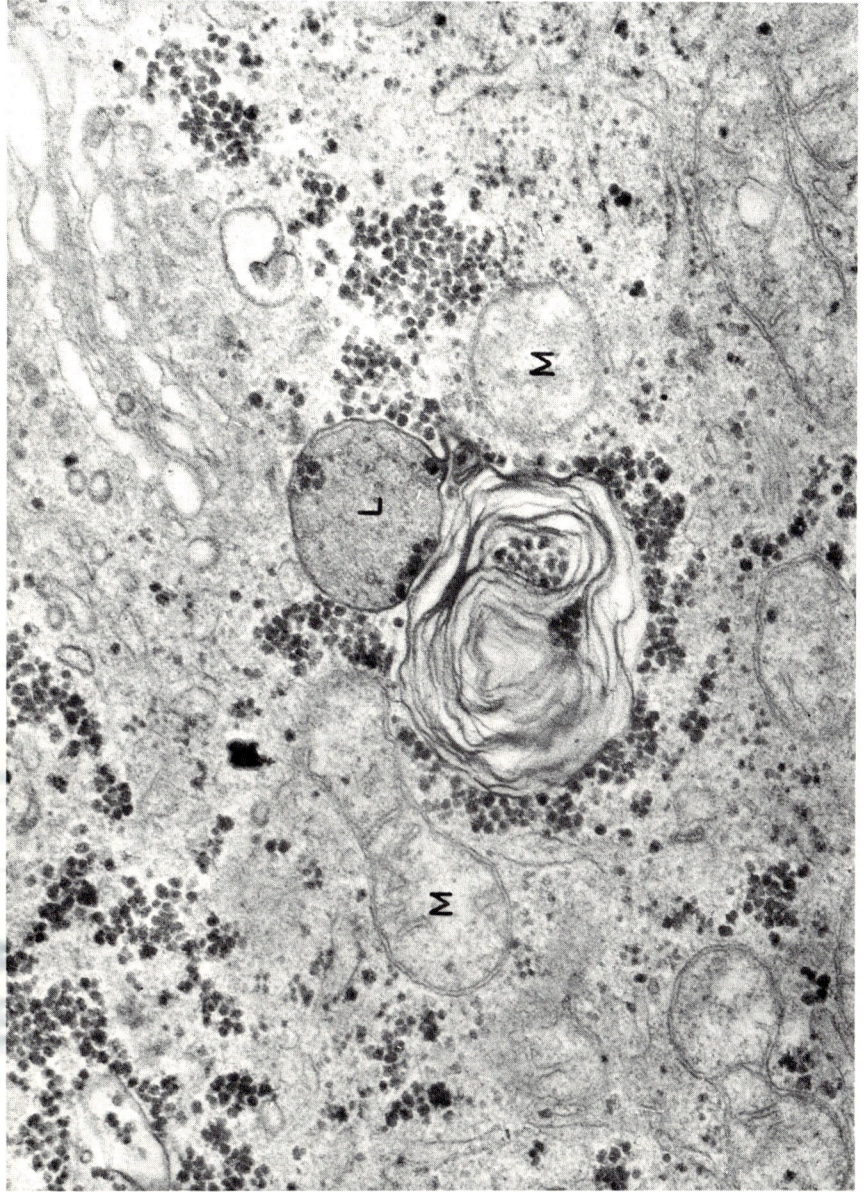

Fig. 17-7. Cytoplasm of endometrial epithelium in advanced luteinic phase. A large myelin body is seen with glycogen bodies around it and within it. A lysosome, *L*, shows glycogen bodies in it and mitochondria, *M*, show degenerative change. ×15,000. (Reprinted with permission from C. Sirtori, 1969. Some ultrastructural aspects of human uterine physiopathology: cyclosenility, senility, antisenile, and antitumoral therapies, and contraceptive drugs. Proc. Int. Symp. Ob. and Gyn. Carlo Erba Foundation, Milan.)

REFERENCES

The Female Reproductive System

Bloch, S. and E. Flury, 1957. Untersuchungen über Klimakterium und Menopause an Albino-Ratten I Mitteilung Gynaecologia, vol. 143, pp. 255–263.
Boot, L. M. and O. Mühlbock, 1954. The ovarian function in old mice. Acta physiol. pharmacol. neerl., vol. 3, pp. 463–464.
Eckstein, P., 1955. The duration of reproductive life in the female. *In* Old Age in the Modern World. Report of IIIrd Congress of the Internat. Assn. of Gerontology, 1954, Livingstone, Edinburgh and London, pp. 190–199.
Engle, E. T., 1944. The menopause, an introduction. J. Clin. Endocr., vol. 4, pp. 567–570.
Krohn, P. L., 1955. Tissue transplantation echniques applied to the problem of the aging of the organs of reproduction. *In* Ciba Foundation Colloquia on Aging, vol. 1, pp. 141–161. Boston: Little, Brown.
Thung, P. J., L. M. Boot, and O. Mühlbock, 1956. Senile changes in the oestrous cycle and in ovarian structure of some inbred strains of mice. Acta Endocr., Copenhagen, vol. 23, pp. 8–32.
——— 1957. Senile amyloidosis in mice. Gerontologia, vol. 1, pp. 259–279.
Sirtori, C., 1969. Some ultrastructural aspects of human uterine physiopathology: cyclosenility, senility, antisenile and antitumoral therapies, and contraceptive drugs. Proc. Int. Symp. Ob. and Gyn. Carlo Erba Foundation, Milan.
Woessner, J. F., Jr., 1963. Age-related changes of the human uterus and its connective tissue framework. J. Geront., vol. 18, No. 3, pp. 220–226.
Zondek, B. and S. Ascheim Das Hormon des Hypophysenvorderlappens. 1. Testobjekt zum Nachweis des Hormons. Klin. Wchnschr., vol. 6, pp. 248–252.

The Male Reproductive System

Engle, E. T., 1938. The relation of the anterior pituitary gland to problems of puberty and of menstruation. *In* The Pituitary Gland, Monograph of the Society for Research in Nervous and Mental Disease. Baltimore: Williams & Wilkins, vol. 15, pp. 298.
——— 1939. Sex and Internal Secretions. Edited by E. Allen. Baltimore: Williams & Wilkins.
——— 1947. The testis biopsy in infertility. J. Urol., vol. 57, pp. 789–798.
——— 1952. The male reproductive system. Chapter 27, pp. 708–729 *in* Cowdry's Problems of Ageing, 3rd ed., A. I. Lansing, ed., Baltimore: Williams & Wilkins.
Faller, A., 1941. Histochemische Untersuchun über das Vorkommen von Ascorbinsäure im Hoden und Nebenhoden von Ratten verschiedener Lebensalter. Z. mikr.-anat. Forsch., vol. 49, pp. 333–358.
Hamilton, H. B. and J. B. Hamilton, 1948. Aging in apparently normal men. I. Urinary titers of ketosteroids and of alpha-hydroxy and beta-hydroxy ketosteroids. J. Clin. Endocr., vol. 8, pp. 433–452.
Jenkins, R. H., C. L. Deming, and G. Van Wagenen, 1934. Comparative histology of prostate and prostatic urethra in man, monkey and rat from the viewpoint of the anatomical origin of prostatic hypertrophy. New Eng. J. Med., vol. 211, pp. 569–571.
Kirk, E., 1948. The acid phosphatase concentration of the prostatic fluid in young, middle aged, and old individuals. J. Gerontol., vol. 3, pp. 98–104.
Korenchevsky, V., 1961. Physiological and pathological aging. Basel: S. Karger, p. 514.
———, S. K. Paris, and B. Benjamin, 1953. Treatment of senescence in male rats with sex and thyroid hormones and desoxycorticosterone acetate. J. Geront., vol. 8, p. 415.
Lynch, K. M. and W. W. Scott, 1950. The lipid content of the Leydig cell and Sertoli cell in the human testis as related to age, benign prostatic hypertrophy and prostatic cancer. J. Urol., vol. 64, p. 767–776.
Moore, C., 1936. Responses of immature rat testes to gonadotropic agents. Amer. J. Anat., vol. 59, pp. 63–88.

——— 1939. Sex and Internal Secretions. Edited by E. Allen, Baltimore: Williams & Wilkins.

Moore, R. A., 1936. The histology of the newborn and pre-puberal prostate gland. Anat. Rec., vol. 66, pp. 1–9.

——— 1936. The evolution and involution of the prostate gland. Amer. J. Path., vol. 12, pp. 599–624.

——— 1936. Morphology of prostatic corpora amylacea and calculi. Arch. Path., vol. 22, pp. 24–40.

Moore, R. A., 1952. Male secondary sexual organs. Chapter 26, pp. 686–707 *in* Cowdry's Problems of Ageing, 3rd ed., A. I. Lansing, ed., Baltimore: Williams & Wilkins.

Oiye, T., 1928. Statische und histologische Hodenstudien. Mitt. u. allg. Path. u. path. Anat., vol. 4, p. 393.

Stieve, H., 1930. Männliche Genitalorgane. *In* von Mollendorf, Handbuch der Mikro. Anat. des Menschen VII/2, II.

Teilum, G., 1953. Estrogen production of Sertoli cells in the etiology of benign senile hypertrophy of the human prostate: testicular "lipid cell ratio" and estrogen-androgen quotient. Personal communication.

18
The Endocrine Glands

Thyroid Gland

Dogliotti and Nizzi-Nuti (1933) studied the thyroid glands of 53 individuals, 9 of whom were 70 years of age or older. They found reduction in size of follicles with a corresponding decrease in amount of colloid, and an increase in the volume of the epithelial cells, with increase in their content of granules. They concluded that there is actually an increased degree of activity of the thyroid gland in old age.

Again, in 1935, they reported a study of the thyroids of 50 aged subjects, upholding their earlier conclusions. Speaking of the findings of an early investigator on the subject (Clerc, 1912), they said: "Alterations of regressive or degenerative type, such as described by Clerc in advanced age, have been observed by us at any period of life, when the cause of death was a serious infectious or exhausting disease." (Clerc had listed as characteristic senile changes: degeneration of epithelial cells, with fat and pigment appearing in them, a "concentration" or solidification of the colloid, and a more or less noticeable increase in connective tissue.)

Rice (1938) believed that definite criteria for normality at different ages can be established. Garau (1938) saw evidence of increased activity in old age in the thinner colloid in numerous follicles with high epithelial cells. Such thin colloid, according to his concept, is colloid recently formed. This increased activity in places, he considered the result of degeneration of follicles elsewhere, which are "smothered" or submerged by the proliferating connective tissue.

Nolan (1938) reported on an extensive series of autopsies (725 cases). He found in this series that as age increased the number of "normal" glands decreased, and he said: "Fibrosis is relatively more common with increase in age." His tables, however, subdivided his cases on the basis of pathological condition without giving the age of the individuals. Ghigi and Reggiani (1939) in a study on constitutional types, age, and the thyroid, found in older persons a tendency to a stabilized condition of the thyroid, numerous larger follicles with "static colloid" and a lining of squamous epithelium. They found also an increase in interfollicular masses of epithelial cells.

In one of the earlier studies on age changes in the thyroid in animals other than man, Lowe (1930) studied the thyroid of sheep. The condition of the secretion in which it appears solid and probably more or less "static" she designated as the "colloid condition." She found this to be the ordinary condition in her older animals, together with, in some, a thickening of the interfollicular connective tissue.

Andrew and Andrew (1942) found that the thyroid gland in young, middle-aged, and senile mice shows marked qualitative and quantitative differences. These differences are greater between middle-aged and senile animals.

The thyroids of young animals show relatively small follicles, cuboidal epithelial cells, and an almost homogeneous colloid present in large, though varying amounts (Fig. 18-1). There is a fairly large amount of interfollicular material, consisting chiefly of immature adipose tissue. In the middle-aged animals the follicles are larger on the average, with low cuboidal or squamous epithelium, and a homogeneous colloid which fills the lumen. There is little interfollicular connective tissue.

In senility there is a great increase in the fibrous connective tissue between the follicles (Figs. 18-2 and 18-3). There is atrophy of some follicles, hyperdistension of others. The colloid presents an extremely varied picture (Fig. 18-4). Many follicles lack stainable colloid. Other contain traces, sometimes in clumped granular form. Other show definite stratification, with solid, fissured colloid in the center, and thinner, finely vacuolated colloid at the periphery.

It should be noted that while the C57 Black strain showed these consistent age changes, Loeb (1941) indicated that there is a great difference in the incidence and to a less extent in the time of initiation of such changes. The sclerosis, or fibrosis, according to him is absent in the CBA strain, rare in the Buffalo strain,

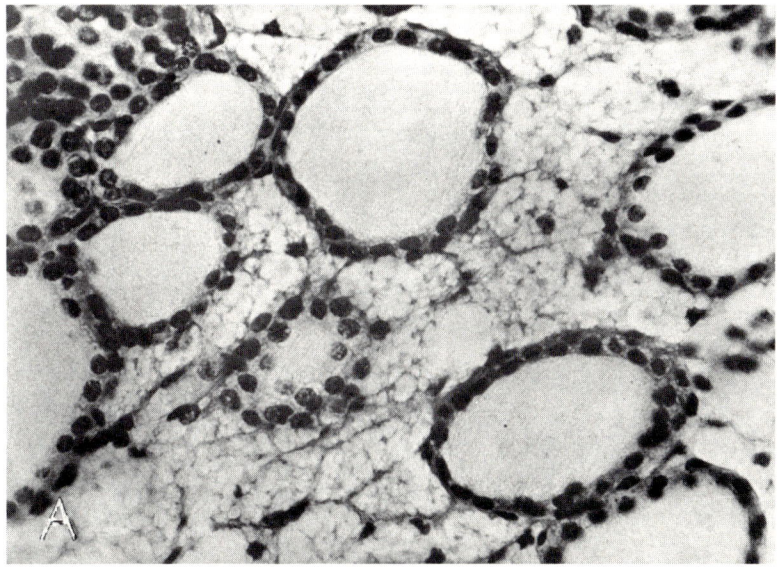

Fig. 18-1. Thyroid gland of a young female mouse, C57 Black, 41 days of age. A finely vacuolated colloid fills the follicles. The epithelium consists of fairly high cuboidal cells. Interstitial material consists of immature or multilocular adipose tissue. ×470. (After Andrew and Andrew, 1942. Reprinted with permission of the American Journal of Pathology.)

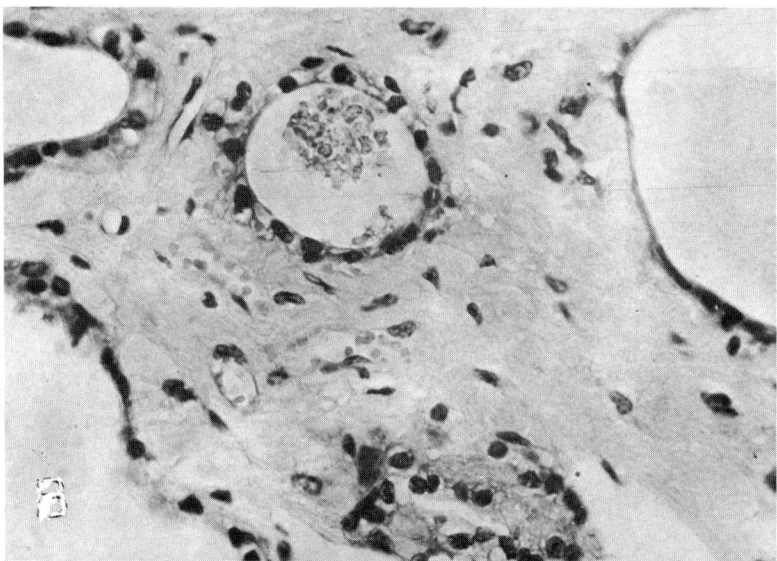

Fig. 18-2. Thyroid gland of a senile female mouse, C57 Black, 596 days of age. The interfollicular material is abundant and consists of dense fibrous connective tissue. The follicles are only partly filled with colloid, which is remarkably variable in aspect. ×470. (After Andrew and Andrew, 1942.)

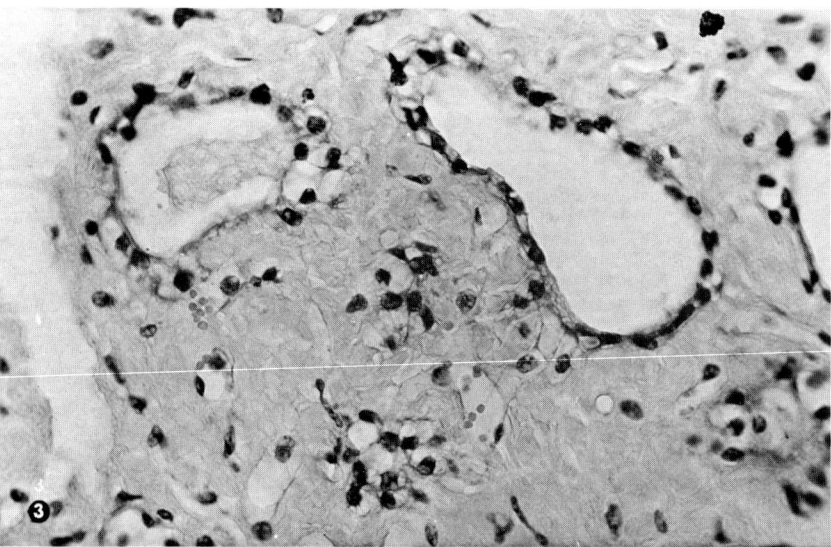

Fig. 18-3. Thyroid gland of a senile female mouse, C57 Black, 595 days old. The follicles are irregular in shape and often lack stainable colloid. Large amounts of dense, fibrous connective tissue are present and the epithelium appears to be degenerate. ×470. (After Andrew and Andrew, 1942.)

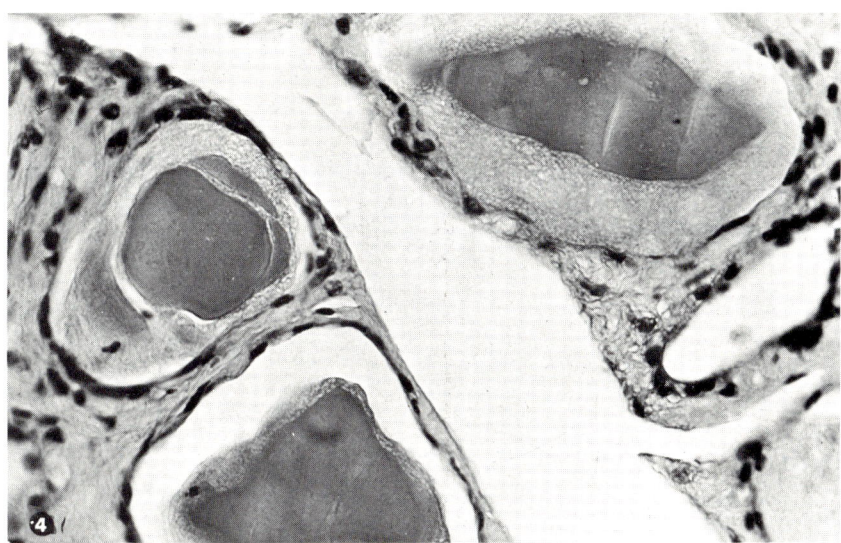

Fig. 18-4. Thyroid gland in senile mice. Thyroid gland of a senile female mouse, C57 Black, 693 days of age. The colloid in this field presents a very solid appearance in the central portion of the lumen of each follicle, a condition which may be due to stasis. × 470. (After Andrew and Andrew, 1942.)

intermediate in A and D strains while in the C_3H and especially in the C57 strain it is an almost constant accompaniment of aging.

It is true, however, that distension of the follicles and increase of connective tissue have been reported as present in senility in the guinea pig (Blumenthal, 1945); the golden hamster (Spagnoli and Charipper, 1955); the rat (Korenchevsky et al., 1950), and again in the mouse (Blumenthal, 1955). For the human thyroid similar changes have been described by Kurimoto (1950) and Mustacchi and Lowenhaupt (1950).

Important functional data might be expected to be available where such an influential regulatory gland is concerned. Other than the generally lower metabolic rate in senility, however, direct relations of function and structure have been difficult to show. Wilansky et al. (1957) have found a decreased uptake of I^{131} by the thyroid in aging as well as a decrease in its turnover rate. The amount of circulating protein-bound iodine, however, which is the active form, was not decreased in the older animals.

An extensive study on the weight of the thyroid glands of dogs was made by Haensly, Jermier, and Getty (1964). Animals of known age, ranging from 1 day to 15 years were studied. There were 171 subjects. After 9 weeks of age the average relative weights decreased sharply from 0.15 Gm. per kilogram of body to about 0.07 Gm per kilogram at *2 years of age*. From this time to the fifteenth year, however, the average relative weights fluctuated *randomly*. Since this study does not consider *type of tissue* in the gland, it is difficult to make an anatomical interpretation of the findings related to weight.

Senile changes have been described not only in mammals but also in birds (Payne, 1949, 1952). This investigator gave details of peculiar age changes in the thyroid gland of the fowl, by which vesicles appeared in the cytoplasm, increased in size, and fused to form larger spheroids. That such types of change may be fundamental seems probable from the fact that Payne discovered them in the old fowl in basophils of the pituitary gland and in cells of the adrenal cortex as well as in the thyroid gland.

Pituitary Gland

Cooper (1925) described the vascularity of the anterior pituitary gland to be at its maximum at puberty. Little change was seen until the age of 40 or 50, after which a decrease occurred.

The individual gland cells in very young subjects seemed well-spaced, while later they became more densely packed. The basophils were present in greatest numbers in older persons. Cooper found also an increase in amount of connective tissue in old age and, in the posterior lobe, an increase of "golden pigment."

Concretions, described as to location and probable source by Shanklin (1946, 1948) have been observed by a number of investigators at various ages, including early childhood but their relationship to aging does not seem to have been clearly elucidated.

The question of changes in absolute and relative weight of the pituitary gland with age has been treated in the literature with somewhat conflicting conclusions. Parson (1935) found little change in size and weight between 20 and 80 years. There was, however, a slight decrease in weight of the pituitary in females over 60 years of age. This author describes an invasion of the posterior lobe by basophil cells which seem to come from the anterior lobe or the pars intermedia. Tessauro (1952) referred to the invasion by basophils, stating that it began at about 10 years of age and reached a peak at about 40 years.

Wolfe *et al.* (1938) found little influence of age on the relative proportions of the cell types in the anterior pituitary of the rat. They did find an increase in amount of colloid in the alveoli. In virgin female rats Wolfe (1943) found a decrease in numbers of eosinophils and an increase in chromophobes. The changes came chiefly at a relatively early part of the life span. In animals over 17 months of age nearly a third of the anterior lobes showed adenomata. He found a continuous decrease in mitoses, so that by 6 months it was difficult to find even one.

The description of pituitary adenomata by Saxton (1941) emphasized multicentric origin of the nodules of chromophobe cells, which he indicated contained many vacuoles filled with lipoid.

The reticular tissue of the anterior pituitary of the rat, according to Lansing and Wolfe (1942), shows an increase in number and thickness of fibers but with no apparent change of these fibers into collagenous ones. They did find groups of atypical hypertrophic cells occurring in the anterior pituitary. These varied from small ones containing only a few cells up to larger ones a millimeter or more in diameter while in two animals they occupied approximately one half of the lobe. The cells were chiefly of chromophobe type but included some eosinophils.

In the mouse pituitary Blumenthal (1955) found only a small increase of connective tissue in old age but a decrease in size of the gland cells. He found smaller numbers of mitotic figures with advancing age until they were no longer seen in the male after 21 months. In the female they were almost absent by 22–26 months but there seemed to be a "spurt" again between the 27th–31st months, an advanced age for the mouse.

Ultrastructural aspects of the mouse pituitary were considered by Weiss and Lansing (1953). They found an enlargement and vacuolation of mitochondria and a fragmentation of the endoplasmic reticulum, together with increasing irregularity of the nuclear membrane.

Payne, in his extensive studies on the pituitary of the fowl (1942, 1944, 1946), found a gradual loss of the chromophobe cells, while in very old birds nearly all of the surviving cells showed nuclear and cytoplasmic evidence of degeneration. The mitochondria tended to change into vesicles, and in aging capons particularly these vesicles fused to form much larger bodies, which he called chondrospheres. Mitochondrial changes occurred earliest in the capon, next in the non-castrated rooster, and latest in the hen.

Using the electron microscope, Payne (1965) in a later paper, revised some of these conclusions, particularly the concept that the vesicles are derived from mitochondria. He found lipid accumulation occurring in the gonadotrophs (Fig. 18-3) of aging and of castrate fowls, but mitochondria and secretory granules play no part in this accumulation. Much of the lipid, perhaps all of it, occupies cavities of the endoplasmic reticulum.

We have mentioned the changes occurring in the salmon after spawning. Robertson (1957) showed that in the salmon pituitary at this time there is a decrease in number of cells, an increase in connective tissue, and a vacuolization of basophils.

For the golden hamster, Spagnoli and Charipper (1959), found an increase in numbers of basophils in older females and found some decrease in numbers of acidophils in older males. Some increase in density was found, with a loss of regularity of the reticular network. Degranulation of many cells seemed to have occurred, vacuolation was more common in the basophils, and some nuclei were pycnotic.

With the complicated functional aspects and histological features of the anterior pituitary gland, it is difficult to draw histophysiological conclusions from the available data. It appears that enough observations have been made on this structure at different ages to show that it does undergo important changes, knowledge of which should be useful in future studies in the growing field of experimental gerontology.

Adrenal Glands

The Rat

As a part of the Wistar Institute project on senile changes, under the leadership of Dr. Edmond J. Farris, Yeakel (1946, 1947) described histological aspects of the aging process in the adrenal glands of the rat. Hyperplastic changes in the cortex and medulla were found in glands of rats 700 days old or more (Figs. 18-5

200 *Man and Mammals*

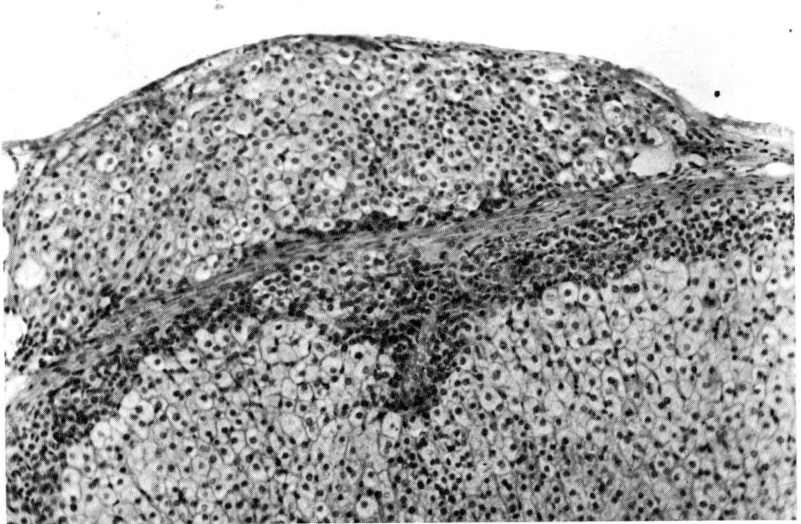

Fig. 18–5. Adrenal gland of a senile Albino male rat, 1002 days of age. Hyperplasia of the cortex is shown, with a large mass of epithelial cells growing on the *external surface* of the capsule. ×125.

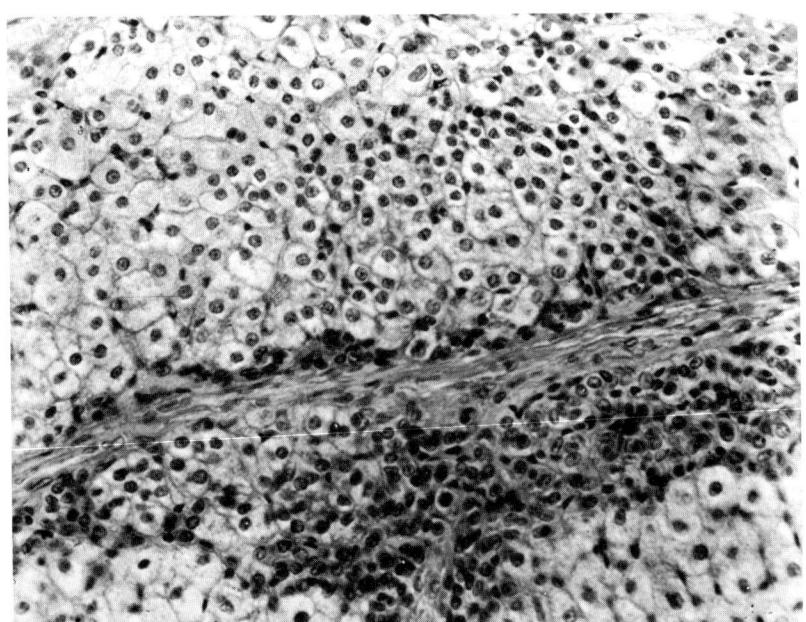

Fig. 18–6. Detail of a portion of Fig. 18–3. The hyperplastic epithelial tissue does not resemble closely any of the normal layers of the cortex. Note the crowded condition of cells in immediate association with the capsule and its extension. ×250.

and 18–6) and were more frequent in males. Hyperplasia was either diffuse or nodular. In a number of cases it was associated with dilatation of the sinusoids.

Absolute and relative weights of the adrenal glands were found to be much greater in aged rats. Yeakel also studied relative volumes of cortex and medulla at different ages. The cortex had its greatest relative volume at 50–100 days of age in both sexes. By 300 days it had decreased greatly in males, but at this age it remained high in females. The relative volume of the cortex diminished still more in older rats, both male and female. However, in animals of 1000 days measures of the volume had to be considered inaccurate because of pathological alterations. In such glands nodular hyperplasia of the cortex and great vascular dilation were seen, together with medullary hyperplasia.

A study of the connective tissue of the adrenal glands of 80 virgin female rats, ranging in age from 10 to 884 days, was made by Dribben and Wolfe (1947). They found, with advancing age, a progressive increase in the width of the capsule with a gradual transformation of reticular fibers of its original inner half into collagenous fibers. In the zona glomerulosa a thickening of the reticular fibers occurred but there was very little transformation. The changes in the zona fasciculata and zona reticularis were similar to those in the glomerulosa but were most apparent in the reticularis. A progressive thickening occurred in the fibers in the adrenal medulla and many delicate reticular fibers were replaced by fibers of collagen but these were still of fine caliber. An interesting finding was the appearance of epithelial cell aggregates within the substance of the capsule, forming capsular nodules, between the ages of 540 and 884 days.

Jayne (1953) described cellular changes in the adrenal, including loss of cytoplasmic lipids with, however, fatty degeneration in some cases, hyperchromasia and pycnosis of nuclei, and obscuring of cell outlines and of zonation. The capsule became thickened, more fibrous and hyaline, and in very old rats showed signs of shrinkage and degeneration. Mitochondria in both cortex and medulla showed fragmentation and in some cases apparent vesiculation. This appeared to be associated with formation of age pigments.

Later, Jayne (1957) studied the cortex histochemically, using various methods. He described the distribution of the Schultz-positive material (esters and cholesterol), the Schiff-positive material, and the Sudanophil material. He found that the greatest intensity of reaction for all of these was reached at about 3 months of age and then remained at a constant level until about 1 year. In animals over a year and a half old cellular degeneration caused a decrease in ability to secrete the fatty or lipoidal substance. As the reaction for such substance decreased, the brownish pigment appeared.

The Mouse and the Hamster

Blumenthal (1955) found a progressive increase in diameter of the gland in the mouse until about the end of the first year.

Blumenthal described increase of connective tissue, at first of the capsule, then extending through the fasciculata, and finally forming a definite layer between cortex and medulla. Cortical cells were found to be hypertrophic and even to form giant multinucleate forms, often containing brown pigment.

A histologic, histochemical, and ultrastructural study of the adrenal gland was undertaken on female mice of the inbred strain, C57BL/10 (Samorajski and Ordy, 1967). Organs from animals of 4, 8, 20, and 30 months were studied.

There was an age-related increase in the number of cells containing pigment. In old mice many pigment cells coalesced to form "multi-nucleated giant cells" with very large pigment bodies, up to 5μ in diameter.

The submicroscopic structure of adrenal lipid pigment is variable (Fig. 18-7) and these authors describe five major configurations. There are no ultrastructural nor histochemical features of the pigment itself which are distinctive of advanced age in the mouse.

There have been some prominent age changes described for the adrenal of the hamster (*Cricetus auratus*) Meyers and Charipper (1956). After 1 year of age

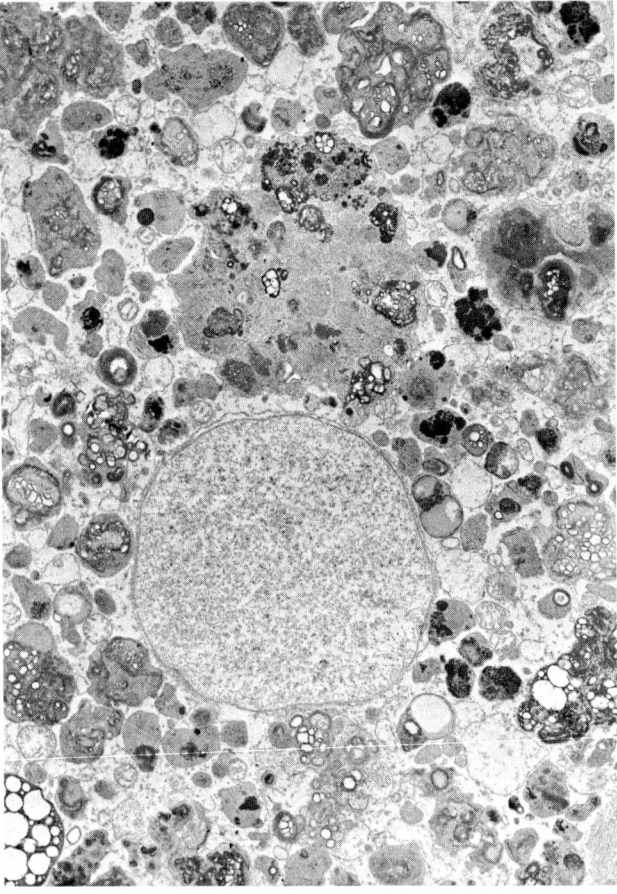

Fig. 18-7. Low power electron micrograph of a pigment cell in the adrenal gland of a 4-month-old female mouse. A variety of membranous, vacuolated and homogeneous pigment granules appear in the cytoplasm. The majority of the mitochondria appear to be swollen. (Reproduced with permission, after Samorajski and Ordy, 1967.)

hemorrhagic regions make an appearance and increase as age progresses. They seem to begin in the zona reticularis and to spread toward the zona fasciculata. According to these authors, degenerative and "generative" changes may be occurring simultaneously. The generative change means a proliferation of either normal or abnormal cortical cells. The degenerative change means atrophy with loss of parenchyma which is replaced by a coagulum type of connective tissue.

The accumulation of pigment is another prominent feature of aging in the adrenal of the hamster. It is deposited first in the inner part of the gland, both in reticulo-endothelial cells and in parenchyma. This pigment does not contain iron and its origin and nature are not clear. It seems to contain a fatty substance which blackens with osmic acid. The hamster adrenal is one which is normally poor in lipids, differing in this way from that of the mouse and rat.

Adrenal of Man

In the medulla of the human adrenal gland the chief age change is an increase of connective tissue which forms an increasingly coarse meshwork around the groups of parenchymal cells.

In the cortex, cells with lipochrome pigment are seen as early as puberty but increase with age, until in very old subjects masses of pigment cells are conspicuous. Other changes in the cortex include nodular hyperplasia and the appearance of aberrant cells.

Welch and Cannon (1968) have reported on the development of the zonular patterns in the human adrenal gland. They studied 58 autopsy specimens, ranging in age from 1 month of gestation to 69 years of age. They give many details on the structure of the "fetal cortex," for which they suggest a new name "the transient zone." This, it seems, would be more in keeping with the nomenclature of the zona glomerulosa, zona fasciculata, and zona reticularis, which we have discussed above in relation to changes during senescence. Coarse granules surrounding a homogeneous mass are seen in the cells of the transient zone as early as the 18th week of fetal life and persist until birth. For the first few months of postnatal life this material decreases in amount and very little is left by the time of establishment of definitive zonation.

It may be of significance that pigment occurs in adrenal cells which are themselves degenerating or are near degenerating cells, as in the inner portion of the zona reticularis.

Adrenal of a "Wild" Animal

An unusual opportunity to study a 13-year-old female swamp lynx, which showed clear signs of aging, was used by Pflugfelder (1967). He concentrated especially on the endocrine glands.

A marked cortical hyperplasia of the adrenals was found by which masses of glomerular zone cells broke through the fibrous capsule and grew outside as well as inside. The thyroid gland showed a conspicuous fibrillary change in the bases of the follicle cells.

The ovary, even at this age, still showed numerous primordial follicles as well as many instances of follicular atresia. Corpus luteum cells still occurred in large numbers.

Islets of Langerhans were reduced in numbers.

In the anterior lobe of the hypophysis there were large intercellular spaces, the presence of which Pflugfelder (1967) attributes to cell loss.

Pineal Gland

In the human pineal gland progressive calcification is a well-documented change associated with aging (Kitay and Altschule, 1954). Histochemically, the human pineal gland shows the presence of hydroxindole-O-methyl transferase, histamine-N-methyl transferase, and monamine oxidase, a fact which would indicate a capacity to inactivate histamine and serotonin and to synthesize melatonin. Such capacities apparently do not vary with age.

Bondareff (1965) described the ultrastructure of the pineal gland of young (9 months) and old (25 month) Sprague-Dawley rats. In the old animals he found an increase in connective tissue and an increase in numbers of various pleomorphic dense bodies in the cytoplasm of cells of the pineal parenchyma. There was no significant alteration in the relative distribution of granular vesicles in postsynaptic sympathetic nerve endings within the gland.

REFERENCES

Andrew, W. and N. V. Andrew, 1942. Senile involution of the thyroid gland. Amer. J. Path., vol. XVIII, pp. 849–863.

Blumenthal, H. T., 1955. Aging processes in the endocrine glands of various strains of normal mice: relationship of hypophyseal activity to aging changes in other endocrine glands. J. Geront., vol. 10, pp. 253–267.

Bondareff, W., 1965. Electron microscope study of the pineal body in aged rats. J. Geront., vol. 20, no. 3, pp. 321–327.

Bourne, G. H. and E. P. Jayne, 1961. The Adrenal Gland, Chap. 19 in Structural Aspects of Aging, edited by G. H. Bourne. New York: Hafner, pp. 305–324.

Charipper, H. A., A. Pearlstein, and G. H. Bourne, 1961. Aging changes in the thyroid and pituitary glands. Chap. 16 in Structural Aspects of Aging, edited by G. H. Bourne. New York: Hafner, pp. 263–276.

Clerc, Edouard. Die Schilddrüse im hohen Alter vom 50 Lebensjahr an aus der norddeutschen Ebene und Küstengegend sowie aus Bern. Frankfurt. Ztschr. f. Path., 1912, 10, 1–19.

Cooper, Eugenia R. A., 1925. The Histology of the More Important Human Endocrine Organs at Various Ages. London: Oxford University Press.

Dogliotti, G. C. and G. Nizzi-Nuti, 1935. Thyroid and senescence. Structural transformations of the thyroid in old age and their functional interpretation. Endocrinology, vol. 19, pp. 289–292.

Dribben, I. S. and J. M. Wolfe, 1947. Structural changes in the connective tissue of the adrenal glands of female rats associated with advancing age. Anat. Rec., vol. 98, pp. 557–585.

Haensly, W. E., J. A. Jermier, and R. Getty, 1964. Age changes in the weight of the thyroid gland of the dog from birth to senescence. J. Geront. vol. 19, no. 1, pp. 54–56.

Jayne, E. P., 1953. Cytology of the adrenal gland of the rat at different ages. Anat. Rec., vol. 115, pp. 459–484.

Korenchevsky, V., S. K. Paris, and B. Benjamin, 1950. Treatment of senescence in female rats with sex and thyroid hormones. J. Geront., vol. 5, pp. 120–157.

Kurimoto, T., 1950. Histological changes of endocrine organs in old age. Folia endocrinol. Japonica, vol. 26 (1/2), p. 5.

Lansing, W. and J. M. Wolfe, 1942. Changes in the fibrillar tissue of the anterior pituitary of the rat associated with advancing age. Anat. Rec., vol. 83, pp. 355–365.

Meyers, M. W. and H. A. Charipper, 1956. A histological and cytological study of the adrenal gland of the Golden Hamster (*Cricetus auratus*) in relation to age. Anat. Rec., vol. 124, pp. 1–25.

Payne, F., 1942. The cytology of the anterior pituitary of the fowl. Biol. Bull., vol. 82, pp. 79–111.

——— 1943. The cytology of the anterior pituitary of broody fowls. Anat. Rec., vol. 86, pp. 1–13.

——— 1944. Pituitary changes in aging capons. Anat. Rec., vol. 89, p. 563 (Abstract).

——— 1946. The cellular picture in the anterior pituitary of normal fowls from embryo to old age. Anat. Rec., vol. 96, pp. 77–91.

——— 1949. Changes in the endocrine glands of fowl with age. J. Geront., vol. 4, pp. 193–199.

——— 1965. Some observations on the anterior pituitary of the domestic fowl with the aid of the electron microscope. J. Morph., vol. 117, no. 2, pp. 185–200.

Pflugfelder, O., 1967. Untersuchungen an einem senilen Sumpfluchsweibchen (Lynx rupelli Brandt) unter besonderer Berücksichtigung der inkretorischen Organe. Zeitschr. f. Alternsforsch., vol. 20, pp. 45–54.

Rasmussen, A. T., 1928. The weight of the principal components of the normal male adult human hypophysis cerebri. Amer. J. Anat., vol. 42, pp. 1–27.

Robertson, O. H., 1957. Pituitary degeneration and adrenal tissue hyperplasia in spawning salmon. Science, vol. 125, pp. 1295–1296.

Samorajski, T. and J. M. Ordy, 1967. The histochemistry and ultrastructure of lipid pigment in the adrenal glands of aging mice. J. Geront., vol. 22, no. 3, pp. 253–267.

Saxton, J. A., 1941. The relation of age to the occurrence of adenoma-like lesions in the rat hypophysis and to their growth after transplantation. Cancer Res., vol. 1, pp. 277–282.

Shanklin, W. M., 1946. The development and histology of pituitary concretions in man. Anat. Rec., vol. 94, pp. 597–613.

———, 1948. On the presence of calcific bodies, cartilage, bone, follicular concretions and the so-called hyaline bodies in the human pituitary. Anat. Rec., vol. 102, pp. 469–491.

Spagnoli, H. H. and H. A. Charipper, 1955. The effects of aging on the histology and cytology of the pituitary gland of the Golden Hamster (Cricetus auratus) with brief references to simultaneous changes in the thyroid and testis. Anat. Rec., vol. 121, pp. 117–139.

Tessauro, G., 1952. La senescence de l'hypophyse. Comptes Rendus de la Société Françoise de Gynecologie, vol. 22 (4): 188–214, 1952.

Weiss, J. and A. I. Lansing, 1953. Age changes in the fine structure of anterior pituitary of the mouse. Proc. Soc. Exp. Biol., N.Y., vol. 82, pp. 460–466.

Wilansky, L. G., S. Newsham, and M. M. Hoffman, 1957. The influence of senescence on thyroid function: functional changes evaluated with I^{131}. Endocrinology, vol. 61, pp. 327–336.

Wolfe, J. M., 1943. The effects of advancing age on the structure of the anterior hypophyses and ovaries of female rats. Amer. J. Anat., vol. 72, pp. 361–383.

———, W. B. Bryan, and A. W. Wright, 1938. Amer. J. Cancer, vol. 34, p. 352.

Yeakel, F. H., 1946. Changes with age in the adrenal glands of Wistar Albino and Gray Norway rats. Anat. Rec. (Abstr. no. 59), p. 525.

——— 1947. Medullary hyperplasia of the adrenal gland in aged Wistar Albino and Gray Norway rats. Arch. Path., vol. 44, pp. 71–77.

19
The Sense Organs*

The Eye

Fibrous Coat

The sclera of the eye is primarily a layer of tough fibrous collagenous tissue. In older human subjects this tissue appears to be infiltrated by varying amounts of adipose tissue. Some of the blood vessels within it show a sclerotic change, varying with the degree of general vascular change in the body. Areas of hyalinization and calcification may be seen but the variation in this regard is great.

The bulk of the cornea is an anterior extension of the fibrous sclera, covered anteriorly by epithelium of the bulbar conjunctiva, resting on Bowman's membrane, and posteriorly lined by a low cuboidal epithelium, resting on Descemet's membrane.

One of the conspicuous signs of aging in man is the appearance of the arcus senilis, or senile ring, an opaqueness just within the circumference of the cornea and separated from its edge or limbus by a narrow clear zone. It may begin as zones of opacity in the upper or lower segments and increases in extent until the complete ring is formed. Forsius (1954) found that none of his ophthalmic patients over 50 years of age were without some arcus senilis, while in 40 per cent of males over 60 there was a complete and marked arcus, with a figure of about 30 per cent for women of that age.

The peripheral edge of the arcus senilis corresponds to the termination of Bowman's membrane and is more sharply limited than the inner edge. Microscopically, a diffuse infiltration of cholesterol esters and droplets of fat can be observed in Bowman's membrane, within the fibrous substantia propria, and in Descemet's membrane. This infiltration, however, also occurs in the iris and ciliary body where its gross manifestation is not marked. Local vascular change, allowing increased diffusion of certain substances from the blood vessels, may be important in the etiology of the senile arc; Forsius (1954) found considerably higher levels of serum cholesterol and of total serum lipids in persons who developed the arc relatively early.

The so-called Hassall and Henle warts are another common lesion of the cornea in later years. They appear to originate as local thickenings near the

---

* Special thanks are due to Dr. Michael Lashmet, of the Department of Ophthalmology of Indiana University, for assistance with the illustrations in this chapter.

periphery of Descemet's membrane, generally being less than one tenth of a millimeter in diameter. Very large numbers of them may cause slight interference with vision by altering the permeability characteristics of the posterior epithelium of the cornea and permitting aqueous humor to enter the substance of the substantia propria.

The conjunctiva, smooth and transparent in younger individuals, tends to become roughened and partially opaque in old age. Microscopically, the epithelium is thickened and shows a tendency to keratinization. The lamina propria may show areas of hyalinization and elastic tissue proliferation, and fragmentation into short segments. A non-inflammatory lesion, the "pinguecula" may appear as a yellow triangular mass on both sides of the cornea (Fig. 19-1). It may be the result of long-continued trauma by dust and wind, or of the sclerosis of blood vessels, or a combination of factors.

Obstruction of glands may lead to concretions in the conjunctiva on the inner side of the lids, which appear as yellowish-white patches. They consist of dammed-up secretion and cast-off epithelial cells and may be a site of calcification. The resulting hardened elevations can cause damage to the cornea.

Iris and Ciliary Body

The iris shows several types of change with age. The stroma throughout this structure often becomes hyalinized. The blood vessels frequently show hyalinization of the adventitia, making distinction of their boundaries difficult. Narrowing of the lumen, however, is seldom noted. Hyalinization of stroma between the

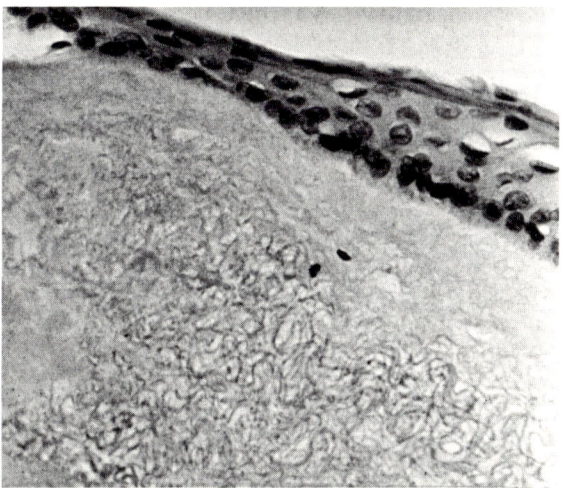

Fig. 19-1. A section through a pinguecula, a non-inflammatory lesion of the conjunctiva. A thickening of the epithelium and a modification of the underlying connective tissue are seen. This condition may be a result of continued trauma and, while often seen in older persons, is not necessarily associated with senility. This subject was 52 years of age. Hematoxylin and eosin. ×350.

pigmented epithelium and the sphincter muscle causes formation of a raised ring at the edge of the pupil, leading to the condition of "senile miosis" in which pupillary dilatation becomes difficult or impossible.

A "fading" of the iris often is observed in older persons. Pigment seems to diffuse peripherally, causing the appearance of patchy areas just outside the pupil, and making its border ill-defined.

It is thought that small masses of the diffusing irideal pigment, together with cellular debris including minute portions of lens capsule, are carried by the aqueous humor into the network of uveoscleral fibers, become trapped there, and help to set up that obstruction which leads to glaucoma.

There is a difference of opinion in the literature as to the role of age changes in the ciliary body in bringing about the loss of power of accommodation so frequently seen. The smooth muscle does show some atrophic change, but it has been said that there is no decrease in its power (Van der Hoeve and Flieringa, 1924) and that loss of accommodation is due entirely to changes in the lens (vide infra). Rones (1938), on the other hand, believed that the deposition of fat within and between smooth muscle fibers, as observed by him, must interfere with the functioning of the ciliary muscle.

The epithelium on the ciliary processes shows occasional fatty infiltration between its two layers and also hyperplasia of its outer, non-pigmented layer. Elevations of the hyperplastic epithelium project into the vitreous humor and may

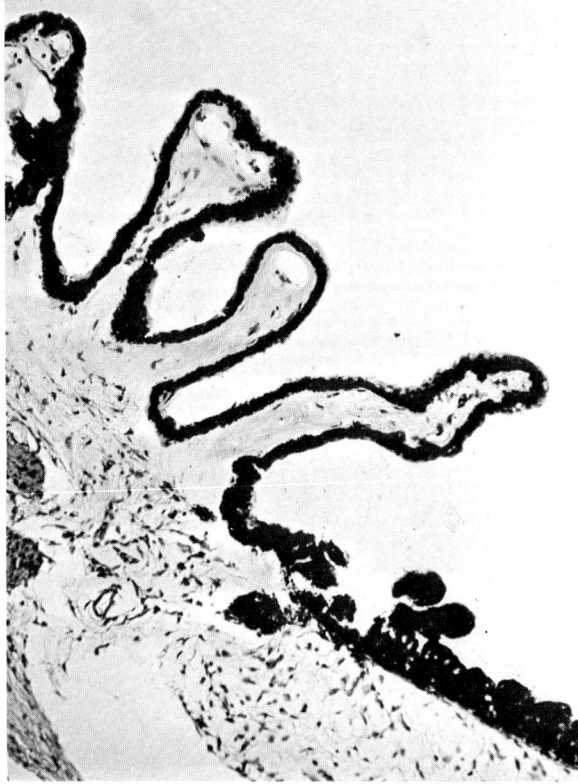

Fig. 19–2. Hyalinization of the stroma of the ciliary processes, a common finding in old age. The hyalinized connective tissue has a homogeneous aspect with relatively few nuclei. 75-year-old subject. Hematoxylin and eosin. ×170.

undergo fusion with subsequent cystic degeneration. This process may cause the iris to bulge forward or even lead to detachment of the retina. The stroma of the ciliary processes in old age often shows varying degrees of hyalinization (Figs. 19–2 and 19–3) and calcification.

The Lens

Cataract, a very common disease of the eye, is actually an especially marked phase of an aging change in the lens. Cataract, in some degrees is seen in over 90 per cent of persons by age 70 (Trevor-Roper, 1955). It may be either one of two types: (1) nuclear cataract in which the center of the lens is chiefly involved, and (2) cortical cataract in which the layers nearer to the surface of the lens are altered.

In nuclear cataract a diffuse central opacification eventually involves the entire nucleus. The fibers are compressed and finally undergo a coalescence to form one homogeneous nuclear mass.

In cortical cataract, an edematous condition causes the fibers to be forced apart and creates regions of opacity. Lens fibers break down into eosinophilic globules and mix with albuminous fluid. The nucleus is not involved at first but later it may sink to the lower portion of the capsule and even be seen floating about

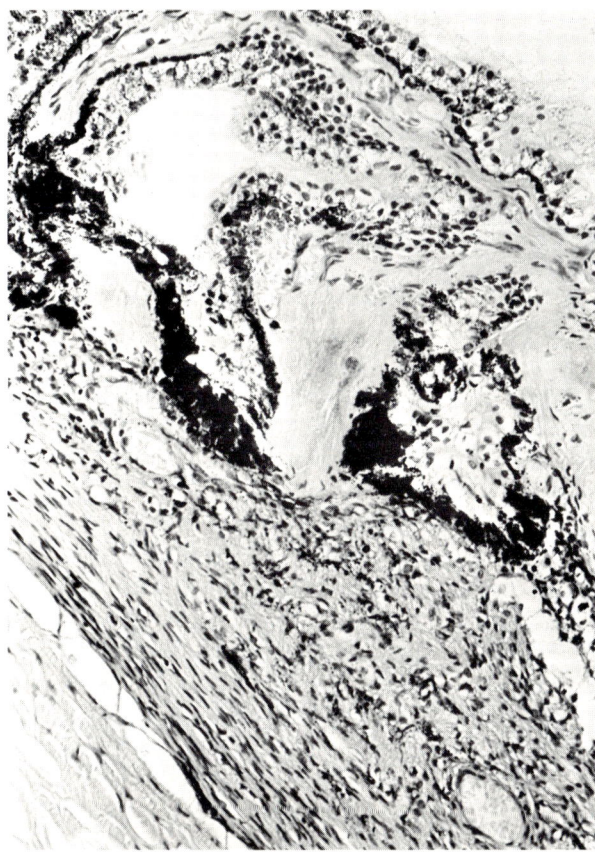

Fig. 19–3. Calcification of the stroma of the ciliary processes. The calcium deposits are deeply basophilic. 81-year-old subject. Hematoxylin and eosin. ×170.

when the head moves. Escape of fluid from the lens in this condition may bring about serious reactions such as uveitis.

The basic age change in the lens is an increasing sclerosis of the fibers of the central part or nucleus and an increasingly closer apposition of the peripheral fibers. The resulting loss of flexibility of the lens interferes with its assumption of a more convex form for close vision; hence presbyopia, or far-sightedness develops.

Friedenwald (1952) has described a relative increase of sodium as compared with potassium, and a loss of glutathione and water, in the lens.

Choroid Layer

Marked changes in the vessels occur in the choroid coat with advancing age. These consist of atherosclerosis or arteriolosclerosis or a combination of the two and can be seen in cases in which such changes are not present in the retina. Often blood vessels which are not sclerosed undergo a compensatory dilatation.

The capillaries of the choriocapillaris layer frequently show thickening of the walls, with partial or complete shutting off of the lumen. "Senile circumpapillary choroidal atrophy" is a condition in which these changes in arteries and capillaries are marked around the optic disc (Duke-Elder, 1940).

Nervous Layer: Retina

Age changes in the choroid coat are not only of intrinsic importance but may have great effect on the outer layers of the retina which are dependent for their nutrition on the blood vessels of the choroid coat. The pigmented epithelium, the outermost layer of the retina, often suffers and shows areas of depigmentation and degeneration. Senile macular degeneration, a degeneration of the outer retinal layers which leads to loss of central vision, seems to be due to sclerosis of arterioles of the choroid near the macular region of the retina, which has no direct blood supply of its own and is wholly dependent on the choroid.

A special phenomenon of aging is seen commonly in relation to Bruch's membrane, where "colloid bodies" develop. It is believed that an abnormal secretion by the pigment epithelium may lead to the formation of these round bodies which contain a homogeneous substance which in later stages may show concentric layers. They apparently seldom cause impairment of vision unless they fuse to form larger structures which may press upon the retina.

A cystic degeneration of the peripheral portions of the retina (Fig. 19-4) usually commences by the beginning of the sixth decade. Spaces develop in the outer plexiform layer and in the outer and inner nuclear layers and come to anastomose to form connecting channels. The immediate cause of the change may be vascular, but this is by no means certain. The important effect, besides visual impairment, is a tendency to detachment of the retina.

Ophthalmoscopic examination of the eyes of the elderly shows the arteries and arterioles to be paler, more narrow, and straighter. Some variation in diameter may be seen. The straightening appears to be due to fibrotic change of the wall and the variations in diameter to localized endothelial proliferation.

The Sense Organs   211

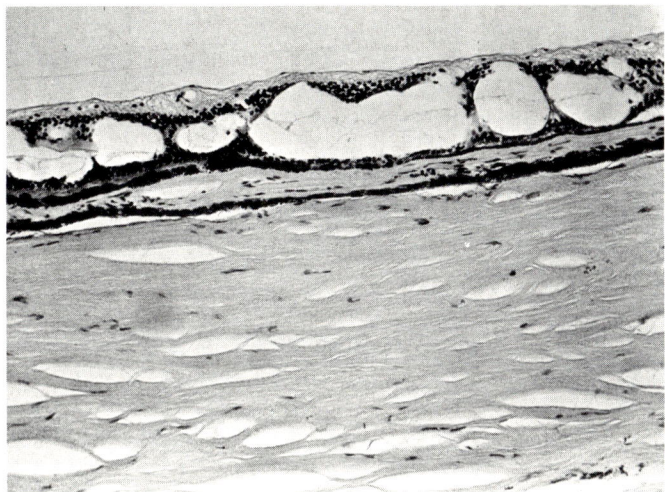

Fig. 19-4. Cystic degenerative change of the retina. The nuclear and plexiform layers are severely affected here. An 82-year-old female subject. Hematoxylin and eosin. ×110.

The distribution of terminal branchings of the central artery, it will be recalled, is to the inner layers of the retina where capillaries can be observed. Atheromatous degeneration of the vessel and its branches may proceed to vascular occlusion with an ischemia and necrosis of those inner layers. Cells of the nerve fiber, ganglion cell, inner plexiform, and inner nuclear layers may undergo degeneration. These layers then are invaded by phagocytes which remove the cellular debris. The final result of occlusion may then be an atrophy of the three or four inner layers of the retina.

Atheromatous change of the retinal vessels on the arterial side can also lead to changes in the veins. At places where the artery and vein cross each other, a hyperplasia of venous endothelium is particularly apt to develop. This may be of such an extent as to occlude much of the drainage to the central vein and thus cause serious retinal damage. Stasis of blood may lead to minute aneurysms which in turn rupture and cause hemorrhages which disturb the retinal architecture. The damage to the circulation may be remedied over a period of time by absorption of extravasated blood and formation of new channels, often with recanalization of the larger venous branches which have been blocked. The damage to vision, however, is seldom much offset by these changes.

The Optic Nerve

There has been little description of age changes in the optic nerve at least as uncomplicated by arterosclerotic change. It is believed that primary age change is relatively slight and not of great significance for function. Where some atrophy of axons occurs on other than a vascular basis, as for instance from changes in the ganglion cells of the retina, the degenerate fibers are replaced by fibrous and glial tissue.

When atrophy within the optic nerve is due to atheromatous change in the blood vessels supplying it, glial proliferation does not occur and after a time the glia itself undergoes degeneration, with formation of cavities which grow by fusion. The spread of this process leads to "Schnabel's cavernous atrophy."

Areas of atheromatous change in the central artery of the retina generally are scattered. They usually show an intact endothelium over foam cells, deposits of hyaline or calcified material, or hemorrhagic extravasations, hyperplasia of subendothelial tissue, and fibro-sclerotic changes of the outer coats.

According to Lyle (1957) the cavernous atrophy of Schnabel may also be brought about by pressure on the nerve due to an atheromatous plaque in the ophthalmic or internal carotid arteries, with the degeneration extending gradually in both directions from the region of pressure where it has been initiated.

"Sand bodies," or corpora arenacea (Figs. 19–5 and 19–6), usually lie in the sheath of the nerve. They grow by accretion of layers or laminae, always show calcification, and vary greatly in size.

Corpora amylacea, found in various parts of the nervous system with advancing age, are seen also in the intracranial part of the optic nerve and in the optic chiasma. They are spheroid or ovoid refractile bodies of homogeneous appearance and of variable size, generally $10-20\mu$ in diameter. It is believed that they develop from products of degeneration of glial tissue. While they generally occur among the nerve axons in the optic nerve, they may be seen in the sheath of the nerve or in the fibrous septa.

The Eyelids

In the eyelids of younger persons there are considerable amounts of adipose tissue and elastic connective tissue underlying the skin. With advancing age such tissue becomes diminished and the skin is thrown into minute wrinkles, the well-known "crows' feet." The skin itself shows the histological features which we have described as characterizing aging skin in general: reduction in number and prominence of rete pegs, thinning of epidermis, and patchy hyalinization of dermis with some degeneration of collagenous tissue. Here, as on other parts of the face, areas of keratosis may appear which are "pre-cancerous" lesions and in some cases develop into squamous cell carcinoma.

A lesion of the eyelids occurring frequently in older women is a formation of yellowish plaques along the inner angles of the lids. It is called Xanthelasma palpebrarum. The plaques contain large foam cells which are shown histochemically to contain phospholipids and cholesterol rather than the neutral fat characteristic of normal cutaneous adipose tissue (Lever, 1954).

Patients with primary hypercholesteremic xanthomatosis often show this condition of the lids, but persons with no rise in blood cholesterol also frequently present it.

The Ear

It is a well-established fact, recognized for many years, that a high tone hearing loss occurs with advancing age (Covell, 1952; Bunch, 1929; Bunch and Raiford, 1931; Ciocco, 1932).

The Sense Organs 213

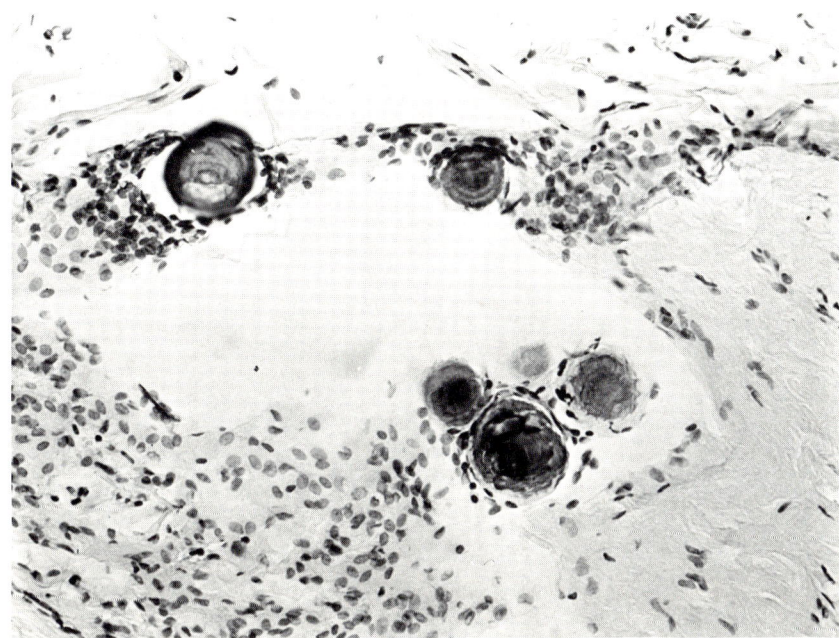

Fig. 19–5. Corpora arenacea or "sand bodies" in the sheath of the optic nerve. These may begin to appear at a relatively early age. Subject 50 years of age. Hematoxylin and eosin. ×170.

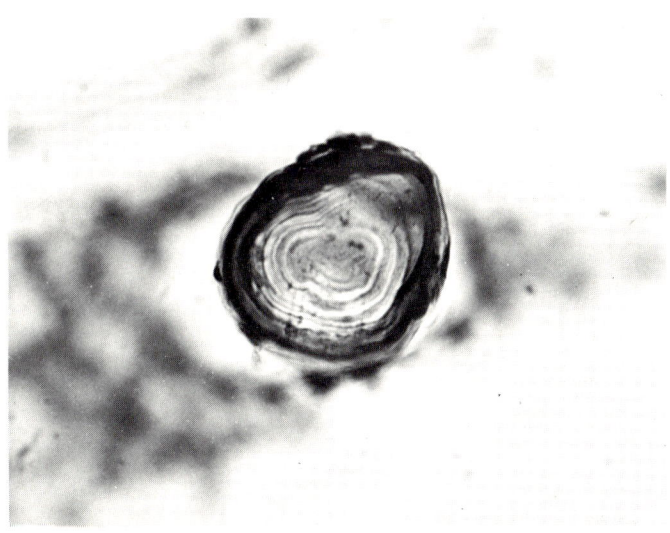

Fig. 19–6. Detail of a "sand body." The concentric lines of development are seen. Same subject and stain as in Fig. 19–5. ×425.

Anatomical changes of the inner ear were described by many early investigators. A recognition, however, of the great susceptibility of the delicate Organ of Corti and of the end organs related to equilibrium makes the descriptions of rather doubtful value.

Studies of hearing loss in large population samples were made by Webster, Hines, and Lichtenstein (1950) and earlier by Steinberg, Montgomery, and Gardner (1940). The results of these studies indicate that with increasing age there is, in general, an increased absolute hearing loss, dispersion, and skewness.

While the causes of increased hearing loss are varied and may concern any part of the hearing apparatus, the most common one after 45 years of age is the degeneration associated with the Organ of Corti, and the ganglion cells of the cochlear ganglion. This degeneration extends from the basal portion into the lower and, at times, the middle turn of the cochlea.

Crowe, Guild, and Polvogt (1934) in a study of 79 subjects, described degenerative changes in the basal turn and found that they were correlated with loss of hearing for high tones, on the basis of examinations made during the last illness of the subjects. The changes in histological structure, however, appear to have been similar in kind and degree to the "normal" changes accompanying aging.

Covell (1952) was of the opinion that human material generally is unsuitable for studies on the very delicate hair cells and their adjoining supporting cells because of the danger of producing artefacts, no matter how rapidly the temporal bone could be removed and immersed in fixative. He felt that it is probable that only material in which an anesthetized animal is perfused with the fixative is of real value in the study of details of structure of the Organ of Corti.

Rasmussen (1940) made a quantitative study of the cochlear and vestibular divisions of the VIIIth cranial nerve from human subjects who had a history of normal hearing. He found a decrease in number of fibers of about 7 per cent in the cochlear division in persons 44–60 years of age as compared with younger persons. There was a decrease also in the number of fibers in the vestibular division.

According to Covell (1952) there is considerable degeneration of spiral ganglion cells and their dendrites (the cochlear nerve fibers) of the basal turn in aging subjects who have shown marked hearing loss for high tones. For lesser degrees of hearing loss, the changes are more difficult to ascertain and danger of misinterpretation of artefactual change is greater.

Stria vascularis

The stria vascularis is thought to be the source of the endolymph which fills the cochlear duct and other parts of the membranous labyrinth. Its cells are chromophile and chromophobe. The chromophile cells show a "foot" anchored to a blood vessel, in a manner similar to that seen for astrocytes and probably indicating a secretory function for this type. The chromophobe cell may be simply a precursor of the chromophile, or a resting form.

According to von Fieandt and Saxen (1937) definite degenerative changes occur in the stria vascularis in advanced age in human subjects.

The Temporal Bone

Osteoporosis of the temporal bone was found to occur with advanced age (Nager, 1947) and is most extensive in the outer portion but eventually does involve the inner ones. Mayer (1930) found that the temporal bone shows microfractures which seem to be due to increased brittleness of the bone in the aged. Guild (1942) noted a decreased stainability of the bone matrix.

The question as to a specific relation between age and degree of pneumatization of the temporal bone appears to be a difficult one to answer. Apparently the process definitely slows as maturity is reached but many factors, such as inflammation, Eustachian tube occlusion, and fibrosis in the mucous membrane, may influence it.

External Auditory Canal

The hairs at the entrance of the external auditory canal become coarse in the aged male as they do in the nostrils and on the eyebrows. An increase in numbers of elastic connective tissue fibers in the fibrocartilaginous part of the canal was described by Bonatti (1950). Zorzoli (1948) found dilatation and irregularity of contour in the ceruminous glands of old men. The epithelium tends to change from a columnar to a low cuboidal or even squamous type. Diminution in size of the external aperture may be due to sagging of the skin caused by degeneration of cartilage in the wall of the canal.

The Tympanic Membrane

In older persons the substantia propria appears thinner, while sclerosed fibers make an appearance around the umbo and at the periphery (Zanzucchi, 1938).

Eustachian Tube

The Eustachian tube appears to share in some of the general atrophic changes of the pharynx although just what proportion of change is intrinsic and what is dependent on changes in neighboring structures is hard to elucidate. The cartilage of the wall does undergo some calcification in old age (Terracol, Corone, and Guerrier, 1949) but the extent of the process varies.

The Middle Ear

Although many details of the variations in weight, position, and joints of the ossicles have been described (von Békésy, 1949), and although the distribution of elastic tissue (Davies, 1948), and the structure of the stapedius and tensor tympani muscles (Byrne, 1938) have been studied in detail, nothing can be said as yet with any certainty concerning age changes in these components of the middle ear. This region, with its very specific topographical arrangements, by which a type of point-to-point comparison of portions of the bones, ligaments, and muscles could be made, would seem to offer a fruitful field for future investigation of anatomical changes with age.

REFERENCES

The Eye

Duke-Elder, Sir W. S., 1941. Textbook of Ophthalmology. Vol. III, St. Louis: C. V. Mosby Co. 3470 pp.
Forsius, H., 1954. Arcus senilis corneae; its clinical development and relationship to serum lipids, proteins and lipoproteins. Acta ophthal., Kbh., vol. 32 (42), pp. 1–78.
Friedenwald, J. S., 1952. The eye. *In* Cowdry's Problems of Aging, edited by A. I. Lansing, 3d Ed., Baltimore: Williams & Wilkins, pp. 239–359.
Lever, W. F., 1954. Histopathology of the Skin. London: Pitman Medical Publishers, p. 258.
Lyle, D. J., 1957. Arteriosclerotic Optic Atrophy. Proc. Roy. Soc. Med., vol. 50, pp. 937–942.
Rones, B., 1938. Senile changes and degenerations of the human eye. Amer. J. Ophthal., vol. 21, pp. 239–255.
Van Der Hoeve, J. and H. J. Flieringa, 1924. Accommodation. Brit. J. Ophthal., vol. 8, pp. 97–106.
Trevor-Roper, P. D., 1955. Ophthalmology, London: Lloyd-Luke, 468 pp.

The Ear

Békésy, G. V., 1949. The structure of the middle ear and the hearing of one's own voice by bone conduction. J. Acous. Soc. Am., vol. 21, pp. 217–232.
Bonatti, M., 1950. Ricerche sul componente elastico della cute del condotto uditivo esterno nelle varie eta nell'uomo. Oto-rino-laring. ital., vol. 18, pp. 439–454.
Bunch, C. C., 1929. Age variations in auditory acuity. Arch. Otolaryng., vol. 9, pp. 625–636.
———, 1931. Further observations on age variations in auditory acuity. Arch. Otolaryng., vol. 13, pp. 170–180.
——— and T. S. Raiford, 1931. Race and sex variations in auditory acuity. Arch. Otolaryng., vol. 13, pp. 423–434.
Byrne, J. G., 1938. Studies on the Physiology of the Middle Ear. London: H. K. Lewis and Co., Ltd.
Ciocco, A., 1932. Observations on hearing of 1,980 individuals: a biometric study. Laryngoscope, vol. 42, pp. 837–856.
Covell, W. P., 1952. The Ear. Chapter 10, pp. 260–276 *in* Cowdry's Problems of Aging, 3rd ed., A. I. Lansing, ed. Baltimore: Williams & Wilkins Co.
Crowe, S. J., S. R. Guild, and L. M. Polvogt, 1934. Observations on the pathology of high-tone deafness. Bull. Johns Hopkins Hosp., vol. 54, pp. 315–380.
Davies, D. V., 1948. A note on the articulations of the auditory ossicles and related structures. J. Laryng., vol. 62, pp. 533–535.
Fieandt, H. von and A. Saxen, 1937. Pathologie und Klinik der Altersschwerhörigkeit. Acta oto-laryng. Suppl., vol. 23, pp. 1–102.
Guild, S. R., 1942. The ear. *In* Problems of Aging, edited by E. V. Cowdry, 2d Ed., Baltimore: Williams & Wilkins.
Mayer, O., 1930. Über die Entstehung der Spontanfrakturen der Labyrinthkapsel und ihre Bedeutung für die Otosklerose. Ztschr. f. Hals-, Nasen-u. Ohrenh., vol. 26, pp. 261–279.
Nager, F. R., 1947. Pathology of the labyrinthine capsule and its clinical significance. *In* Nelson's Loose-Leaf Medicine of the Ear, edited by E. P. Fowler, New York: Thomas Nelson and Sons.
Rasmussen, A. F., 1940. Studies of the VIIIth cranial nerve of man. Laryngoscope, vol. 50, pp. 67–83.
Steinberg, J. C., H. C. Montgomery, and M. B. Gardner, 1940. Results of the World's Fair Hearing Tests, J. Acous. Soc. Am., vol. 12, pp. 290–301.

Terracol, J., A. Corone, and Y. Guerrier, 1949. La Trompe d'Eustache. Paris: Masson et Cie., Editeurs.
Webster, J. C., H. W. Himes, and M. Lichtenstein, 1950. San Diego County Fair Hearing Survey. J. Acous. Soc. Am., vol. 22, pp. 473–483.
Zanzucchi, D. G., 1938. Sulle modificazioni delle fibre elastiches della membrana timpanica in rapporto coll'eta. Arch. ital. di otol., vol. 50, pp. 203–224.
Zorzoli, G., 1948. Ricerche sulla morfologia delle ghiandole ceruminose dell'uomo nelle varie está. Arch. ital. anat. e embriol., vol. 53, pp. 117–129.

20
The Nervous System

The nervous system has a position of special importance in relation to considerations of the aging process because of two factors: (1) changes in this communicating and governing system will affect many organs in other systems and lead to disturbance of precisely balanced functions, and (2) the neurons, the units of the nervous system, differ from the majority of types of cells in that in postnatal life they generally do not reproduce, and thus as individual cells or groups of cells, they are not replaced if they degenerate or die before the death of the organism.

The Meninges

The coverings of the brain and cord, at least in man, tend to show certain alterations with age. The leptomeninges (arachnoid and pia mater) present thickenings and adhesions to one another and to the surface of the brain and cord. The arachnoid villi which project into the venous sinuses of the dura mater often show hypertrophy. Calcium deposits occur frequently in the meninges of old persons.

Brain

Gross Changes

Gross changes in the human brain include some atrophy, with narrowing of the cerebral gyri and widening and deepening of the sulci between them, accompanied by loss of weight of the brain as a whole. It is not possible to say how constant such changes are. The color of the gray matter is darker in many older brains than in younger ones.

Arteriosclerotic changes vary greatly in their intensity and extent in different parts of the brain in persons of advanced age; while such arterial change is the rule, there are many brains of old persons in which it is minimal. The question as to whether senile alterations occur in neurons and neuroglial tissue in such brains is, of course, of great importance and will be discussed later.

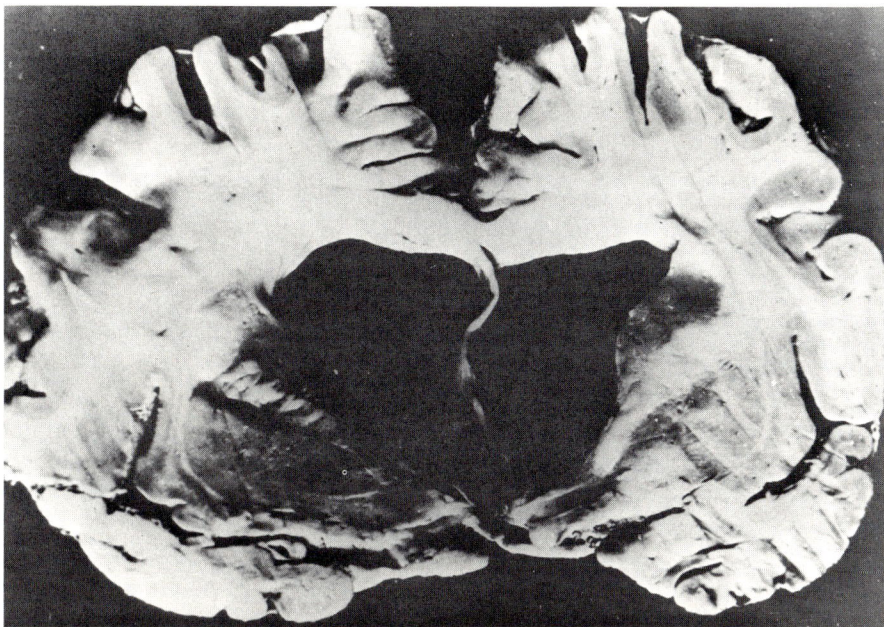

Fig. 20-1. Cross section of the brain of a subject with senile dementia. There is expansion of the lateral ventricles and atrophy of the cerebral cortex. Lesser degrees of cerebral atrophy are common in the "normal" brain at advanced ages. (After Andrew, 1956.)

In the senile cerebrum, the lateral ventricles of the hemispheres are often increased in size. In senile dementia such an enlargement may be marked (Fig. 20-1).

Microscopic Changes

The nerve cells are placed in the category of fixed "postmitotics"—cells that have undergone their last division and which are "fixed" in their condition of inability to divide for the remainder of their lives (Cowdry, 1952).

Why is it that such highly specialized cells should be far longer lived than many of the less specialized cells? While we cannot point to the specific organization within the nerve cell which makes such long life possible, we can see the need of continued existence of such cells throughout the life of an animal, for they are the cells that retain the individuality of the organism and receive and conserve the patterns of behavior in the developing animal or human being. The impulses flow along the pathway of the nerve cell processes and, when repeated over and over again, eventually make us what we are. Were these cells to undergo the great turnover that occurs in cells such as those, for instance, of the blood or epidermis, we probably could not retain the individuality of mind and spirit in the control of our material bodies.

Do nerve cells die a natural death or do they continue to exist as long as the organism of which they are a part? Since the latter part of the past century studies have been made which indicate that there are changes in these cells that tend to lead toward a decrease in their functional ability and eventually toward the death and disappearance of many of them before the death of the organism.

The study of the death of nerve cells in the aging nervous system and their decrease in number from a quantitative standpoint is not at all easy. Indeed, there have been conflicting results in relation to the numbers of nerve cells present at various stages in several parts of the nervous system. Nevertheless, evidence at the present time shows that this loss not only is present but apparently is considerable in many regions. In certain groups of cells, as in the Purkinje cells of the cerebellum (Fig. 20–2), it is seen readily.

Difficulties encountered in such quantitative studies are due to the problems of counting nerve cells from section to section and others are due to the great variation from individual to individual which necessarily exists in a biologic phenomenon such as the aging process. Naturally it is impossible to compare anatomically the nervous system of any one individual at different times throughout its life history, so that the factor of individual variation always enters into such studies.

Brody (1955, 1960) studied the cells in various regions of the human cerebral cortex at different ages. He finds a "patchy appearance," due to areas deficient in cells in the older cortex. These areas were scattered throughout the regions studied and were particularly evident in the superior temporal gyrus. Actual counts show a considerable decrease in number of cells—as high in some areas as one-third or more of the original number. The cells in general are spaced more widely in senile than in young adult brains. The picture, then, in many areas of the cortex is such that even low power microscopic examination permits a differentiation of sections from senile and young persons, the technical preparation of both having been as similar as possible.

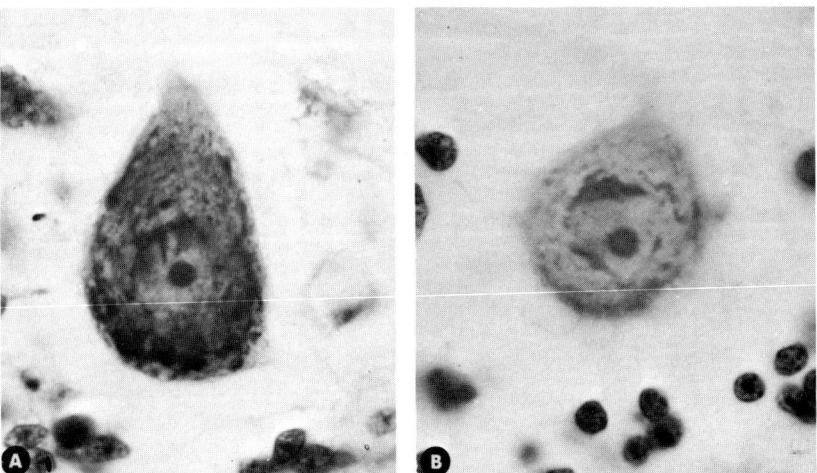

Fig. 20–2. Purkinje cells of human subjects. (A) 53-year-old man, showing an abundance of deeply staining Nissl substance. (B) 78-year-old woman, cell with well-rounded outline but scanty Nissl substance. Note that the small deeply staining nuclei appear similar in *A* and *B*. Cresyl violet. ×1495. (After Andrew, 1956.)

The observation, however, of the changes which lead to the loss of nerve cells is not as difficult. Cells undergoing these changes are present in all older individuals and indeed can be seen to arise in some cases at a fairly early age. If we use the comparison of a community, we may think of one in which there is no longer any production of new individuals. In such a community viewed from time to time there will be an increasingly large number of old individuals and individuals showing the infirmities of age. Gradually, various members of the community will begin to die out. The same is true in nerve cells—we see an increasing number of cells undergoing the changes of age and an increasing number that are reaching the stage of final degeneration.

A number of investigators have given attention to the question as to whether or not age changes are seen in the nerve cells in man. Indeed, several papers had described changes as of a "disintegrative" and "atrophic" nature before the beginning of the present century. Among these were Hodge (1894) and Robertson and Orr (1898). Andrew (1938) and Andrew and Cardwell (1940) have described degenerative changes of Purkinje cells and cells of the cerebral cortex in old age in the human subject. Truex (1940) has shown a very definite process of degeneration and loss of cells in the human trigeminal ganglion with advancing age.

Riese (1946) found considerable degeneration in a group of brains from 18 subjects ranging from 77 to 107 years of age, but these were "mental cases." Kuhlenbeck (1954) found degenerative changes throughout the cerebral cortex of the rat, the changes being very similar to those described for man.

Degenerative changes in the nerve cells of various parts of the human brain have been described by Oskar and Cecile Vogt (1946). These workers and their disciples in more recent years have contributed a large amount of new information in this field of which we shall have more to say later.

A term commonly used in referring to the change in size and shape which comes about in the nerve cell body with advancing age in many parts of the nervous system is "shrinkage." Frequently, this is manifested in a conspicuous way. Where the normal or "young form" of the cell is well known and characteristic, as for example in the case of the Purkinje cell of the cerebellum with its pear-shaped body, shrinkage often leads to a great change from the original appearance. In the pyramidal cells of the cerebral cortex the process of shrinkage also often leads to a clearly evident change in form.

In many kinds of nerve cells the question of whether there is a decrease in size during the process of aging is a difficult one to settle and few really quantitative studies have been done. The complicated shape of the cells, as in the case of the large motor cells of the ventral horn of the spinal cord, increases the difficulty. Among cells of both autonomic and sensory ganglia, decrease in size of the cells appears as they degenerate and often appears to be a more or less rapid process, or even to be brought about by the actions of the satellite neuroglia cells.

Change in size of the cells in the nervous system is evidenced by an increase in the size of the pericellular lymphatic space; the cell which was originally of such volume as to leave very little of such space about it, shrinks inward and leaves a space which on section appears as a large, white, vacant cavity surrounding the cell body. There is some question, of course, as to how much of this space is "real," that is, existing in the living condition and how much is due to fixation change;

but in many cases, such as in the inferior olive of the medulla oblongata and in the dentate nucleus of the cerebellum, the pericellular spaces in old individuals particularly are so large and constant as to leave little doubt of their reality. Also, the larger the space the more shrunken and degenerative the contained cell appears. Spaces appear to be present which have even been completely cleared of the detritus of the dead cell which once occupied them.

Decrease in size is not the only size change which may be a part of the degeneration of the nerve cell body. In cases where accumulation of relatively inert materials such as pigment and fat, of which we shall say more later, are an accompaniment of the degenerative process, they may cause an actual increase in the size of the cell body or of parts of it. Such is true in the case of the accumulation and increase in size of droplets of fat in certain ganglion cells. Such also seems to be the case in some instances of accumulation of pigment. This, of course, is not an increase in the actual living substance of the cell.

Nuclear differences between young and old animals, while they were not universally present, still are criteria of aging in a general sense (Andrew, 1936, 1938, 1939, 1956, 1959; Andrew and Cardwell, 1940). Thus, it is noted in the Purkinje cells and in the pyramidal cells of the cerebral cortex, and also in the ganglion cells from the trigeminal ganglion, that whereas the differentiation of the nucleus from the cytoplasm seen at low power is very sharp in the young animals, it becomes less so with advancing age and in senile animals the borderline is often difficult to discern at lesser magnifications. Studies of the tissues of human subjects showed a similar difference but the exceptions were more marked; for instance, in the cerebellum and to some extent in the cerebral cortex of individual aged human subjects we have seen many cells with very clear vesicular nuclei standing out from the general cytoplasm. Nevertheless, as a generalization concerning change of the nucleus with age, the statement of this decrease in differentiation between nucleus and cytoplasm holds true. It is due actually to two factors: (1) the general decrease in the amount of Nissl material in the cytoplasm, and (2) a tendency to a greater basophilia of the "nuclear sap" in the nucleus of the senile nerve cell. The nucleolus in the nucleus of the nerve cell generally is paler in the old animal and sometimes shows interesting changes, some of which may be degenerative or show signs of exhaustion and others which may indicate, as we shall mention later, reactive defensive changes.

Since many nerve cells are in the process of degeneration we may expect to find nuclei in various stages of actual death and dissolution. Dolley (1911) described in senile dogs cells which seem to lack a nucleolus or even an entire nucleus, such cells being extremely distorted and staining atypically. In regard to size, there are no consistent figures for the relationship of the nucleus to the cell body with advancing age. General observations would indicate that in many cells it is smaller in the old animal than in younger animals for any given group of cells. In others again, perhaps as a reactive process, it may be unusually large. The shape of the nucleus in the senile nerve cell is more frequently "atypical" in the sense of being elongated, somewhat irregular or angular in outline, and occasionally lobated. Again, there are many cells in which the well-rounded shape of the nucleus is retained.

In old age the position of the nucleus within the cell may be greatly altered. This seems to be due, however, not to any intrinsic tendency of the nucleus to shift position but rather to the accumulation of the inert materials in the cytoplasm. Thus, accumulations of pigment in the apical portion of the cell may force its nucleus further toward the basal portion while accumulations in the basal portion will force a centrally located nucleus apically. This is true especially in the case of the pyramidal cells. In the case of the accumulation of vacuoles of fat, as shown clearly in the trigeminal ganglion, the nucleus is gradually pushed off far to one side just as in the case of the development of a fat cell. Here, the nucleus is also distorted in shape and may be seen as a crescent-shaped body.

One of the cytoplasmic components long recognized to be of great importance in the nerve cell is the Nissl material. The changes which this material undergoes under various physiologic and pathologic conditions has made it serve as a good criterion of the condition of the neuron. As is well known, in conditions such as poliomyelitis and other diseases, the Nissl substance may disappear almost entirely from the body of the nerve cell. On section of the nerve, cutting the axons of a particular group of nerve cells, there often is clear-cut chromatolysis or dissolution of this material, a fact which makes it possible to follow the course of the groups of axons. We might well expect, then, that in a fundamental change such as the aging process, changes would occur also in the Nissl substance in many of the neurons of older animals. This is the type of change which does not seem to be specific for a particular group of cells but which appears to be widespread or even universal in the nervous system, both central and peripheral.

It is a striking change, since in many of the cells in youth, as in the large motor cells of the spinal cord, in the Purkinje cells, or in the pyramidal cells of the cerebral cortex, the Nissl flakes are so conspicuous. In senile animals, cells of this type may be almost completely lacking in Nissl substance. On the other hand, there will always be some which have retained a fairly large amount of Nissl material. According to our observations, Nissl flakes may be retained in approximately the same size and architectural pattern as in youth in some cells but their basophilia is greatly decreased as though there might be a gradual "fading out" (Fig. 20-3). This we have felt to be of particular interest since the publication of the most recent findings with the electron microscope on the fine structure of the neuron. Palay and Palade (1955) are of the opinion, and show convincing evidence for their view, that the Nissl substance is composed of at least two separate components, only one of which, a minute particulate component, is responsible for the basophilic property which it exhibits. It seems, then, that the architectural structure of the flakes could be maintained while the basophilic component, the extremely minute granules scattered in the second component, a system of fine canals, were decreased in amount.

The relationship between the amount of Nissl material and the presence of inert substances—for the Nissl material is not inert—is of interest. In general, the amount of Nissl material seems to decrease as the pigment increases. On the other hand, cells may be seen in which there is a large accumulation of pigment in one portion with well-preserved Nissl granules in the other portion. In such cells, however, Nissl granules are decreased in number.

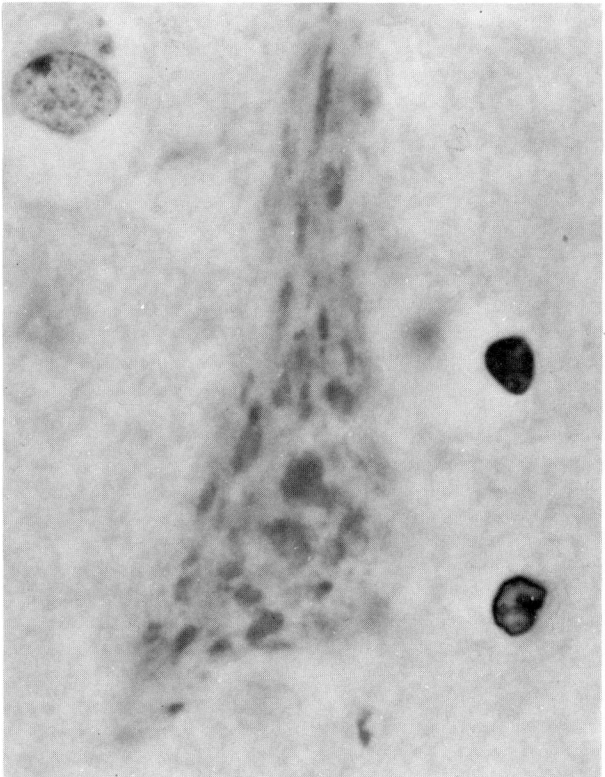

Fig. 20-3. Large pyramidal cell of cerebral cortex of a 78-year-old woman. In this particular cell the Nissl "bodies" have retained their size and shape but have less staining capacity, as though a partial loss of their basophilic component had occurred. Cresyl violet. ×1495.

The pigment, which may be considered a result of the aging process in nerve cells, is known as lipofuscin (Fig. 20-4). This lipofuscin ranges in color all the way from a relatively light golden brown through various shades of brown toward dark brown and indeed sometimes almost black in appearance. There are certain groups of nerve cells that contain a pigment throughout all or almost all of the postnatal life in man. This material is generally considered to be entirely distinct from the age pigment. It is thought to be melanin. It is found particularly in the cells of the substantia nigra and of the locus coeruleus.

Pigment accumulation in the nerve cells of dogs and hogs seems to be independent of sex and breed but is clearly associated with aging (Whiteford and Getty, 1966). The distribution of pigment, when it occurs, also varies with age. The youngest animals do not show pigment. The pattern in relatively old specimens varies in dissemination of the pigment granules. Perinuclear patterns are more common in 7-8-year-old animals while in 12-13-year-old ones polar aggregations predominate. In some nuclei, however, the pattern is fairly constant and

The Nervous System 225

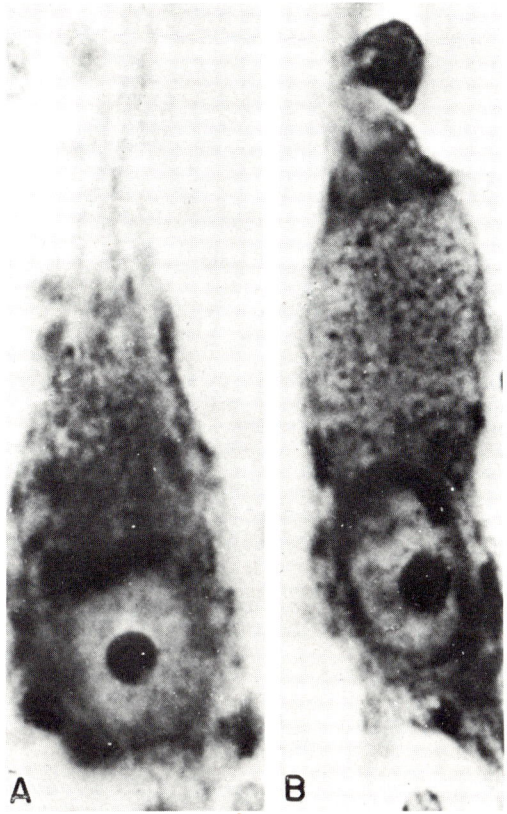

Fig. 20-4. Large pyramidal cells of layer V of the motor cortex. (A) 30-year-old subject. No pigment visible. (B) 81-year-old subject, showing a large mass of pigment which has altered the shape of the cell. Cresyl violet. ×2000. (Courtesy of Professor Oskar Vogt.)

without relation to age. Thus the cochlear ganglion shows the perinuclear pattern almost exclusively.

Histochemical studies on age pigment of the human autonomic ganglion cells tend to confirm earlier observations on the difference in the nature of the aging pigment from the pigment found in some localized groupings of cells. The pigment of senescence does not give the reactions that are typical of melanin, such as the bleaching by strong oxidizing agents, e.g., potassium permanganate, hydrogen peroxide, and sodium hydroxide. Rather, the age pigment is resistant to this type of change. A number of other histochemical tests also serve to distinguish the two pigments. It may be pointed out, however, that Levi (1946) states that lipofuscin, like all lipochrome pigments, consists of a nucleus of melanin surrounded by a lipoid component, so that some genetic relationship between these pigments is at least conceivable.

The best evidence at present is that lipofuscin is of the nature of a ceroid, similar to the insoluble pigment found in cirrhotic livers. While this pigment generally appears in the form of granules, it has been our experience to see it also in what appear to be vacuolar formations. In fact, the amount, appearance, and distribution of the pigment vary according to the specific type of nerve cell at any given age, at least in man.

The accumulation of lipofuscin pigment in nerve cells (Fig. 20-5), like many of the other changes which we describe as occurring with advancing age, are not to be thought of as specific changes. In other words, pigment accumulation can and probably does occur in a variety of conditions other than old age; in fact, such accumulation is one of the common "non-specific" changes in pathologic states. This does not make the gradual accumulation during the aging process any less important for the study of that process, particularly as the time of its appearance and the degree of its accumulation seem to be rather constant for special groups of nerve cells even when pathologic conditions do not complicate the picture.

Truex (1940, 1942) found accumulation of fat in the trigeminal ganglion both in lower mammals and in man. This may, of course, have to do with its later accumulation in the cytoplasm or such accumulation may be independent of the development of fat within the nucleus. In any event, nerve cells from a number of localities in man may show vacuoles of fatty material accumulating with advancing age. Thus, among the large pyramidal cells there are always some of these cells degenerating through fatty change in individuals of advanced age (Vogt, 1955). While the picture of fatty change, with the presence of large, frequently coalescing vacuoles, is quite different from that of the accumulation of the lipofuscin pigment,

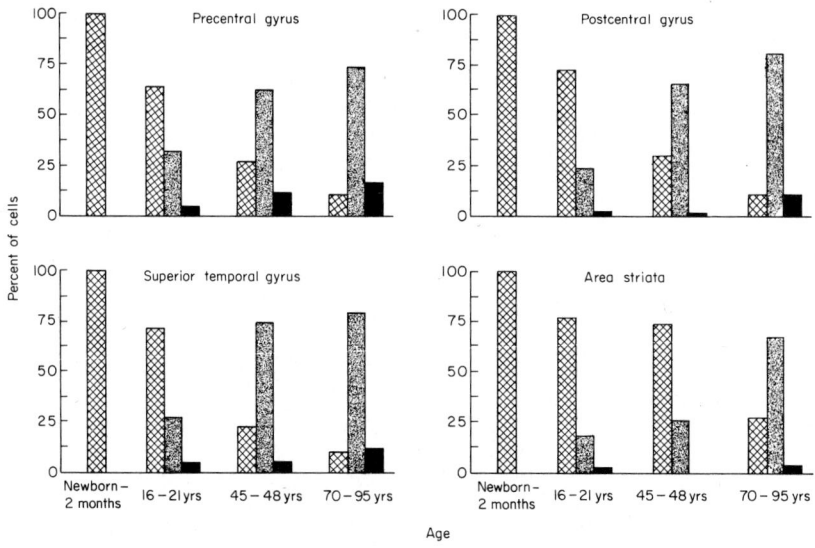

Fig. 20-5. Distribution of lipofuscin ("age pigment") in the cells of layer III in four regions of the cerebral cortex, from birth to 95 years. The block patterns represent three groups of cells: cross-hatched, those with no pigment; stippled, those with scattered pigment; and solid black, those with congregated pigments. (Reproduced with permission from Brody, H., 1960.)

there may well be some definite connection between the two which has not yet been elucidated. In fact, accumulation of fat in the nerve cell is one of the more conspicuous of the degenerative changes. According to Vogt and his disciples (see particularly Buttlar-Brentano, 1954), certain groups of cells in the hypothalamus never show true fatty degeneration but do show lipofuscin accumulation (Nucleus tuberomammilaris), although other groups in close proximity to them (Nucleus basalis) do show such degeneration as well as the pigment.

Many of the early investigators of nerve cells described fine fibrils coursing through the cytoplasm of the cell. They have been described in cells of many invertebrates, of vertebrates, and of man.

The type of neurofibrillary change known as "Alzheimer's disease" originally was supposed to be related only to presenile dementia but later indications are that it is fairly widespread in senile brains of human beings in general. It consists in an hypertrophy and complication of the pattern of the neurofibrillae, often with formation of whorls and baskets of these elements. Sosa (1952) made a study of the deposition of lipochrome (lipofuscin) pigment and correlated it with a peculiar change in the neurofibrillae which he described as neurofibrillar degeneration. This degenerative process involves a dissociation and agglutination of neurofibrils and ends with a total dissolution of these elements. The course of the change runs parallel with the accumulation of the lipofuscin pigment.

Andrew (1939a) found the Golgi apparatus in the Purkinje cells of young mice to occur as a large, well-developed reticular structure composed of coarse threads, while in senile animals it was represented by a mass of argentophilic granules of varying shapes and sizes, distributed irregularly in the cell. The susceptibility of the Golgi apparatus to change was shown in experimental studies by Sulkin and Kuntz (1948), in which the reticular structure was altered to a granular one as a result of long electrical stimulation and by induced hypertension. Not only did the apparatus become granular in many cells but in many others it seemed to vanish entirely.

Mitochondria of the nerve cell has been a somewhat vexing problem, partly because of the difficulty of the technical procedures to demonstrate them. In fact, for a number of years during the beginning of this century there was considerable doubt in the minds of many investigators as to whether or not the nerve cells in adult animals possessed such elements. Recently several papers have appeared concerning the mitochondria of the nerve cell in relation to age. Hess and Lansing (1954) studied the spinal ganglion cells in a group of guinea pigs ranging from newborn to senile. Their work was done with the electron microscope. They found some mitochondria which seemed to be "degenerating" at all ages examined, and concluded that the condition of the mitochondria does not serve as a good criterion of the stage of the aging process in nerve cells. Andrew (1955, 1956) studied the nerve cells from several regions of the central nervous system, including the Purkinje cells of the cerebellum, the pyramidal cells of the cerebral cortex, and the large motor cells of the ventral horn of the spinal cord in the mouse, using pedigreed animals of the C57 Black stock. His findings include a fairly consistent type of change from a predominance in the younger animals of the filamentous or long rod-like type of mitochondrion in these types of cells toward a short rod and granular type, frequently even spheroidal or of bead-like form, in the cells of the senile mice (Figs. 20-6A,B).

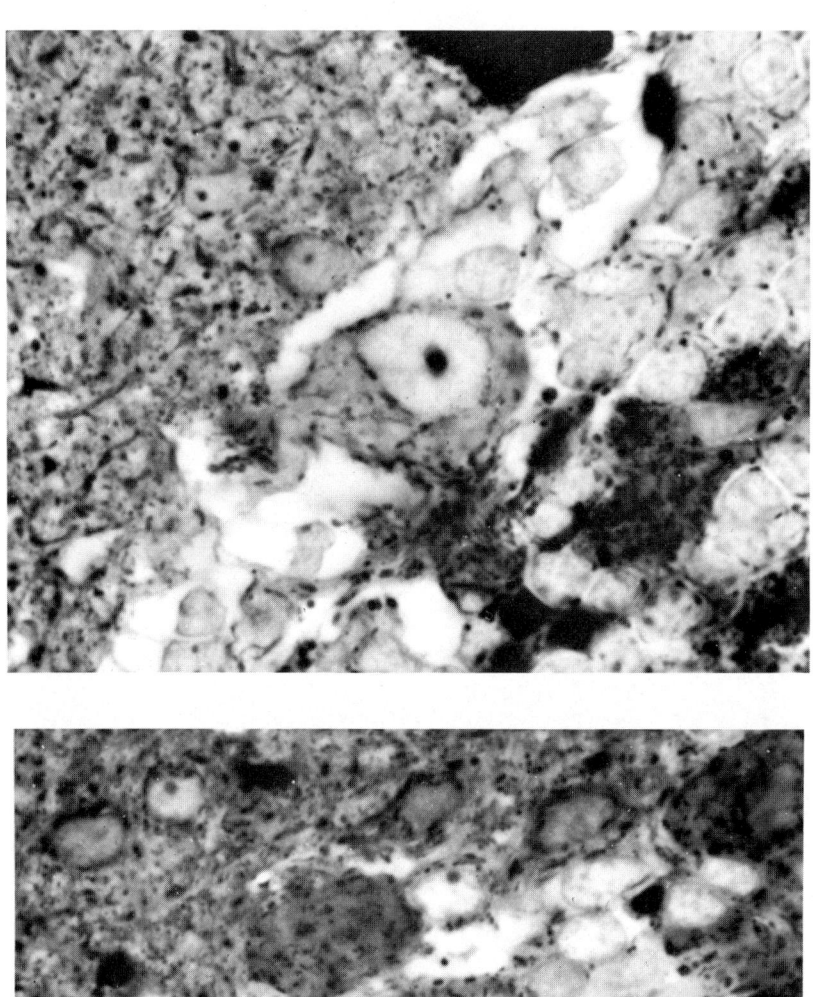

Fig. 20–6. Mitochondria in Purkinje cells. (*top*) Young male mouse, C57 Black, age 6–10 months. The mitochondria are generally long and filamentous but include some thicker rods. Molecular layer above and to left; granular layer with its mitochondria-rich glomeruli, below and to right. Regaud's fixation, acid fuchsin stain. ×1260. (*bottom*) Old female mouse, C57 Black, age 19 months. The mitochondria appear chiefly as short rods and granules, although some filaments occur. Preparation of specimen and orientation of layers are as in preceding figure. ×1260.

The Research of Oskar Vogt and His Disciples

Some of the most significant work on age changes in the human brain has been done by Oskar Vogt and his disciples. Their studies were carried out on a collection of some 500 brains in serial section, information on which was carefully documented at the Institute for Brain Research at Neustadt, Germany. As a result of studies on many areas of the brain in this large group of specimens, Vogt concluded that degenerative changes in the nerve cells are not, in general, due to changes in the vessels of the brain. In many instances neuronal age changes were observed in areas and in brains in which arteriosclerosis was lacking or at a minimum.

Work with this excellent collection also led to the finding by Buttlar-Brentano (1954) of certain singular characteristics of the nerve cells of the hypothalamus, in particular of the nucleus supraopticus and nucleus paraventricularis, in relation to age. In the first place, she found no evidence in this part of the brain of any destruction of cells. In the second place, no accumulation of age pigment occurs in these cells. Some changes, such as decrease of Nissl substance, probable diminution in size of cell body, and some degree of vacuolation, occur in a part of the cell population. But the outstanding feature in this part of the brain is the occurrence of certain phenomena which can be interpreted as reactive or defensive in nature. These changes are of several distinct kinds: (1) some of the cells increase tremendously in size to become what may well be called "giant cells," with a surface area 8 to 10 times that of these cells in younger individuals; (2) a number of cells become binucleate or multinucleate, thus presenting an increased area of nuclear membrane for exchange with the cytoplasm; (3) the nucleolus, which has for many years remained as a single element, undergoes division and up to 6 nucleoli can be found in a single nucleus; and (4) there is in some nuclei a hyperchromatosis which is not a pycnotic phenomenon but one apparently due to an increase in the amount of ribonucleotide. These "reactive" phenomena would appear to place the cells in which they occur in a better position to survive the metabolic vicissitudes of old age. It seems that such phenomena would be most likely to occur in parts of the nervous system that are vital to continued life of the whole organism and in which defensive measures against the onset of age-related degeneration could have evolved.

Spinal Cord

Studies specifically on the nerve cells and neuroglia of the spinal cord on human and lower animal material indicate a considerable degree of change, both quantitative and qualitative.

Sander (1900) examined 30 spinal cords from human autopsies ranging in age from 51 to 87 years. He found definite changes both in gray and white matter in all of his specimens. Neurons showed degeneration which he described as pigmentary in character. In his opinion, the Nissl bodies dissolve themselves into a yellow-brown pigment, first appearing near the axon hillock and gradually replacing the nucleus and other cell constituents. The cell is progressively converted into a bag of pigment, which may completely disappear, leaving an empty space. After the age of 85, scarcely a normal ganglion cell could be seen.

Harms (1924) studied spinal cords of senile dogs and found a definite preponderance of pigment, loss of regularity of the cell outline, and the loss of the sharp demarcation between the nucleus and cytoplasm found in younger animals. He also speaks of numerous "phagocytes" among the ganglion cells.

Fleugel (1927) studied cords of six persons over 80 years of age, dying either of cerebral hemorrhage or pneumonia, and found a definite increase in the glial cells in gray matter. Pigment accumulation was present in every neuron, the accumulation being more in the motor cells. Gliosis was present and had a predilection for the dorsal and lateral white columns.

Critchley (1931) in his review of the literature dealing with the neurology of old age said that neuroglial proliferation was constantly present and always in the area of myelin degeneration, e.g., the marginal area, perivascular areas, and the lateral and posterior columns. The ganglion cells showed intracellular pigment accumulation and a dark-staining and eccentrically placed nucleus with an angular and shrunken nucleolus.

Stankiewitsch, in 1934, found pigment deposition in the nerve cells of spinal cord and dorsal root ganglia, and vacuoles in the nucleoli of some of these cells in a woman aged 70 years.

Andrew (1941) studied the cytological changes in spinal cord and other parts of the nervous system of mice of different age groups. He found that the large ventral horn cell of the young animals (40-day age) had a clear nucleus and a deeply staining nucleolus in the center. Considerable amounts of Nissl substance as large blocks were present throughout the cytoplasm. No pigment was found in the neurons of this age group. In the middle-aged animals fewer cells had large amounts of Nissl substance, and some even showed vacuolization and pigmentation of the cytoplasm, the latter being present in less than 5 per cent of cells. The nuclei took a light basophilic staining in contrast to the clear appearance presented by those of the younger animals. In the senile animals (595 to 699-day age) the Nissl picture, though uniform in the individual members of this group, differed remarkably from that of the other groups. He found that the majority of the cells were hypochromatic, some almost completely lacking in the Nissl substance. But there were always some cells which retained Nissl material in normal amount, size, and architecture, although the basophilia was greatly decreased as though there were a gradual ' fading out." The nuclei of all of the cells were basophilic, nucleoli pale and often rather poorly differentiated from the nucleoplasm. Many neurons had vacuolation and shrinkage of the cytoplasm. Pigment accumulation was a prominent feature, almost all cells being affected, the majority having half or more of their volume filled with pigment.

Bailey (1953) studied spinal cords of 100 patients, representing each decade of life from first to tenth, and removed according to the method described by Kernohan in 1933. He observed that the pigment atrophy began in the third decade and became a constant finding by the fifth decade of life. He considered the sixth to tenth decades together as the findings were much the same in all cases. There appeared to be a progressive increase in the degree of pigment atrophy with advancing age. He is of the opinion that one does not find a single spinal cord after the fourth decade of life without pigment atrophy, and scarcely a normal ganglion cell after the seventh to eighth decade.

Bari and Andrew (1964) compared the spinal cords of two groups of rats, a "young adult" group of 162 to 240 days (6 animals), and a "senile" group of 900 to 1100 days (8 animals). The cords were in serial sections and those of the young and old rats were stained simultaneously in the same slide-trays, with 0.5 per cent cresylecht violet.

They found a definite loss of ventral horn neurons, ranging from 17.8 to 38.3 per cent in the senile cords (Fig. 20-7). The neurons present in the older cords showed decrease in amount and degree of basophilia of the Nissl substance, increased basophilia of the karyoplasm, and paler and vacuolated nucleoli. In addition, the cell bodies showed shrinkage and accumulation of pigment. Some of them seemed to be clearly in process of degeneration.

A statistically significant increase in amount of satellitosis (from 0.599 ± 0.025 satellite cells per unit area in the young to 0.686 ± 0.02 per unit area in the old) was found. On the other hand, a *decrease* in numbers of non-satellite neuroglial cells has been found.

There seemed to be no correlation in these cords between the vascularity in the ventral horns and the amount of loss of neurons. Senile plaques were found in the older group in both gray and white matter, more abundantly in the latter.

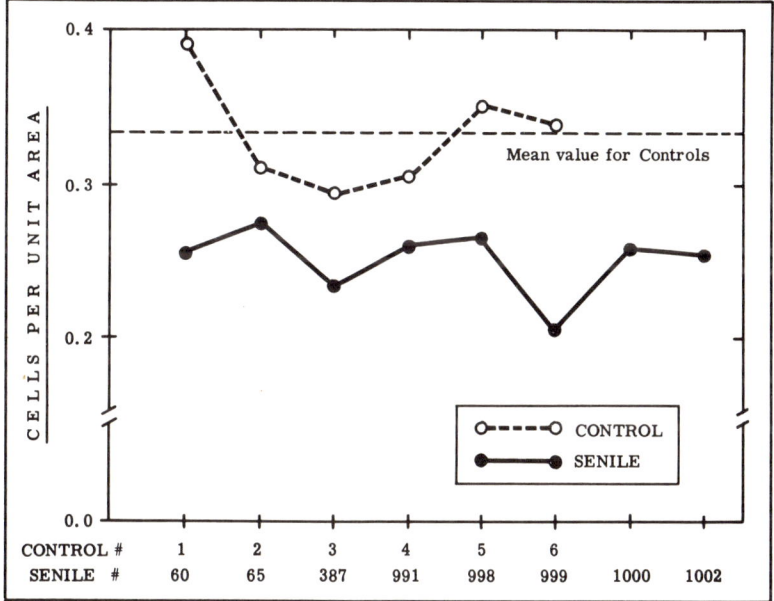

Fig. 20-7. Difference in the number of ventral horn neurons (showing nucleus) *per unit area* in the spinal cords of Wistar Institute Albino rats. A "unit area" is 6400 square μ and 40 unit areas were surveyed to obtain the counts. The upper, discontinuous line connects white circles each of which is a "control" or young rat, five being 162, one 240 days of age. The lower, continuous line connects dark circles, each of which is a senile rat. These are identified by number, as "60, 65, 387, etc." Three were 900 days, four 1000 days, and one 1100 days old. (After Bari and Andrew 1964.)

The meninges of the spinal cord of the old rats showed a general thickening with patchy calcification of the dura mater, some localized proliferation of cells of the arachnoid, also with calcific foci, and a slight thickening of the pia.

Dendrites and Axons

We turn now to a consideration of the processes of the nerve cell, the dendrites and the axons. It will be recalled that of these, the axon is the more specialized and that the dendrites partake more of the characteristics of the cell body. Nissl material and even Golgi apparatus extend partway into the dendrites. Mitochondria, on the other hand, are found both in the axon and dendrites.

In regard to quantitative studies on the numbers of axons, or actually the number of nerve fibers in given nerves or nerve roots, some of the same difficulties occur as in relation to the quantitative studies on the number of nerve cells at different ages. Corbin and Gardner (1937) have shown a decrease in the number of dorsal root fibers with advancing age after the third decade in man.

There appears to be an increasing complexity of form, with the appearance of new processes, i.e., short dendrites scattered over many parts of the surface of the cell, with increasing age. This, to begin with, is not a "senile" phenomenon, since it may begin fairly early in life. Conti (1946) has shown that in the sympathetic ganglion cells of the heart this complication in form occurs in early adult life, while after 50 years it seems to cease and in fact there appears to be a slight regression in volume of the cell and in richness of the processes in old age. A morphologic complication of the nerve cells is reported in the submucous and intermuscular plexus of Meissner and Auerbach in the gastro-intestinal tract of man, the process apparently continuing up through middle age, according to Borsello and Cavazzana (1917). Levi (1946) cites a number of Italian workers who have found increasing complexity with age in the processes of cells of the autonomic system. In some cases this increasing complexity seems to be correlated with the degree of accumulation of pigment. Thus, for example, in the ciliary ganglion where no pigment accumulates even at a late age, the number of prolongations remains constant even in advanced age.

In a number of mammals as in the ox, sheep, and the goat, many "polydendritic" cells are found in old specimens. Such cells have been particularly well demonstrated in the fascia dentata. There is a marked new formation of delicate short branches arising from all parts of the surface of the cell body. At least part of this formation of new processes seems to be concerned with an increase in the surface of the neuron, which may help to compensate for the poor supply of tissue fluid or blood in the senile animal. On the other hand, according to Truex (1940), cells with aberrant processes in the trigeminal ganglion probably represent degenerate neurons. Such atypical cells may be found here in varying numbers at all ages, even in normal ganglia; however, they are more common in old age. A curious phenomenon of end-bulb formation on the ends of certain of the dendritic processes also is seen in senility, according to Truex (1940). Andrew (1956) has described a thickening and vacuolization in dendrites of some Purkinje cells in aged mice (Fig. 20-8).

The Nervous System   233

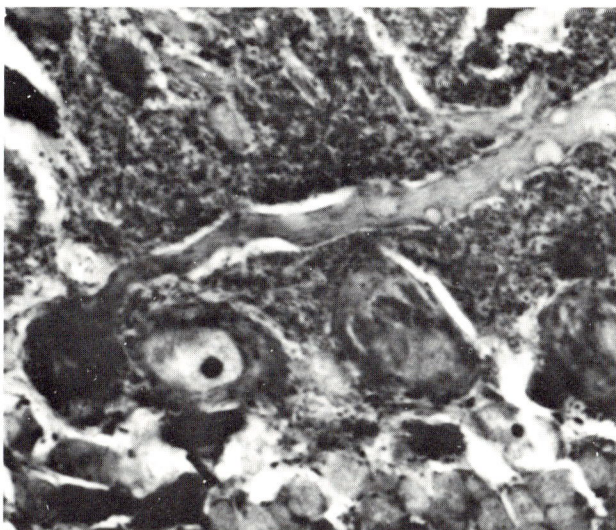

Fig. 20–8. Aberrant dendrite of a Purkinje cell in a senile male mouse, C57 Black strain, 634 days of age. The dendrite, proceeding from the small dark cell on the left, is seen to be hypertrophied and to contain numerous vacuoles. Regaud fixuation, acid fuchsin stain. ×1260.

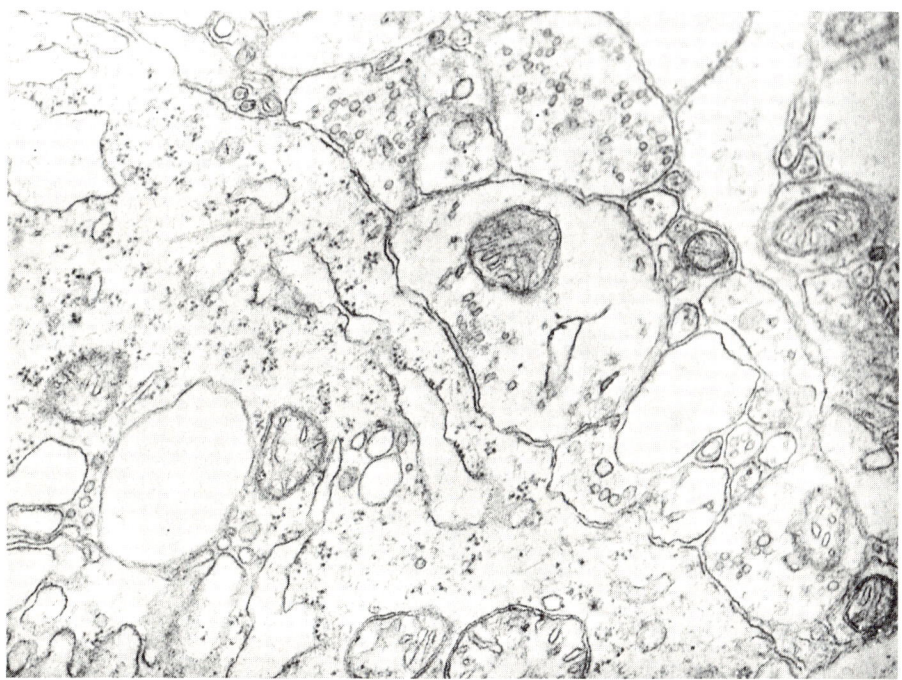

Fig. 20–9. Part of the body of a nerve cell in the motor cortex of the cerebrum of a senile rat. The cytoplasm is pale, with relatively few ribosomes. Cavities of the endoplasmic reticulum are dilated. ×29,200. (Courtesy of Dr. N. I. Artiukhina.)

Changes with Age in Fine Structure of Nerve Cells

Andrew (1961) found in nerve cells of the brains of senile mice a decrease in ribosomes and a tendency to invagination of the nuclear membrane. He found no apparent change in the numbers of "dark" or chromophilic neurons with advancing age, such cells occurring in about the same numbers in young adult and in senile mice. He found the lipofuscin pigment to show two components when studied with the electron microscope. One is a vacuolar one, the other an amorphous solid which varies in electron density.

Artiukhina (1966) reported on fine structure changes in the nerve cells and in the neuropil of motor and visual cortex of the rat, using animals of 18, 24, and 30 months. She found the neuropil in older animals presented large numbers of dilated glial processes which often seem to surround the cell bodies of the neurons. Her findings with the electron microscope agree with earlier light microscope studies on some of the common changes with age in nerve cells and with the statements by Andrew (1961) concerning fine structure, in relation to decrease in number of ribosomes (Fig. 20-9) and to the appearance of invaginations of the nuclear membrane (Fig. 20-10). In her classification of nerve cells in the brains of old rats, she points out that there are always some neurons which appear similar to those generally seen in young rats.

Artiukhina (1969) also has given evidence of formation of the aging pigment, or lipofuscin, in the lysosomes and Golgi vesicles of the nerve cells.

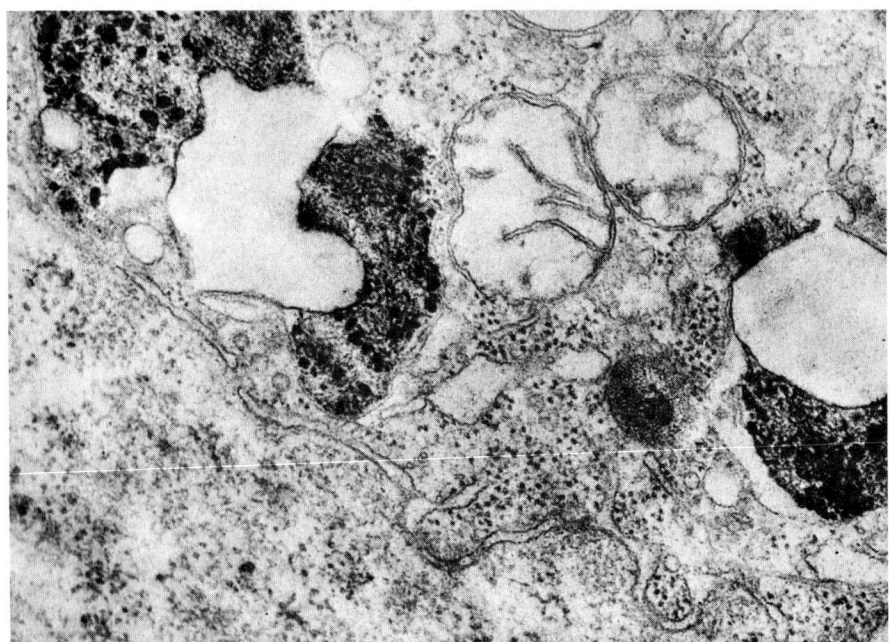

Fig. 20-10. Part of the body of a nerve cell in the motor cortex of the cerebrum of a senile rat. The nuclear membrane, along the lower portion of the field, shows many irregularities. There are several large dark pigment bodies, each with a vacuolar component, and smaller mitochondria. ×29,200. (Courtesy of Dr. N. I. Artiukhina.)

Division of Nuclei of Neurons in Senility

The amitotic division of nuclei in the Purkinje cells of the mouse appears to be a characteristic feature of the senile brain (Figs. 20-11 and 20-12). Studies on the frequency of such division in other parts of the nervous system and in other vertebrates should be investigated.

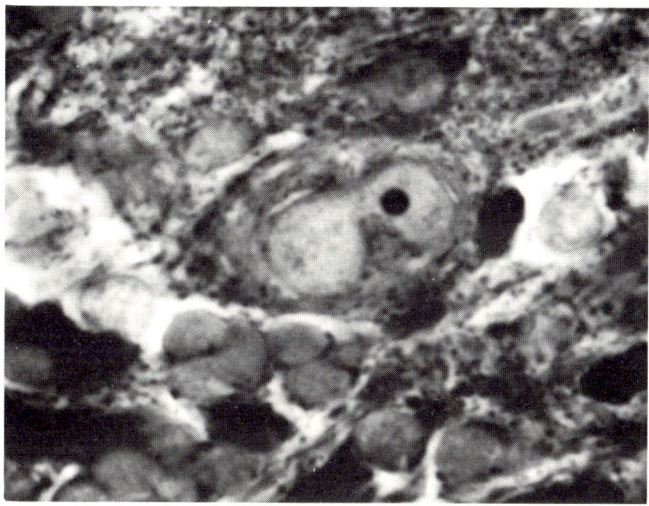

Fig. 20-11. Amitotic division of the nucleus in a Purkinje cell of a senile mouse, a 634-day-old C57 Black male. A narrow neck connects the major portions of the nucleus and a prominent nucleolus is seen in the part on the right. Nucleolar division precedes nuclear division. ×1800.

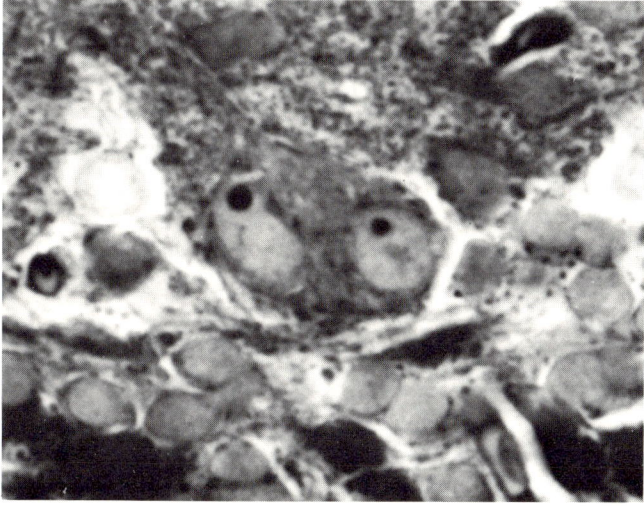

Fig. 20-12. Binucleate Purkinje cell in a senile mouse, a 634-day-old C57 Black male. A prominent dark nucleolus is seen in each nucleus. ×1800.

REFERENCES

Andrew, W., 1936. The Nissl substance in the Purkinje cell in the mouse and rat from youth to senility. Ztschr. f. Zellforsch., vol. 11, pp. 583–603.

——— 1937. The effects of fatigue due to muscular exercise on the Purkinje cells of the mouse, with special reference to the factor of age. Ztschr. f. Zellforsch. u. Mikr. Anat., vol. 27, pp. 534–554.

——— 1938. The Purkinje cell in man from birth to senility. Ztschr. f. Zellforsch., vol. 28, pp. 292–304.

——— 1939a. The Golgi apparatus in the nerve cells of the mouse from youth to senility. Amer. J. Anat., vol. 64, pp. 351–375.

——— 1939b. Neuronophagia in the brain of the mouse as a result of inanition, and in the normal aging process. J. Comp. Neurol., vol. 70, pp. 413–425.

——— 1956. Structural alterations with aging in the nervous system. J. Chronic Dis., vol. 3, no. 6, pp. 575–596.

——— 1959. The reality of age differences in nervous tissue. J. Geront., vol. 14, no. 3, pp. 259–267.

——— 1961. An electron microscope study of age changes in nerve cells, with particular reference to chromophilia and to accumulation of pigment in man and laboratory animals. J. Geront., vol. 16, no. 4, p. 388.

——— and E. S. Cardwell, 1940. Neuronophagia in the human cerebral cortex in senility and in pathologic conditions. Arch. Path., vol. 29, p. 400.

Artiukhina, N. I., 1966. Some age features of the cortical ultrastructure of the animals (in English). In Electron Microscopy. (Proc. VIth Int. Congr. for Electron Microsc., Kyoto, Japan), vol. II, pp. 475–476.

Borsello, P. L. and P. Cavazzana, 1917. Sull'aspetto microscopico dei plessi di Auerbach e di Meissner lungo i vari tratti dell'intestina dell'vomo in rapporto all'eta. Soc. ital. Biol. Sper., vol. 23, pp. 598–602.

Buttlar-Brentano, K., 1954. von: Zur Lebensgeschichte des Nucleus basalis, tuberomammillaris, supraopticus and paraventricularis unter normalen und pathogenen Bedingungen. J. f. Hirnforsch., vol. 1, p. 337.

Cammermeyer, J., 1963. Cytological manifestations of aging in rabbit and chinchilla brains. J. Geront., vol. 18, no. 1, pp. 41–54.

Conti, G., 1946. Sulle trasfomazion: Struttural; dei neuroni dei gangli simpatici del cuore dell'vomo in rappor to all'eta. Soc. ital. Biol. Sper., vol. 22, pp. 603–605.

Corbin, K. B. and E. D. Gardner, 1937. Decrease in number of myelinated fibers in human spinal roots with age. Anat. Rec., vol. 68, pp. 63–74.

Dolley, D. H., 1917. Further verification of functional size changes in nerve cell bodies by use of the polar planimeter. J. Comp. Neurol., vol. 27, pp. 299–324.

Hess, A., 1955. The fine structure of young and old spinal ganglia. Anat. Rec., vol. 123, pp. 399–424.

Hodge, C. F., 1894. Changes in ganglion cells from birth to senile death. Observations on man and honeybee. J. Physiol., vol. 17, p. 129.

Inukai, T., 1928. On the loss of Purkinje cells, with advancing age, from the cerebellar cortex of the Albino rat. J. Comp. Neurol., vol. 45, pp. 1–33.

Kuhlenbeck, H., 1954. Some histologic changes in the rat's brain and their relationship to comparable changes in the human brain. Confinia Neurologica, vol. 14, pp. 329–342.

Palay, S. L. and G. E. Palade, 1955. The fine structure of neurons. J. Biophys. and Biochem. Cytol., vol. 1, p. 69.

Robertson and Orr, 1898. The normal histology and pathology of the cortical nerve cells (especially in relation to insanity). J. Ment. Sc., vol. 44, p. 729.

Samorajski, T., J. R. Keefe, and J. M. Ordy, 1964. Intracellular localization of lipofuscin age pigments in the nervous system. J. Geront., vol. 19, no. 3, pp. 262–276.

Solyom, L., H. E. Enesco, and C. Beaulieu, 1967. The effect of RNA on learning and activity in old and young rats. J. Geront., vol. 22, no. 1, pp. 1–7.

Sosa, J. M., 1952. Aging of neurofibrils. J. Geront., vol. 7, pp. 191–195.

Spiegel, A., 1928. The degenerative changes in the cerebellar cortex in the course of the life cycle of C. cobaya. Zool. Anz., vol. 79, pp. 173–182.

Sulkin, N. M. and A. Kuntz, 1948. The golgi apparatus in autonomic ganglion cells and peripheral neuroglia and its modification following stimulation and induced hypertension. J. Neuropath. Exp. Neurol., vol. 7, p. 154.

Truex, R., 1940. Morphological alterations in the gasserian ganglion cells and their association with sensecence in man. Amer. J. Path., vol. 16, pp. 255–268.

——— and R. W. Zwemer, 1942. True fatty degeneration in sensory neurons of the aged. Arch. Neurol. and Psychiat., vol. 48, pp. 988–995.

Vogt, C. and O. Vogt, 1946. Aging of nerve cells. Nature, vol. 158, p. 304.

Wilcox, H. H., 1951. Changes accompanying aging in the brains of guinea pigs. J. Geront., vol. 6 (suppl.), no. 3, p. 168.

part IV

Conclusion

21
An Overview

Lessons Taught by a Comparative Study of Senescence

In the Preface to this book we used the phrase "the very probable value of the comparative approach." Have the descriptions of anatomical change in the chapters on animal and human structure justified such an estimate of the worth of comparative study?

The case of the Protozoa alone seems to us to have shown the need for such study even within that one great phylum of the single-celled animals. It was natural for many early investigators to think of single-celled organisms, reproducing over and over again simply by division of a mother-cell into two daughter cells, as potentially immortal and not as subject to senile change as man and other higher organisms. Yet if one makes the comparative study within the phylum, the group Suctoria will show a life cycle curiously similar to the ones with which we are more familiar. Here a process occurs within the "mother" by which a portion of its cytoplasm and nucleus go to constitute an embryo which then develops and leaves the body as a free-swimming ciliated larva. Soon it settles down and metamorphoses into a typical adult suctorian. The "mother" after a certain term of reproductive life undergoes changes of senescence which we have described in Chapter 1. Thus by comparative studies within the Protozoa some species have been found which do present the process that is the subject of our study. More than this, the very changes within the protozoan cell, the accumulation of pigment and the increasing irregularity of the nuclear membrane, are very similar to changes in various cells of the body of man and higher animals.

Comparative studies appear to show us, on the other hand, that considerable numbers of invertebrates may not be subject to a process of senescence and may, in this sense, be potentially immortal. Among such animals are the sea anemones (Chapter 2).

Study of the rotifers, among the smallest and lowest of the Metazoa, has contributed another example of senescence in a short-lived organism. The early investigators already had described such important changes as pigment accumulation with age in the digestive gland, the mastax and the gut of rotifers, and some possibly valid data on ultrastructural changes have been described.

The information obtained on senescence in arthropods offers a rich field for consideration. The very specialization of these animals in ways quite different from the chordates makes comparisons more difficult. There is, for example, nothing in the vertebrate body that can be compared, *sensu strictu*, with the fat body of an insect, an organ in which important changes occur with age. Where comparisons can be drawn, however, as in the question of a decrease in population of neurons in aging animals, arthropods and vertebrates, significant similarities are seen. Further work on the details of senescent changes in arthropods on which much genetic and ecological data are available, such as Drosophila and Apis, should be of much importance.

Knowledge of comparative longevity dispels the common concept that a relatively short life span is a necessary concomitant of the arthropod anatomical and physiological makeup. A lobster, if not as long-lived as man, is far more so than many mammals and may well live out its two score and ten! Anatomical studies on such long-lived arthropods through their entire span may prove highly rewarding in the future. It is an important question, surely, whether the determinant factors for length of life of the individual members of a given species are the same among invertebrates and vertebrates. We have seen (Chapters 6–8) that size and relative rate of metabolism seem to be related to longevity in the vertebrates. Whether any similar relation exists, say, in the Crustacea where great size differences occur, is open to discussion and to further study.

The comparative study of aging and of the anatomical changes that go with it in the vertebrate series shows us what seem, at least at first sight, to be surprising discrepancies as to how clearly defined the process of senescence is, not only from class to class but also among the members of an individual class. In the cold-blooded vertebrates (poikilotherms) the discrepancies are chiefly between the species which have indeterminate growth as compared with those in which determinate growth occurs. There is a serious question whether we can speak of an actual process of senescence in the fishes or reptiles in which growth seems to continue indefinitely; while in those in which it does terminate at a rather definite time in the life history, organ and tissue changes, similar to some extent to those described for warm-blooded animals (homoiotherms) can be seen in a number of instances.

Among the fishes, the migratory forms such as the salmon seem to offer a model for a biological deterioration which can be considered as a very rapid "aging" process. While this has many of the features of what we may consider as "natural" aging, in its later stages at least the organic deteriorative sequences become complicated by pathological change, including bacterial infection and fungoid infestation. Thus, death itself is difficult to attribute either to natural or pathological change alone but rather, as in many cases in our species, appears to be due to a combination of the two.

Perhaps there are better examples of natural senescence than rapid aging in the migratory salmon, yet it seems probable that where senescence can be clearly demonstrated in various forms of invertebrate and vertebrate life, the role of other organisms in the destruction of the senescent individual is important. Such an individual is, of course, likely to become the prey of the ever-present predators in the wild. But perhaps of equal or greater importance would be the enormous numbers of fungi, bacteria, and viruses. It seems likely that some of these organisms are already present in the tissues of higher organisms and ready, when opportunity presents itself in the vulnerable senescent, to spread beyond the limits of safety for that aging organism and to weaken functions already impaired.

Features of Aging Common to Many Animal Forms

The bewildering variety of animal life might in itself be a discouraging influence in a search for common features in the senescence which occurs in organisms of such diverse size, structure, way of life, and duration of life. Yet the efforts of scientific inquiry in many fields have seen their most significant triumphs in enabling one to arrive at concepts common to all varieties. In the past century the Cell Theory is the great example of such a triumph. In the present century perhaps the further knowledge of the basic similarity of the organelles within the cell, as studied with the high resolving power of the modern electron microscope, rivals the new understanding of the nucleic acids as a triumph in this acquisition of the concept of unity.

If senescence is, indeed, a fundamental process of life, the search for common features in it through the animal kingdom would seem to have a sound basis. This is not the same as the effort to create a theory to "explain" aging which has been done many times in the past and is being done by many at the present time. It is, rather, a hard look at "hard data" of a morphological nature and a comparison of such data from one phylum, one class, and one species to another.

Certainly many of the cells of higher organisms from old individuals look very much like those from young individuals. Such cells, however, while they are part of an older individual, are themselves of no greater "age" than similar cells in the body of a young individual.

Anatomical Changes and the Current Theories of Aging

The breadth of the problem of aging and perhaps the mystery that seems to be connected with a process whose effect is to clear away the members of each succeeding generation of organisms, continues to lead to attempts to explain the basic causes. Thus, general "theories" of aging will be found more or less in vogue at different periods in the development of knowledge. At the present time there are several such broad theories, each of which has its proponents and its opponents.

Briefly, certain of these theories, with special regard to the relation of anatomical evidence to their validity are: (1) the gene mutation theory; (2) the chromosomal theory; (3) the collagen deposit theory; (4) the wear and tear theory; and (5) the auto-immune theory.

The Gene Mutation Theory

A comparative study of average longevity shows it to be specific, to a large extent, for each type of animal. It seems that it is not a matter of phyla, classes, or other large divisions showing particular characteristics but rather of each *species* presenting its own longevity value. There appears, then, to be a *genetic* basis for lifespan. The problem may be more complicated, however, since within a given class, as the Mammalia, the question of size, in turn linked with that of metabolic rate, is important. The C57 Black mouse has a maximum longevity of somewhat over 2 years while man lives some 30 times as long, but *lifespan* itself may be dependent on other differences between these two mammalian organisms.

The gene mutation theory states that during the life of an animal or man spontaneous somatic mutations occur in cells of various tissues. The accumulation of cells with altered structure and function would lead, with advancing age, to malfunctions of the tissues and organs which eventually bring about the manifestations of senility.

A few years ago there was considerable excitement concerning the fact that radiation, an important mutagenic agent, shortens the lifespan of many organisms, and the concept was advanced that the effects of radiation might actually be an acceleration of the process of somatic mutation, and thus an acceleration of the process of senescence. A significant contribution to this concept was made by Curtis and Gebhard (1958) who employed chemical mutagens to bring about a shortening of the lifespan of experimental animals.

From the anatomical standpoint, it is not yet possible to gather *direct* evidence relating to the gene mutation theory. There is evidence, however, that the number of aberrations in chromosomes is in direct proportion to the number of mutations occurring. Curtis (1966) believes this evidence to be valid enough to indicate that when mitotic figures in regenerating liver show an increased number of chromosome abnormalities, as they do in older mice, rats, and dogs, an increased number of somatic mutations can be inferred to be present.

From the standpoint of the visible structure of body cells, it seems to us that there is considerable evidence of marked change, especially in the nuclei. We have called attention to these changes in individual cells, apparently randomly distributed, in the liver and salivary glands (Chapter 15) and in the kidney (Chapter 16). These descriptions relate to parenchymal epithelial cells but the cardiac muscle (Chapter 11) also shows such altered nuclei. It seems probable that further histological observation, a process often neglected or bypassed in recent years, would bring out examples of such cellular change in other tissues.

While not direct evidence for the gene mutation theory, these nuclear changes seen in cells not in mitosis, seem to add some further indirect evidence in favor of the theory.

The Chromosomal Theory

The increase in number of aberrations of chromosomes in mitotic figures of older animals might in itself, aside from gene mutations as such, mean the production of aberrant, malfunctioning cells. It is interesting that this increase appears to develop in pure-bred dogs, which have a lifespan about six times that of mice, at a rate only about one-sixth that seen in mice.

The functional relation of the genes to the chromosomes as such is still obscure in many ways. For example, present theory should indicate that the chromosomes in cardiac muscle nuclei would carry genes for determination of hair color—and a number of other genes which would seem to bear little or no relation to the function of the particular type of body cell. If this is true, chromosomal aberrations might involve both useful and relatively useless genes in particular groups of somatic cells. Yet it is difficult for this author to believe that a large part of each chromosome in a somatic cell has no rôle to play in the life of that cell.

The Collagen Deposit Theory

Various factors, including stress, are said to lead to deposit of collagen, both fibrous and amorphous. The gradual accumulation of this protein substance, which apparently is reabsorbed at a very slow rate or not at all, interferes with the supply of oxygen and nutriments to the cells.

Verzár (1957) has noted that changes in collagenous fibers are probably an important factor in senile change. His classic studies on heat shrinkage of the fibers of the rat tail-tendon have led to the conclusion that the rate of such shrinkage can be taken as a reliable measure of physiological aging of the organism.

The anatomical descriptions in the preceding chapters have shown many instances of fibrosis in senility. Among the more striking ones are those seen in some of the endocrine organs, in the reproductive organs, and in the muscles. Fibrosis often seems to occur at the expense of the essential functioning tissue of an organ; but whether it is a primary process inimical to such tissue, or whether it is a means of replacing such tissue which has degenerated, is not clear.

We have seen in this book that elastic as well as collagenous tissue increases in amount in old age in many parts of the body. It does not, however, seem to do so to the extent of the collagenous tissue nor to replace any considerable bulk of "functioning" tissue, such as glandular parenchyma or muscle.

The collagen deposit theory, it seems to us, is much harder to apply through the great range of animals other than vertebrates than is either the somatic mutation or chromosomal aberration one.

The Wear and Tear Theory

The wear and tear theory is an old and perhaps a natural one. The many examples of the products of man's ingenuity which have a "lifespan" or period of usefulness limited by the wear and tear upon them are always before us. The life of over-exertion or stress, with inadequate rest, is pointed out as a deterrent to longevity, and the phrase "a short life but a merry one" is familiar.

A more scientific statement of the idea relates longevity to metabolism. According to Rubner (1908) there is a direct proportion, such that each animal, and perhaps each cell, has a particular amount of metabolic energy the rate of expenditure of which will determine its length of life. Many examples of evidences for Rubner's hypothesis, and of later refinements of it, might be given. Recently Johnson et al. (1961) kept rats in a cold environment and found that a considerable increase in their metabolic rate was accompanied by a decrease in their lifespan. The classical studies by McCay et al. (1939), extending the lifespan of rats by decreasing the amount of food consumed, is an example of the reciprocal effect on longevity.

Anatomical support for the wear and tear concept is difficult to assess. It is interesting that we still use the term "wear and tear" pigment (Abnutzungspigment) for the yellow-green lipofuscin which appears in the nerve cells with advancing age (Andrew, 1952, 1956). If accumulation of pigment is indeed an evidence of wear and tear on the cells, there are many examples of it in various organs (thyroid gland, Clerc, 1912; pituitary, Cooper, 1925; adrenal, Meyers and

Charipper, 1956; Jayne, 1957). Increase in amounts of pigment in smooth, cardiac, and skeletal muscle also have been cited in this book (Chapter 11). New knowledge concerning the nature of age pigment has been furnished by Samorajski, Ordy, and Keefe (1965) (Fig. 21-1).

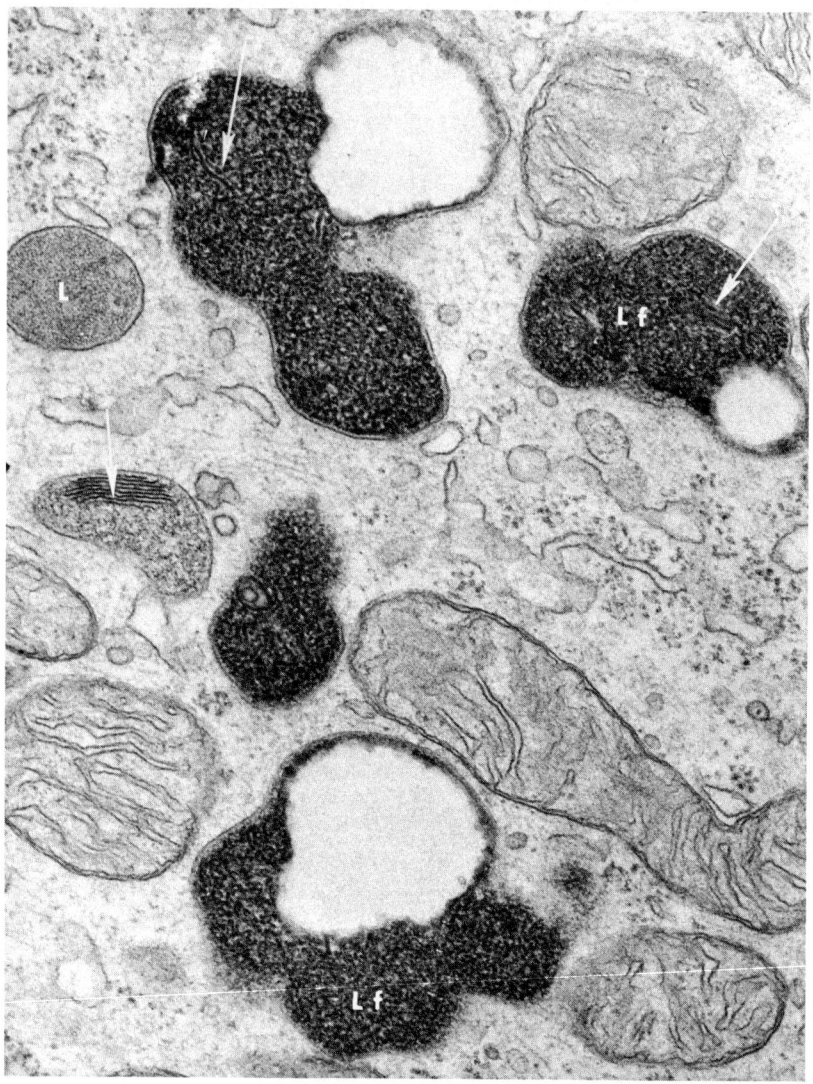

Fig. 21-1. Electron micrograph of a dorsal ganglion cell in a senile mouse, 24 months of age. Early stages in development of lipofuscin granules (Lf), as well as lysosomes (L) are seen. A single unit membrane surrounds each lysosome and each lipofuscin pigment complex. A lamellated residual body is shown in one lysosome and striated bands in two pigment complexes (arrows). Lead hydroxide. ×60,000. (Reprinted from Samorajski, Ordy and Keefe: The fine structure of lipofuscin age pigment in the nervous system of aged mice. J. Cell Biology, 1965, vol. 26, no. 3, pp. 779-795.)

The recent discovery (Andrew, 1969) of a marked increase of pigment in the liver in amphibians of advanced age further extends the significance of pigment accumulation during senescence.

In the invertebrates, including even the Protozoa, pigment accumulation has been described as a phenomenon of aging (Chapter 1).

There are examples in the animal kingdom of wearing down of hard organs, such as teeth, claws, and beaks, in old age, without adequate repair, but how to identify anatomical signs of wear and tear in tissues and cells is a question difficult to answer.

The Auto-Immune Theory

Every cell in an individual is immunologically similar to every other cell in that individual. If, however, cells within an organism undergo variation from its genetic pattern, as may occur through somatic mutation, then autoantigens may be produced by either the cytoplasm or the nuclei of such cells. These cells or the tissues of which they are a part then take on a character foreign to that of their surroundings and the body will react to them, much as it would to tissue transplanted to it from another individual. If complete rejection is not possible, other manifestations of the reaction will become evident.

Walford (1964) has discussed in detail the theory of the role of the auto-immune reaction in senescent change. He also has indicated that when organisms carry cells or tissues other than their own, their lifespan will be shortened.

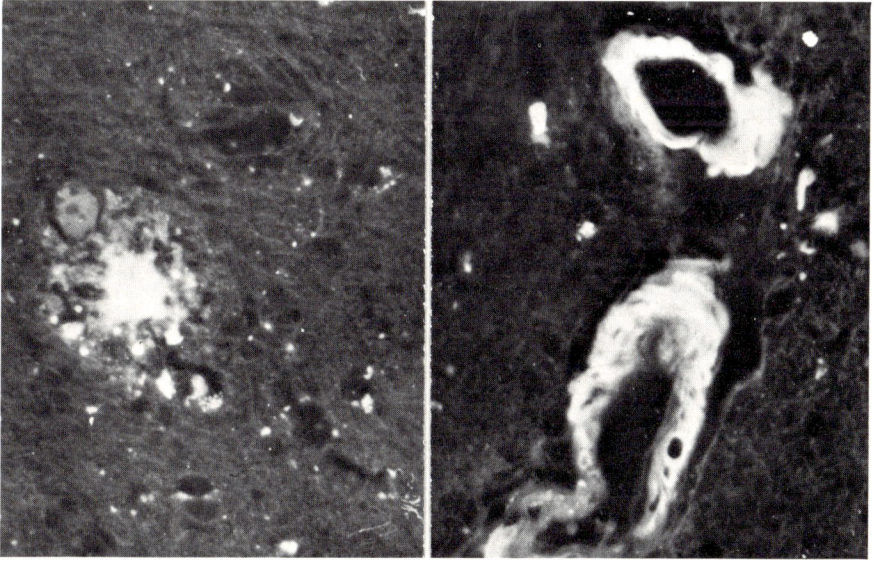

Fig. 21-2. Specific fluorescence in senile brain after staining with thioflavine-T. In this 85 year old human subject the brain parenchyma, blood vessels of the brain, and the spleen gave highly positive reactions. (A) Senile plaque. (B) Small artery of brain. (Reprinted from Walford, R. L. and J. R. Sjaarda. 1964, J. Geront., vol. 19, pp. 57–61.)

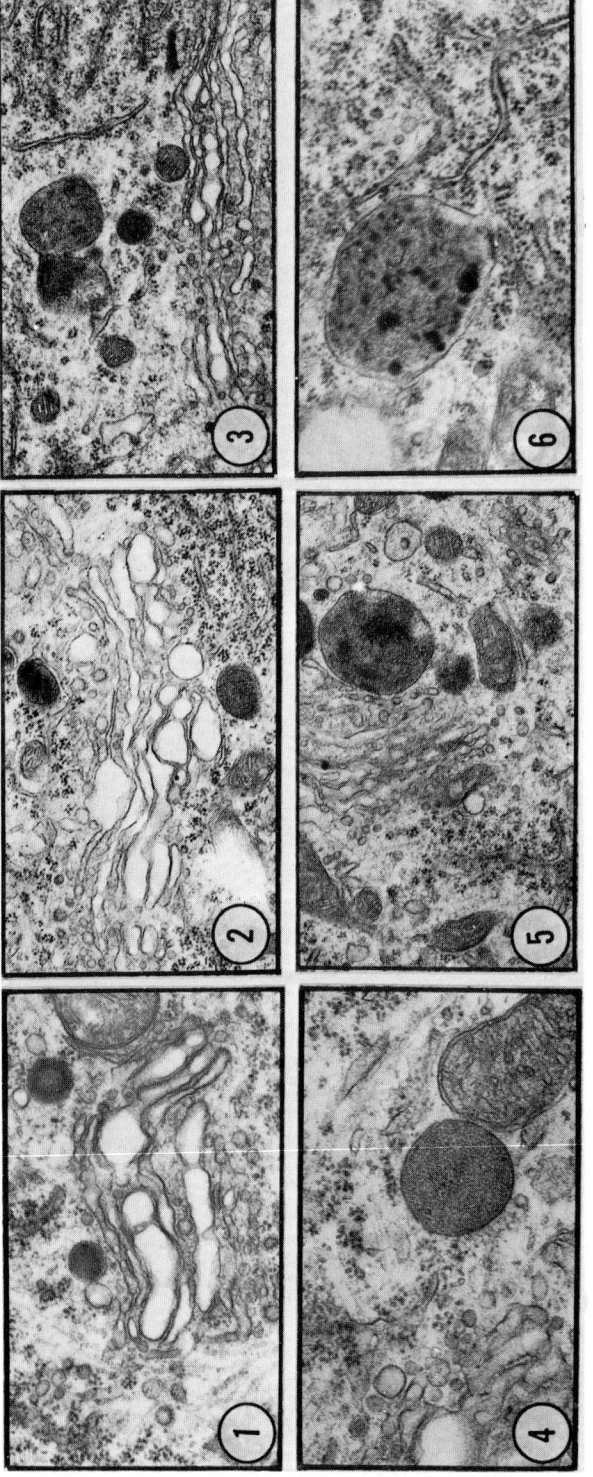

Fig. 21-3. Probable sequence of events in the deposition of pigment in the cytoplasm of nerve cells, according to Sekhon (1969). Stages 1–3 inclusive show granules of increasing size in the vicinity of the Golgi apparatus. In stages 4–6 the pigment masses are larger and an increasing heterogeneity is seen. Magnification of '1–2,' × 24,000; of '3,' × 16,800; '4,' 33,000; '5,' 16,800; and '6,' 24,000. (Courtesy of Dr. L. S. Sekhon.)

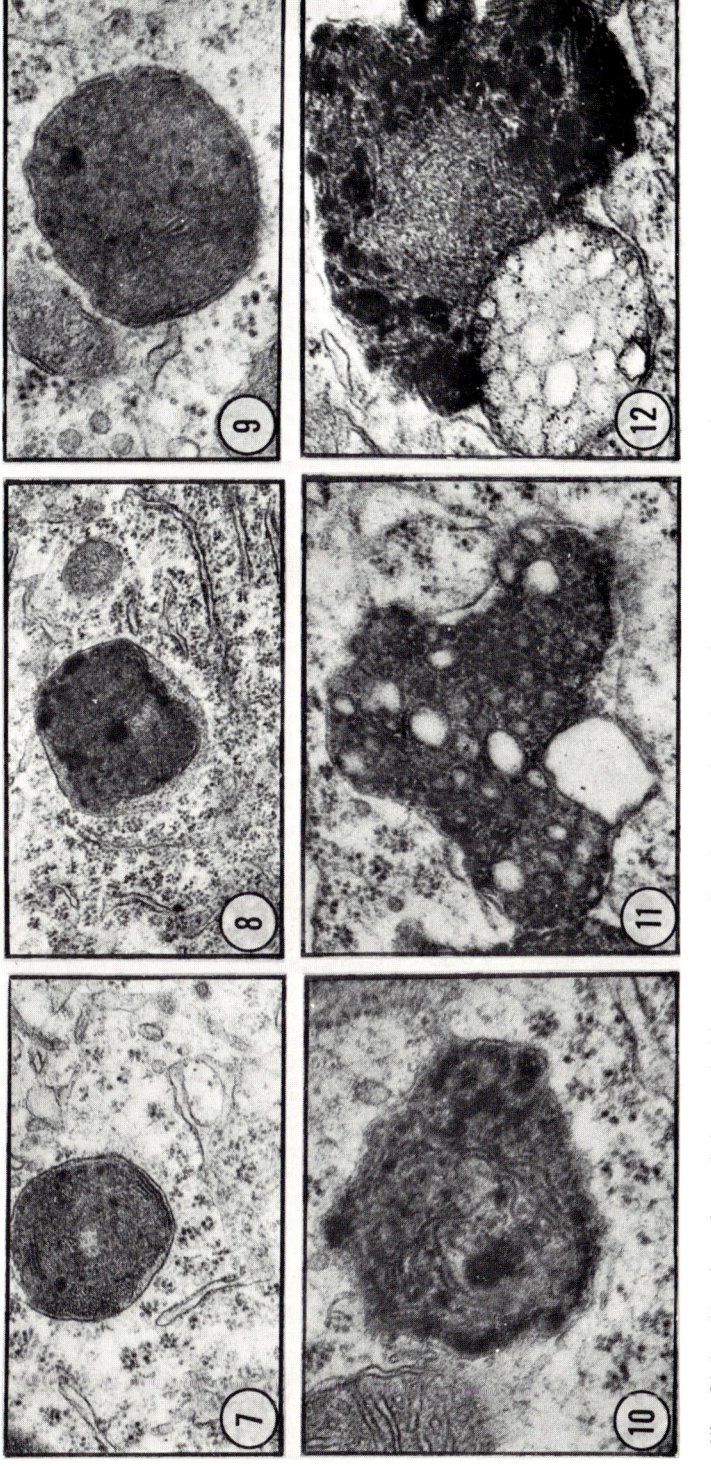

Fig. 21-4. Continuation of the probable sequence in pigment deposition. (Stages 7–12 inclusive). 10–12 show the development of marked heterogeneity with the vacuolar element becoming very prominent. Magnification of '7,' × 33,000; '8,' × 24,000; and of 9–12 inclusive, ×48,000. (Courtesy of Dr. L. S. Sekhon.)

249

An important feature of this theory is that it may provide the basis for explanations of both the "natural biological" process of aging and of a number of its common pathological manifestations (Fig. 21-2). There is a disease hierarchy known as the "auto-immune diseases" which include arthritis in its various forms, muscular dystrophy, probably diabetes mellitus, and even cancer. Characteristics of these auto-immune diseases include: (1) infiltration by amyloid, a homogeneous substance identified as a chondroitin sulfuric acid-protein complex; (2) accumulation of lymphocytes and plasma cells; (3) death of parenchymal cells, and (4) changes in the blood vessels with thickening and hyalinization of their walls and perivascular inflammation.

In human subjects it has been found that islets of Langerhans which have undergone hyalinization will specifically bind fluorescein-conjugated insulin as well as fluorescein-conjugated antihuman globulin (Berns et al., 1964). This was true in both diabetic and non-diabetic groups. The investigators discuss their findings in terms of the concept that maturity-onset diabetes may be an age-related autoimmune disease with an *altered insulin* as the responsible antigen. The hyaline (amyloid) deposits in the islets would seem to be an age-related phenomenon representing a deposit of an *insulin—insulin antibody complex*.

Sekhon (1969) has made an intensive study by electron microscopy of the deposition of pigment in various kinds of nerve cells. He has been able to construct a probable sequence in this process (Figs. 21-3 and 21-4). Similar studies on other tissues such as epithelium, cardiac, and skeletal muscle might be very profitable in ascertaining whether there is a mechanism common to different types of tissue.

All of the changes characteristic of auto-immune diseases have been described in the preceding chapters as occurring in the various organs of the body as types of senile change. It may be, then, that these are the visible reactions to alterations which have separated, as it were, certain cells from the general self-identity of the body.

A Closing Word

In attempting to relate the anatomical features of aging in man and animals to some current theories of the aging process, we emphasize here that we are considering the morphological evidence as only one type of contribution in the consideration of such theories. The rapidly growing knowledge of biochemistry, molecular biology, immunology, and other fields of basic science, as well as the enormous addition of material from new clinical experiences within recent years, must all be taken into careful consideration. The prospects for experimental gerontology, based on data from all of these areas, seem to us to present a vista of great hope and promise for the future. In the important studies which will be carried out in many laboratories, by many individual investigators, and in an increasingly large number of nations, the anatomical knowledge of the gross, microscopic, and ultramicroscopic changes of the aging process will, it is hoped, continue to serve an important rôle.

REFERENCES

Andrew, W., 1952. Cellular Changes with Age. A Monograph in American Lecture Series. Springfield, Illinois: Charles C Thomas, 72 pp.

———, 1956. Structural alterations with aging in the nervous system. J. Chronic Dis., vol. 3, no. 6, pp. 575–596.

——— 1968. The Fine Structural and Histochemical Changes in Aging. *In* The Biological Basis of Medicine, vol. 1, edited by E. E. Bittar and N. Bittar, New York: Academic Press, Chap. 13, pp. 461–492.

Berns, A. W., C. T. Owens, and H. T. Blumenthal, 1964. A Histo- and immunopathologic study of the vessels and islets of Langerhans of the pancreas in diabetes mellitus. J. of Gerontology, vol. 19, no. 2, pp. 179–189.

Clerc, Edouard. Die Schilddrüse im hohen Alter vom 50 Lebensjahr an aus der norddeutschen Ebene und Küstengegend sowie aus Bern. Frankfurt. Ztschr. f. Path., 1912, 10, 1–19.

Comfort, A., 1964. Ageing: The Biology of Senescence. London: Rutledge and Kegan Paul, 365 pp.

Curtis, H. J., 1966. Biological Mechanisms of Aging. Springfield, Illinois: Charles C Thomas.

——— and K. L. Gebhard, 1958. Comparison of life-shortening effects of toxic and radiation stresses. Radiat. Res., vol. 9, p.104.

Cooper, Eugenia, R. A., 1925. The Histology of the More Important Human Endocrine Organs at Various Ages. London: Oxford University Press.

Jayne, E. P., 1963. A histo-cytologic study of the adrenal cortex in mice as influenced by Strain, sex and age. J. Geront., vol. 18, no. 3, pp. 227–234.

Johnson, H. D., L. D. Kintner, and H. H. Kibler, 1963. Effects of 48°F. (8.9°C.) and 83°F. (28.4°C.) on longevity and pathology of male rats. J. Geront., vol. 18, pp. 29–36.

McCay, G. M., L. A. Maynard, H. Sperling, and L. L. Barnes, 1939. Chemical and pathological changes in aging and after retarded growth. J. Nutr., vol. 18, pp. 1–13.

Meyers, M. W. and H. A. Charipper, 1956. A histological and cytological study of the adrenal gland of the Golden Hamster (*Cricetus auratus*) in relation to age. Anat. Rec., vol. 124, p. 1.

Rubner, N., 1908. Probleme des Wachstums und der Lebensdauer. Mitt. d. Gesellsch. f. inn. Med. u. Kinderh. in Wien, vol. 7, pp. 58–81.

Samorajski, T., J. M. Ordy, and J. R. Keefe, 1965. The fine structure of lipofuscin age pigment in the nervous system of aged mice. J. Cell. Biol., vol. 26, no. 3, pp. 779–795.

Sekhon, S. S., J. M. Andrews, and D. S. Maxwell, 1969. Accumulation and development of lipofuscin pigment in the aging central nervous system of the mouse. J. Cell Biol., vol. 43, p. 127a (Abstract).

Verzár, F., 1957. The Aging of Connective Tissue. Gerontologia, vol. 1, pp. 363–378.

Verzár, F. and K. Huber, 1958. Thermic-contraction of single tendon fibers from animals of different age after treatment with formaldehyde, urethane, glycerol, acetic acid, and other substances. Gerontologia, vol. 2, pp. 81–103.

Walford, R. L., 1964. The immunologic theory of aging. Gerontologist, vol. 4, pp. 195–197.

Index

Adipose tissue, *see* Fat
Amphibians, *see* Vertebrates
Arteriosclerosis
  of brain, 218, 229
  in eye, 210
  of heart, 115
  of spleen, 139
Atherosclerosis
  in eye, 210
  of heart, 113
  in man, 68, 112, 113, 117, 118, 210
  in rats, 101, 102, 105, 112, 113
  of vessel walls, 112, 117
Atrophy
  of brain, 218–229
  of eye, 208, 210, 211
  of kidney, 172
  of muscles, 69, 72, 101–108
  of nervous system, 218, 231
  of ovaries, 187, 188
  of rat kidney, 176
  of small intestine, 157–159
  of spinal cord, 231
  of spleen, 139
  of thyroid, 195
  of uterus, 187, 188
  of vagina, 187
Auto-immune theory of aging, 247–250

Birds, *see* Vertebrates
Blood (vascular system), 110–120
  arteries, 110–113
    atherosclerosis of, 112
    calcification of, 111, 112
    of heart, 113
    of lung, 114
    thickening of, 111
  blood-forming tissue, 118
    in rat, 118
  cardiac nodes, fatty infiltration of, 117
  heart, 114–117
    adhesions, 116
    calcification, 116
    collagen infiltration, 115
    corrugation, 115
    endocardial thickening, 115
    fat spots, 115
    hypertrophy, 115
    mast cells, 74
    milk spots, 117
    opacities, 115, 117
    plaque formation, 115–117
    sclerotic changes, 115
  linear regression of hemoglobin content, 120
  quadratic regression of hematocrit and mean corpuscular volume, 120
  veins, 114
  vessel wall, 117, 118
    atherosclerosis, 118
    fraying, 118
    lesions, 118
Bone, *see* Skeleton
Brain, *see* Nervous system

Calcification
  of arteries, 111, 112
  of bone, 98
  of cartilage, 94
  in eye, 209, 212
  in heart, 116
  of pineal gland, 204
  of small intestine, in mouse, 58
Cancer, 68
  of cervix, 188
  of mammary glands, 188
  of uterus, 188
  See also Carcinoma
Carcinoma, 68
  in eye, 212
  of mouth, 150
  of stomach, 157
  See also Cancer
Cartilage, *see* Skeleton
Chromosomal theory of aging, 244
Collagen deposit theory of aging, 245
Cornea, *see* Eye
Cysts, *see* Nodules

Digestive system, 148–170
  epithelium, 159
  esophagus, 157
  large intestine, 101, 159
    diverticulosis in, 101, 159
    fibrosis of, 159
  liver, 161–170
    aging pigment, 168
    fibrosis in, 161, 162
    hyperplasia of, 164
    mitochondrial changes, 168–170
  mouth, 148–157
    carcinoma, 150
    dryness, 148
  pancreas, 159–161
    fat in, 160
    metaplasia of, 159
    in rats, 159–161
  pharynx, 148–157
    fat in, 148, 150, 152, 153, 156
    fibrosis of, 155
    honeycombing, 148
    oncocytes, 148, 153
    in rat, 148–150, 153, 155
    salivary glands, 148–157
    sebaceous glands in, 152, 154
  small intestine, 157–159
    amyloid, in mouse, 158
    atrophy, 157
    calcification, in mouse, 158
  stomach, 157
    carcinoma, 157
    gastritis, 157
    leucocytic infiltration, 157
    polyps, 157

Ear, 212–215
  external auditory canal, 215
    cartilage degeneration, 215
    coarse hairs, 215
    dilatation, 215
    fibrosis, 215
    sagging, 215
  hearing loss, 212, 214
  middle ear, 215
  stria vascularis, 214
  temporal bone, 215
    fibrosis, 215
    inflammation, 215
    occlusion, of Eustachian tube, 215
    osteoporosis, 215
Endocrine glands, 194–204
  adrenals, 199–204
    in hamster, 201–203
    in man, 203
    in mouse, 201–203
    in rat, 199–201
    in wild animals, 203
  pineal, 204
    calcification, 204
    in rats, 204
  pituitary, 198, 199
    of birds, 199
    in hamster, 199
    in mouse, 199
    in rat, 198
    in salmon, 199
  thyroid, 194–198
    in birds, 197, 198
    fat in, 194
    fibrosis of, 194, 204
    of golden hamster, 197
    in guinea pig, 197
    in mouse, 195–197
    pigment in, 194
    in rat, 197
Endomixis
  in Metazoa, absence of, 13
  in Protozoa, 4, 5
Eye, 72, 206–212
  choroid layer, 210
    arteriosclerosis, 210
    atherosclerosis, 210
    atrophy, 210
    thickening, 210
  ciliary body, 206, 207
    atrophy, 208
    calcification, 209
    fat in, 208
    hyalinization, 208
    hyperplasia, 208
  conjunctiva, 207
    fragmentation, 207
    hyalinization, 207
    opaqueness, 207
    "pinguecula," 207
    sclerosis, 207
  cornea, 206
    opaqueness, 206
    senile ring, 206
  eyelids, 212
    carcinoma, 212
    fat in, 212
    hyalinization, 212
    keratosis, 212
    lesions, 212
    plaques, 212
    rete pegs, 212
    wrinkles, 212

Iris, 206, 209
 bulge, 209
 cellular debris, 208
 "fading," 208
 glaucoma, 208
 hyalinization, 206
 iridial pigment, 208
 "senile miosis," 208
lens, 209, 210
 cataract, 209
 presbyopia, 210
 sclerosis, 210
 uveitis, 210
optic nerve, 211–213
 atheromatous change, 212
 atrophy, 211
 calcification, 212
 cavities, 212
 corpora amylacea, 212, 213
 fibrosis, 211, 212
 glial degeneration, 212
 hemorrhage, 212
 hyperplasia, 212
 sclerosis, 212
retina, 209–211
 atheromatous degeneration, 211
 atrophy, 211
 "colloid bodies," 210
 cystic degeneration, 210, 211
 detachment of, 209, 210
 depigmentation, 210
 ischemia, 211
 macular degeneration, 210
 necrosis, 211
 phagocytes, 211
 sclerosis, 210
 spaces, 210
 vascular occlusion, 211
sclera, 206, 207
 corneal opaqueness, 206
 fat, 206
 hyalinization, 206
 sclerosis, 206
 warts, 206

Fat formation, 68
 in adrenals, 201
 in brain, 222, 223, 226
 in eye, 206, 208
 in eyelids, 212
 in heart, 68, 115, 117
 of lymph nodes, 135
 in pancreas, 160
 in rat kidney, 177
 in salivary glands, 148, 150, 152, 153, 156
 in thyroid, 194
Female reproductive system, 186–191
 atrophy, of ovaries, 187
 of uterus, 187, 188
 of vagina, 187
 cancer, of cervix, 188
 of mammary glands, 188
 of uterus, 188
 cyst formation, in *rat* uterus, 187
 of uterus, 188
 fibrosis, of ovary, 186
 of uterus, 186
 of vagina, 186
 keratinization, of vagina, 186
 of vulva, 186
 mammary glands, cancer of, 188
 menopause, 186
 of mouse, 185–187
 ovary, atrophy of, 187
 cancer of, 187
 uterus, atrophy of, 186
 cancer of, 187, 188
 cysts in, 188
 fibrosis of, 186
Fibrosis
 in ear, 215
 in eye, 211, 212
 of heart, 117
 of kidney, 172, 179
 of large intestine, 159
 of liver, 161, 162
 of ovary, 186
 of rat kidney, 180
 of salivary glands, 155
 of spleen, 139
 of thyroid, 194
 in mouse, 195, 196
 of uterus, 186
 of vagina, 186
Fishes, *see* Vertebrates

Hamster
 adrenals of, 201–203
 muscular system, 102, 103
 nail growth, 88
 pituitary of, 199
 thyroid of, 197
Heart, *see* Blood
Hemixis, in Protozoa, 7
Hemorrhage
 of eye, 212
 in mouse adrenals, 203
 in spinal cord, 230

Hyalinization
  of blood vessel walls, 250
  of eye, 206–208, 212
Hyperplasia
  of adrenals, 199–201, 203, 204
    of rat, 199–201
  in eye, 208, 212
  of liver, 164

Invertebrates, 3–39
  Annelida, 21–23
    Lumbricus terrestris, 22, 23
  Arthropoda, 24–33
    Apis mellifica, 32
    Drosphila, 26–30, 33, 242
    Musca domestica, 31
  Ciliates, 3–9
    Ancistruma, 9
    Aspidisca, 9
    chromosomal aberrations of, 4
    endomixis of, 4, 5
    hemixis in, 7
    Paramecium aurelia, 4
    Suctoria, 5–8
    Uroleptus, 4
    Uronychia, 9
  Coelenterata, 12–14
    Campanularia flexuosa, 13, 14, 17
    Cereus pedunculatus, 13
    Hydra, 12
    Pelmatohydra oligactis, 12
  Crustacea, 24–26
    Daphnia longispina, 25
    Daphnia magna, 25
    Entomostraca, 24
  Echinoidea, 37–39
    Echinus esculentus, 38
    Mellita sexiesperforata, 38
    Ophiopholis aculeata, 38
    Ophiura sarsi, 38
    Ophiocten sericeum, 38
    Ophiura texturata, 38
    Psammechinus miliaris, 37
    Stichopus badionotus, 39
    Strongylocentrotus drobachinesis, 38
  Echinodermata, 35–37
    Antedon bifida, 36
    Asterias forbesi, 37
    Asterias rubens, 37
    Asterias vulgaris, 37
    Asteroidea, 37
    Ctenodiscus crispatus, 37
    Heliometra glacialis, 36
    Holothuria floridana, 37
    Paracaudina chilensis, 37
    Pseudarchaster parelii, 37
    Psilaster andromeda, 37
    Stichopus japonicus, 37
  Eumetazoa, 12
  Mollusca, 20–21
    Agrolimax, 21
    Chambered Nautilus, 20
    Crassostrea virginica
      Gmelin, 21, 22
    Helix, 21
    Ostrea iridescens, 22
  Nemathelminthes, 35
    Aschelminthes, 36
  Platyhelminthes, 35
    Planaria maculata, 35
  Porifera, 11, 12
    Adocia alba, 12
    Suberitus carnosus, 12
  Protozoa, 3–9, 15, 241
    Paramecium aurelia, 4
    Tokophyra infusionum, 4
  Rotifera, 13–19
    Chlorella vulgaris, 16
    Epiphanes senta, 15
    Philodina citrina, 16

Kidney, *see* Urinary system

Large intestine, *see* Digestive system
Liver, *see* Digestive system
Lungs, *see* Respiratory system
Lymphocyte, 123–125
  and appendix, 125
  bizarre nuclei, 123
  "seeding," 125
  and spleen, 125, 141
  in thymectomy, 125
  in thymoma, 123
  and tonsil, 125
Lymphoid tissue, 125–157
  in adenoids, 128, 129, 156
  in arsenic poisoning, 125
  in liver, 125
  lymphatic nodules, 126, 127
    in appendix, 127
    in gastrointestinal tract, 127
    in human skeleton, 127
    in respiratory tract, 127
  lymph nodes, 132–137
    cyclic changes, 133
    in rats, 133–135
    shrinking, 133
    in starvation, 132
  pharynx, 128
  spleen, 125, 137–141

arteriosclerosis, 139
atrophy, 139
megakaryocytes, 141
metaplasia, 14
in parotid gland, 126
in rat, 126, 141
in submandibular glands, 126
thymus, 130–132
of birds, 132
of fishes, 131, 132
of frogs, 132
of mouse, 126, 131, 133–137
in tonsils, 128–130, 156, 157

Male reproductive system, 183–186
androgen-estrogen upset, 183
atrophic change, 184
concretions, 184
corpora amylacea, 185
fertility, 183
fibrosis of, 183
prostate gland, 183–185
hypertrophy of, 183
testes and ducts, 185, 186
Mammals, 67–250
bats, 67, 68
cat, 67, 126
cow, 112, 120
elephant, 112, 120
hamster, see Hamster
horse, 67, 112
man, see Man
monkey, 67, 107
mouse, see Mouse
rabbit, 112, 132
rat, see Rat
Man
arteriosclerosis in, 115, 139, 210, 218, 229
atherosclerosis in, 68, 112, 117
cancer in, 68, 150, 157
diverticula in, 101, 159
emphysema in, 68
Mönckeberg's sclerosis in, 103
pneumonia in, 230
presbyopia in, 68
Mast cells, 74
in alimentary tract, 74
in heart, 74
in kidney, 74
in liver, 74
in lung, 74
in rat heart, 104
in rat liver, 74
in rat muscle, 104, 105
in thymus, 74

in uterus, 74
Metazoa, see Invertebrates
Mitochondrial alteration,
in brain, 227
of guinea pigs, 227
in mice, 227, 228
Mollusca, see Invertebrates
Mouse, 119, 124, 131
atheromatous lesions in, 112, 113
calcium deposits, 112, 158
digestive changes, 158
glandular changes, 195, 199, 201–203
influenza in, 144, 145
liver changes, 161–168
nodules, 168
mitochondria, 168
lymphoid tissue, 131
nail growth, 88
nervous system changes, 227, 228, 230, 232–234
reproductive changes, 185–187
skin changes, 118
vessel wall changes, 117, 118
Mouth, see Digestive system
Muscular system, 72, 101–108
in alimentary tract, 101, 157, 158
of diaphragm, 104
of heart, 72, 101, 106–108
in hind limbs, of rat, 104
of larynx, 104
in rats, 102, 104, 105
of skeleton, 72
in uterus, 101, 102
in hamster, 102, 103
Mutation theory of aging, 243, 244

Nervous system, 218–235
brain, 218–229
Alzheimer's disease, 227
arteriosclerosis, 218, 229
atrophy, 218
fat formation, 222, 223, 226
fibrosis, 227
giant cells, 229
hyperchromatosis, 229
pigmentation, 222, 224–227
senile dementia, 219
"shrinkage," 221, 222
spaces, 222
weight loss, 218
cellular fine structure, in rat and mouse, 234
pigment, 234
meninges, 218
adhesions, 218

thickening, 218
spinal cord, 229–233
  atrophy, 231
  hemorrahage, 230
  of mice, 230–233
  pigmentation, 229–231
  pneumonia, 230
  in rats, 231–233
Nodules
  in bone, 126, 127
  in cartilage, 95
  in eye, 210, 211
  lymphatic, 126, 127
  in mouse liver, 168
  in mouse reproductive system, 187
  in prostate, 183
  on skin, 87
  in stomach, 157

Osteoporosis, 97, 98
  in temporal bone, 215

Pancreas, *see* Digestive system
Pigment, 248, 249
  in amphibians, 247
  of brain, 222, 224, 227
  in heart, 246
  in liver, 168
  in mouse, 201, 246
  in nervous system, 222, 224–227, 229–231
  of rat and mouse, 234
  in Protozoa, 247
  in skeletal muscle, 246
  of thyroid, 194
Polyps, *see* Nodules

Rats
  atherosclerotic lesions, 101, 102, 105
  blood forming tissues, 118
  digestive changes, 148–153, 155
  glandular changes, 197, 199–201, 204
  liver changes, 74, 169
  lymphoid tissue, 126, 133–137
  lymph nodes, 133–135
  muscle changes, 102, 104, 105
    in heart, 104
    in hind limbs, 104
  nervous system changes, 231–234
  pancreatic changes, 159–161
  reproductive changes, 183, 187
  skin changes, 73, 82, 85–88
    nail growth, 88
  splenic changes, 139, 141
  tendon changes, 72

Reproductive system, 183–191
  female, *see* Female reproductive system
  male, *see* Male reproductive system
Reptiles, *see* Vertebrates
Respiratory system, 144–146
  lungs
    cancer, 145
    emphysema, 68, 144
    in mice, 144, 145

Salivary glands, *see* Digestive system
Sclera, *see* Eye
Sclerosis
  of bone blood supply, 97
  of eye, 206, 207, 210, 212
  of heart, 115
  of kidney, 172
  of nervous system, 218, 229
  of thyroid tissue, 195
Sense organs, 208–217
  ear, *see* Ear
  eye, *see* Eye
Skeleton, 94–99
  bone, 96–98
    appositional growth, 96
    "brittle" bone, 96
    calcification, 98
    collapse, 98
    curves, 98
    demineralization, 98
    endocrine deficiencies in, 97
    hyalinization, 98
    hyperplasia, 98
    lymphatic nodules, 126, 127
    modeling, 96
    osteoarthrosis, 98
    osteoporosis, 97, 98
    resorption, 96
    sclerosing, of blood vessels to, 97
    softening, 98
  cartilage, 94, 95
    calcification of, 94
    cyst formation, 95
    erosions of joints, 95
    ossification of, 95
Skin, 69, 71–92
  collagen, 90
  cutaneous tag, 88
  cysts, 87
  dermal papillae, 79
  face, 87
  fasciae of body, 89
  forearms, 87
  in guinea pigs, 73
  hands, 87

in humans, 71–92
keratotic lesions, 81, 87–89
in matrix, 92
in mouse, 73
on neck, 88
pigmented nevi, 87
in rat, 73, 82, 85–87
in starvation, 91
telangiectases ("blood spots"), 87
thorax, 88
Small intestine, *see* Digestive system
Spinal cord, *see* Nervous system
Spleen, *see* Lymphoid tissue
Stomach, *see* Digestive system

Thymus, *see* Lymphoid tissue
Theories of aging, 243–250
auto-immune theory, 247–250
cell theory, 243
chromosomal theory, 244
in mouse, 243
collagen deposit theory, 245
gene mutation theory, 243, 244
in mouse, 243
wear and tear theory, 245

Urinary system, 172–181
blockage, of bladder, 183
kidney, 172–181
adhesion, 172
atrophy of, 172
capsular thickening, 172
fibrosis, 172–179
of rat, 173–181
sclerosis, 172

Vascular system, *see* Blood
Vertebrates, 43, 251
amphibians, 43, 50–61
anura, 50, *see also* Vertebrates, Frogs
Bufonidae, 50
Cystignathidae, 50
Discoglossidae, 50
Hylidae, 50
Ranidae, 50
birds, 59–63
house sparrow, 61–62
royal albatross, 63

tits, 61
yellow-eyed penguin, 61
fishes, 43, 44
Alaskan whitefish, 48
Atlantic herring, 44
catfish, 43
eel, 43
fin rays, 43
goby, 44
goldfish, 48
haddocks, 43
lake trout, 48
minnows, 44
mosquito fish, 48, 61
salmon, 43, 45, 46
shark, 43
sturgeon, 43
teleost, 46, 47
tuna, 43
whale shark, 44
frogs, 50, 57
Ambystoma mexicanum, 51; *see also* Anura
Axolotl, 57
Megalobatrachus, 50
Rana pipiens, 56, 57
Triturus pyrrhogaster, 50
man, *see* Man
reptiles, 43, 59–63
crocodiles, 59
lizards, 59
snakes, 59, 60
turtles, 59–61
salamanders, 59–61
Hydromantes platycephalus, 51
snakes, 59–61
Astyanax, 61
Natrix natrix, 60
Vipera aspis, 60
Vipera berus, 61
Toads, 50
Turtles, 59–61
Testudo elephantopus, 60
Testudo nigrita, 60

Wear and tear theory, 245
Worms
flatworms, 35
marine worms, 21
roundworms, 36